Handbook on
Wild and Zoo Animals

A Treatise for Students of Veterinary, Zoology,
Forestry and Environmental Science

Ajit Kumar Santra BVSc & AH, MSc (Dairying), PhD
Associate Professor
Department of Livestock Production and Management
College of Veterinary Science and Animal Husbandry
Indira Gandhi Agricultural University
Anjora, Durg, Chhattisgarh and
Formerly Livestock Specialist, AFPRO, New Delhi

CBSPD

CBS Publishers & Distributors Pvt Ltd

New Delhi • Bengaluru • Chennai • Kochi • Kolkata • Lucknow • Mumbai
Hyderabad • Jharkhand • Nagpur • Patna • Pune • Uttarakhand

Handbook on
Wilk and Zoo Animals

ISBN: 978-93-87085-09-1

©Author

First Edition: 2008
Reprint: 2018, **2024**

Published by **Satish Kumar Jain** and produced by **Varun Jain** for

CBS Publishers & Distributors Pvt Ltd

4819/XI Prahlad Street, 24 Ansari Road, Daryaganj, New Delhi 110 002, India.
Ph: 011-23289259, 23266861   Website: www.cbspd.com
                              e-mail: delhi@cbspd.com

Corporate Office: 204 FIE, Industrial Area, Patparganj, Delhi 110 092
Ph: 011-4934 4934   Fax: 011-4934 4935
                     e-mail: publishing@cbspd.com; publicity@cbspd.com

Branches

- **Bengaluru:** Seema House 2975, 17th Cross, KR Road, Banasankari 2nd Stage, Bengaluru 560 070, Karnataka, India
  Ph: +91-80-26771678/79   Fax: +91-80-26771680   e-mail: bangalore@cbspd.com
- **Chennai:** 7, Subbaraya Street, Shenoy Nagar, Chennai 600 030, Tamil Nadu, India
  Ph: +91-44-26680620, 26681266   Fax: +91-44-42032115   e-mail: chennai@cbspd.com
- **Kochi:** 42/1325, 1326, Power House Road, Opp KSEB, Power House, Ernakulum Kochi 682 018, Kerala, India
  Ph: +91-484-4059061-65,67   Fax: +91-484-4059065   e-mail: kochi@cbspd.com
- **Kolkata:** 147, Hind Ceramics Compound, 1st Floor, Nilgunj Road, Belghoria, Kolkata-700056, West Bengal, India
  Ph: +033-25633055, 033-25633056   e-mail: kolkata@cbspd.com
- **Lucknow:** Basement, Khushnuma Complex, 7 Meerabai Marg (Behind Jawahar Bhawan), Lucknow-226001, UP, India
  Ph: +0522-4000032   e-mail: tiwari.lucknow@cbspd.com
- **Mumbai:** PWD Shed, Gala no 25/26, Ramchandra Bhatt Marg, Next to JJ Hospital Gate no. 2, Opp. Union Bank of India, Noorbaug, Mumbai-400009, Maharashtra, India
  Ph: 022-66661880/89   e-mail: mumbai@cbspd.com

Representatives

• Hyderabad	0-9885175004	• Jharkhand	0-9811541605	• Nagpur	0-8692091830
• Patna	0-9334159340	• Pune	0-9664372571	• Uttarakhand	0-9716462459

Printed at Sanjay Printers, Sahibabad, U.P, India

Dedicated to
those students, doctors, nurses, teachers, scientists,
non-teaching staff, family members and innumerable
friends who saved the author from the clutch of death

PREFACE

There is no denying the fact that our rich natural resources are dwindling rapidly day by day. Wild animals are recklessly killed for their various products. There has already been a warning signal that animals like tiger would be vanished from this country if we do not take appropriate measures at this juncture. A great threat is poised to the existence of wildlife. Majority of recent extinction of plants and animals have been due to man's interference. Therefore, the concern for wildlife is concern of man himself. The wildlife is so fascinating and enchanting that our first prime minister, Pandit Jawahar Lal Nehru once said in his forward note in E.P. Gee's Book 'The Wild Life of India', "It is much more exciting and difficult to shoot with a camera than with a gun." Despite the growing awareness about wildlife conservation, we still have to go a long way in order to achieve the goal. The father of nation, Mahatma Gandhi said, "The greatness of a nation and its moral progress can be judged by the way its animals are treated." Therefore, our endeavour would be to sensitize people about wildlife through various means. Obviously book is one of them.

Ever since the introduction of wildlife in the academic curriculum of Veterinary Science, there has been an avid interest about the wildlife among the student community. Simultaneously, it has expanded the horizon of employment opportunity for the students other than their traditional jobs. Although, veterinarians have been restricted their activities to the health care of wild and zoo animals, the need of the hour, however, is to expand their involvements in other branches of wildlife science. In fact, they should play a cohesive role in the gamut of wildlife including wildlife biology, conservation and man-animal conflict. Therefore, there is a need of thorough understanding of the subject in the present perspectives.

With this background frankly speaking there is no dearth of books on wildlife. Every year people from various sections have poured their knowledge in the arena of wildlife through different forms such as new research findings, writing of books and awareness programmes by socially sensitive organizations. Books available in this field either have been written primarily on game purpose or they are purely meant for the requirement of academic interest. However, looking at the science wildlife so vast and varied in nature, it is a herculean task for an author to incorporate entire aspect within limited pages of a book. Therefore, an effort has been made to incorporate the information on wildlife in such a way so that it serves the purpose of wide readership. This book is an out come of more than ten years rigorous work on collection of information on wildlife. The different chapters have been carefully selected in order to meet the requirement of graduate students of Indian universities. The present book contains three sections with twenty three chapters. The section one has ten chapters dealing with the information about general wildlife. This is extremely important for knowing the present status of wildlife in the country. The section two discusses the biology of wild animals highlighting the distribution, habitat and other salient biological features. The section three on the other hand highlights feeding, housing, management and health care of wild and zoo animals. Each chapter in different sections has been focused with the objective of getting acquaintance with the fundamental and relevant information about wildlife preferably to Indian context within a short space as maximum as possible. Another purpose is to marriage the anecdotes narrated in game books with some recent work available in the literature and need for the competitive examinations so as to give an in-depth exposition to various facets of wildlife. I hope that this book will provide useful and concise information to the students from different fields such as veterinary, zoology, botany and environmental science. This will also help the general readers who have not attended the formal course of biological science. However, the final assessment of the book will be done by learned readers and their valuable suggestions and constructive criticisms would be incorporated in subsequent editions.

Wildlife science is a conglomeration of multidisciplinary subjects. So many people have extended their helping hands in preparing this book, I fear that some of them may be inadvertently skipped. My profound gratitude to Dr. L.N. Acharyo, the veterinarian of international repute in the field of wildlife science, for editing the major portion of this book. Surely without his whole hearted guidance and valuable suggestion I could not venture this kind of work. My students Sandhya and Neha deserve sincere appreciation for their brilliant illustrations. I am grateful to Dr. Sudhakar Jogi, Dean, College of Veterinary Science and Animal Husbandry, Durg for his instrumental role in strengthening of wildlife in the college. Relevant books and literature supplied by Dr. Sandip Tiwari, Dr. S.S. Bargoli, Dr. Nalin Sharma and Dr. S. Roy is worth mentioning. The academic staff from the Department of Livestock Production and Management were very cooperative during my virtually one point programme on relentless writing of this handbook.

My stay at Wildlife Institute of India, Dehradun for library work was extremely warm and joyful. This was possible due to sincere cooperation and help from the Director, erudite faculty members and library staff. Cooperation and help from the library staff from the Indian Veterinary Research Institute, Izatnagar, Ravishankar University, Raipur and College Library, Anjora will never be forgotten. My special thanks to Mr. Suneel Gomber, International Book Distributing Co, Lucknow for his pragmatic vision in publishing the book. I have acknowledged the materials taken from different sources in the respective tables with a full reference lists in the end. However, any omission is considered not to be an intentional one. Finally my wife Sonali and son Aaranyak have a sigh of relief after burn the midnight oil.

31 December, Anjora   **Ajit Kumar Santra**

Contents

gibbon, Macaque and Langur; Order Pholidota-pangolin; Lagomorpha-rabbit and hare; Rodentia- rat and mice, chinchilla, hamster and porcupine; Order Carnivora-cheetah, lion, leopard, tiger, snow leopard, small cat, hyaena, Viverrid, mongoose, bear, Asiatic black bear, sloth bear, Procyonid giant panda, red panda, cavid, wilddog, dhole, Mustelid and otter; Order Proboscidea-elephant; Order Perissodactyla-rhinoceros, wild ass and zebra; Artiodactyla-dear, swamp deer, sambar , cheetal, muntjac, musk deer, chevrotain, Antelopes, blue bull, blackbuck, chinkara, four-horned antelope, yak, mithun, Indian bison, wild buffalo, wild sheep, wild goat, goat antelope, wild pig, camelid, camel, giraffe and hippopotamus.

Turtle- olive ridley turtle, tortoise, terrapin; Crocodile- gharial, marsh crocodile and estuarine crocodile; Snake- Indian python, Russell's viper, Indian cobra, king cobra and common Indian krait.

Ostrich, pelican, flamingo, stork, crane, jungle fowl, peafowl, common quail, pigeon and parrot.

Monotremes- echidna and platypus; Marsupials- kangaroo, wallaby and koala; Insectivores- hedgehog; Chiropterans - fruit bat, insectivorous bat and vampire bat; Primates- Pangolin; Lagomorphs-hare and rabbit; Rodents- Carnivores: Felids-hyaena; Viverrids-mongoose; Bears-Himalayan black bear and sloth bear; Mustelids-otter; Procyonids- Canids- dhole, jacal; Elephant; Perissodactyls-rhinoceros, zebra, wild horse and wild ass; Artiodactyls- wild pig, hippopotamus, mouse deer, giraffe, camelids, deer, cheetal, muntjac, swamp deer, Antelopes, nilgai, blackbuck, chinkara, wild sheep and goat, goat antelope, wild cattle, yak, mithun and wild buffalo.

Chelonians, crocodilians and snakes.

Section 1
GENERAL WILDLIFE

Chapter 1

Wildlife and its Importance

Introduction:

The life first appeared on earth more than 3.5 billion years ago. Scientists believe that the life probably originated from the ocean. Since then, many animals and plants had appeared on earth, invaded new areas and then faced competition from others. As a result, many species became extinct. The earth has spawned about 13.5 million species of living things including plants, microbes and animals. There are about 7,50,000 species of insects, 43,000 vertebrates and 2,50,000 plants. Among vertebrates, there are 4200 species of mammals, 9000 birds, 6300 reptiles, 4200 amphibians, 18,000 bony fishes and 900 jawless fishes and cartilaginous fishes. Some animals and plants occur in a restricted geographical region, while others have adapted to a diverse environment.

India has a rich heritage of wild flora and fauna. With 3,287,263 sq km of land mass comprising 2.4 per cent of world's land area, the country contributes about 8 per cent to the world's diversity of life-forms including 350 species of mammals, 1224 birds, 408 reptiles, 197 amphibians, 2546 fishes, 57,548 insects and 46,286 species of plants. It comprises 11 per cent world's plant species, 7.6 per cent of world's mammal species, 12.6 per cent bird species, 6.2 per cent reptiles, 4.4 per cent amphibian and 11.7 per cent of world's fish species. The country has been recognized by the world community to be one of the world's major centres of crop and livestock origin. It is reported that about 320 species belonging to 48 families and 116 genera of wild relatives of cultivated and domesticated stock are known to have originated from India's wild biodiversity. India is one of the 12 mega biodiversity countries in the world.

Since time immemorial men were interested in catching and keeping the animals for various purposes such as food, worship, work and companionship. The ancient records reveal that wild elephants were domesticated about 4000 years ago. The Mughal emperor Jahangir had 12,000 elephants. The king Soloman had a large collection of lions, antelopes, cheetah, apes, deer, giraffe and crocodiles under his captivity. The findings of Harappa and Mohenjodaro reveal that tigers, elephants and one horned rhinoceros lived in the valley of Indus civilization. Many animals play a major role in man's religious belief and practices. The Bishnoy community, for example, has a religious zeal and fervour for the blackbuck. The great emperor Ashoka, in true sense, was the world's first known conservationist.

Lord Buddha preached his first religious address in a deer park near Sarnath. The long and legendary history also reveals that Mughal emperors had established large tracts of forests for wildlife reserves. Our ancient Rajahs and Maharajahs used to organize great hunting parties for visiting or local dignitaries. For many years, wildlife was the game of many rulers and at the same time, wildlife became the victims of many rulers. It has been reported that many British rulers had shot many tigers during their stay in India. In 1938, a shooting party headed by the then Viceroy of India, Lord Linlithgow, shot as many as 4,273 birds in Keoladeo, Bharatpur as depicted in shooting record inscribed on the pillar near Keoladeo temple. Even today, many wild animals have been trapped, shooted and poisoned by human beings. Many factors such as pollution, reduction of habitat, deforestation and rapid industrialization are responsible for the dwindling wildlife population. As a result, a large number of animals have now become endangered.

However, the wildlife is really a wonder of wonders. It is fascinating to see that squirrels are nursed by a cat; a hunting dog nurses a young deer; elephants become painters; an individual large ape analyzes computer language; some birds nest only at particular periods of the year; the blue-footed booby rears a brood when conditions are favourable and the call of the cuckoo heralds the beginning of monsoon. The wildlife has also many important roles in the improvement of human society. Their contributions to the maintenance of healthy ecosystem, maintaining of gene pool and recreation have been well documented throughout the world. So, our endeavour would be to protect all these fascinating creatures found on the earth. Though, it is ironical that human beings are the destructive agents, they however, have also become the conservationists. Therefore, human progress must continue but it should not be at the cost of our precious wildlife.

Importance of wildlife:

As mentioned earlier that the wildlife has many important roles in the improvement of human society. Therefore, it is meaningful to highlight some important values of wildlife in this text. The values are as follows :-

- **Commercial value**: The wildlife has become an important source of revenue nowadays. Incomes are obtained by selling or trading of wild animals and their products. Meat, fur, wool, bones, horns, nerves, urine, toe nails, claws, musk glands etc are harvested for commercial purposes in many countries. Many wild animals such as deer, wild sheep and goats, birds, reptiles, antelopes etc are considered to be the valuable sources of food. Marine fisheries have potential commercial values as food. The wildlife attracts many domestic and foreign tourists, serving as an important source of income. For example, in Kenya the wildlife tourism has been a major source for earning foreign exchange. The whale watching is now a revenue source in many countries in the world. Today, the whale watching industry generates more than $1 billion in annual revenues worldwide.

- **Biological value**: The biological value of wildlife is concerned with their contribution of productive ecosystems such as pollination, seed dispersal and planting, control of animal and plant population, recycling of nutrients and sanitation through scavenging. The pollination of wild and domestic plants takes place through birds, insects and molluscs. Seeds are transported from one place to another by birds through faeces. The quality of soil is improved by faecal materials and dead animals. Many mammals and birds (e.g. jackals, hyenas, crows and vultures) are scavengers keeping the environment clean. Birds control the population of pests and rodents that damage the agricultural crops. Therefore, the wildlife is considered to be an ecological asset and an indicator of environmental health.

- **Recreation and game value**: Many professionals like photographers and bird watchers are immensely benefitted through their visits in wildlife sanctuaries and national parks. Tourists enjoy after seeing wildlife. These recreational values can not be quantified in terms of money. The physical and mental healths are enhanced through outdoor activities. It has recreational value to those who are involved in fishing or hunting for sport. Trout fishing in Kashmir, for example, attracts many tourists from various parts of the world. The hunting of wild animals and birds still remains an important recreational value among elite in India.

- **Scientific and educational value**: Wild animals are used for various studies. Rhesus monkeys are widely used in biomedical research. How a chimpanzee learns to communicate by sign language and how dolphins talk in their own language - all we can better understand is through scientific studies. Our knowledge on ultrasonic research is based on the research findings on bats. Moreover, we can manage the wild animals and plants better through scientific studies.

- **Aesthetic value**: The aesthetic values of wildlife such as nature studies, bird watching and wildlife photography can not be simply expressed in words. Many paintings, literature, poetry, art and music are based on wildlife. The paintings and carvings in caves and many historical places indicate that how cultural relationship of man with wildlife is as old as human civilization in India.

- **Social and ethical value**: Wildlife has also various social values on our day to day activities. Many wild animals become the object of worships in folklore and legends. Indians worship many trees, animals and rivers for well being of mankind. Tulsi plant *(Ocimum sanctum)* and banyan tree *(Ficus bengalensis)*, for example, have high religious value. Even today, the milk is given to the snakes on the occasion of *Nagapanchmi*.

- **Negative aspect**: However, the wildlife has negative values also. Many a time wild animals cause great damage to agricultural crops and other properties. The carnivores like tiger, lion, panther, wolf etc kill domestic live-stock. Wild animals like elephants cause havoc damage to agricultural crop during their seasonal movements. It is estimated that elephants damage 10,000 to 15,000 houses and 8 to 10 lakh hectares of crops yearly in India. The government spend rupees 10 to 15 crore every year on measures for controlling depredation and payment of *ex-gratia* to the victims of depredation. Depredation by wild elephants throughout its ranges is a common phenomenon. Depredation caused by elephants is normally of four types- Crop raiding, killing or injuring of human being, house breaking, killing of livestock and damaging other properties. For example, depredation by wild elephants in South West Bengal has been well documented. The tangible problems caused by these migratory elephants from Dalma Wildlife Sanctuary, Jharkhand have several dimensions on local people and the habitat. Man-elephant conflict sometimes is turned to forest department - villager conflict. Forests in some areas have been cleared by desperate villagers to deny shelter to the elephants near the village. Agitation, road blockade, gherao and even physical assault on the forest staff have become a routine affair during elephant depredation. Many small mammals and birds damage orchards. Many wild animals are also reservoirs of various diseases.

Chapter **2**

Definitions and Concepts of Some Terminologies Related to Wildlife

Many a time we use some terminologies related to wildlife that need to be clearly defined for better understanding of the subject. Therefore, the following text in this chapter describes the definition and concept of some terminologies.

Wildlife

The present definition of wildlife has been expanded from the past. Initially, it was restricted to only those wild vertebrates that lived in their natural environments. In recent past, the definition of wildlife, however, has received a much wider concept. Wildlife in true sense can be defined as those plants and animals which are not domesticated and are found in their naturally associated environments. The natural environment of a species is referred to that kind of environment which enables the species to use all its adaptations. However, according to the Wildlife Protection (Amendment) Act, 1 April, 2002, the definition of wildlife includes "Any animal, aquatic or land vegetation which forms part of any habitat." The Act has also clearly defined the term wild animal as "Any animal specified in Schedules I to IV and found wild in nature." Therefore, wildlife can be appropriately defined as the plants and animals which are listed in the various schedules and occur in their natural habitats or wild ecosystems as undomesticated form.

Wildlife forms an important component of any natural ecosystem and is considered as mirror of environmental health. Wildlife population occurs in those areas where their basic needs such as shelter, food, water, reproduction and movement are fulfilled. Wild animals are most characteristically expressed by their wild behaviours in the natural environments.

Ecology

The word "ecology" was first coined by the German biologist Ernst Haeckel in 1869. The subject ecology deals with the distribution and abundance of organisms and their interactions among themselves and with the environment. In other words, it is the study of the inter-relationships between plants, animals and their environment.

7

The subject ecology has wide applications in many fields. It plays a major role in the management of wildlife. The wildlife management, in fact, is largely related with the application of fundamental knowledge of ecological science. The study of ecology is, therefore, the prerequisite for the improvement of wildlife resources. It also helps in the scientific management of agriculture, forestry, fishery, grasslands, soil conservation etc.

The subject ecology can be subdivided into the followings:

- **Based on taxonomic affinity:** Based on this principle there are two sub-divisions of ecology-plant ecology and animal ecology. The plant ecology studies the plants, whereas, the animal ecology deals with the animals. There may be specialized ecology such as bacterial ecology and ecology of turtles etc.

- **Based on habitat:** The habitat ecology deals with the descriptive study of different types of habitats and the organisms living there. Freshwater ecology, forest ecology and grassland ecology are classified on this principle.

- **Based on levels of organization:** This is an approach to the classification of ecology based on area, unit of study etc. It is of two types-autecology and synecology. Autecology studies the relation of individual species to its environment. Here, the individual species is studied for details of its geographic distribution, taxonomy, morphology, life cycle etc along with the various ecological factors which might influence the different stages of its life cycle. Synecology, on the other hand, studies the groups of individuals and their interactions in relation to their environment. So, it is based on communities. It deals with population ecology, community ecology, biome ecology etc.

The concept of "applied ecology" is also described by many authors. This is based on the practical applications of ecology. Wildlife management is the major specialized field of applied ecology.

Environment:

The term "environment" is derived from the French word *"environ"* which means "encircle". It is described as the surrounding of an organism in which it lives. It is the whole association of both living and non-living factors, of which, the organism is a part. The environmental factors such as light, temperature, soil, water etc (abiotic factors) have a profound role in the life of an organism. A reciprocal relationship exists between the organism and its environment and any attempt to disrupt this association will eventually lead to an ecological dislocation.

The environmental factors can be classified in many ways. Sometimes all the factors are classified into two groups- direct factors and indirect factors. Direct factors include light, temperature, humidity of air, soil, water, soil nutrients etc. Indirect factors, on the other hand, are soil structure, soil organisms, wind, altitude etc. A short description of some climatic factors is given here:

(a) The light plays a vital role in the development and distribution of various plants and animals. The effects of light on plants for chlorophyll production, overall vegetative development of plant parts and distribution of plants are widely known. It has also important effects on an animal's physiological and behavioural characteristics such as pigmentation, reproduction, development, growth, locomotion and migration. The migration of many birds and larger mammals is initiated by photoperiod. Many smaller animals (e.g. squirrels) store food materials in favourable time for future use and this behaviour is controlled by photoperiod. Another effect of light on animals is the development of pelage and feathers. The swamp deer of central India, for example, has characteristic summer pelage.

(b) The temperature has a profound effect on plants and animals. These include metabolism, reproduction, growth and development, morphology, colouration, migration and behaviour. Many exothermic animals suspend their physiological activities to a large extent during low environmental temperatures. Some animals such as Himalayan bear and polar bear exhibit a phase of hibernation during extreme low temperatures. Turtles and tortoises bury in the pond mud so as to tide over high temperature during summer months. Many desert animals become nocturnal so that they can avoid the heat during day time. Elephants prefer cool dense forests during dry season, while in rainy season they spread out in the open areas. Animals found in the cooler regions have a longer life span than animals living in warmer regions. Another well known effect of temperature on animal is the morphological change. Temperature affects the absolute size of an animal and the relative proportions of various body parts (Bergman' rule). This rule explains that mammals and birds attain greater body sizes in cold regions as compared to warm areas. The polar bear and the kodiak bear are larger than those of their counterparts found in the less cooler temperate zones. However, the mammals like elephants and gaur that are solely found in the tropical parts of the world are exceptions to this rule.

The ear, tail, snout, neck, bill and other extremities of colder animals are relatively shorter than extremities of animals found in the warmer zones (Allen's rule). The arctic fox, for example, has a smaller ear lobe than the desert fox. Some animals found in the warm humid climate are darker in colour than their counterparts living in cool or dry climate (Gloger's rule). A leopard in tropical rain forest of Arunachal Pradesh, for instance, may be darker than animal found in dry region of Rajasthan.

In the light of the above discussion, it can be concluded that the temperature has many profound roles on organisms. Moreover, temperature together with moisture determines the climatic condition of a particular zone.

(c) The amount of rainfall greatly influences the vegetation and animal population of a particular region. Annual rainfall determines the type of vegetation to be found in any region. The evergreen forests, for example, are found in tropical

9

areas where heavy rainfall occurs throughout the year.

(d) The edaphic factors are concerned with the structure and composition of soil along with its physical and chemical characteristics. Soil is one of the most important factors that profoundly influences on organisms and plants. A good soil helps in development of luxuriant vegetations that support a rich animal community. Soil also influences the type of vegetation which ultimately dictates the fauna to be thrived in a particular locality.

(e) The organisms (biotic factors) living in a particular environment influence each other's life through various interactions. The interactions may be through pollination, grazing, symbiosis etc. It may be intra-specific (between the individuals of same species) or inter-specific (between individuals of different species).The interactions among organisms may be beneficial or harmful. In general, inter-specific relationships can be categorized into symbiosis and antagonism. In the former category of relationship one or both partners are benefited and neither is harmed. The later category of relationship is such where at least one species is harmed. Symbiosis again can be classified into two- mutualism and commensalism. In mutual relationship, both the species are benefited. Pollination by animals (bees, moths etc take food and in return cause pollination), dispersal of fruits and seeds (fruits are eaten by animals and seeds in fruits are dropped in the excrement at various places), symbiotic nitrogen fixers (where the bacterium *Rhizobium* forms nodules in the roots of leguminous plants that fixes atmospheric nitrogen and the bacterium lives symbiotically with the host) are some examples of mutualism. The commensalism, on the other hand, is the type of relationship where only one of the partners is benefited and neither is harmed. The vivid example of commensalism is the presence of some microbes (e.g. *Escherichia coli*) in the lower intestine of animal. In antagonistic relationship, one or both the species are harmed during their life period.

Biome:

Biome may be defined as a specific life zone where a major community of plants, animals with similar life forms and environmental conditions exist. In other words, a biome refers to a climatic zone where a group of organisms are restricted by the environment. The examples are savannah, mangrove, swamps etc. A particular biome is the result of a combination of various factors such as temperature, rainfall and soil type. Each biome has specific vegetation. However, no biome is solely restricted to one region. The biomes often merge with no clearly defined borders and sometimes it is very difficult to recognize where one biome ends and another begins. The boundary areas, however, do have distinctive plant communities and the animals from adjoining biomes can move in and out of them. Average annual rainfall and temperature are considered to be the major factors determining the boundaries between two biomes. Plants (producers), animals (consumers) and microbes (decomposers) are the components of a biome. One species interacts with other species in biological community. Biotic interaction may be of inter-specific or intra-specific. In a biotic unit, plants grow and herbivorous animals eat these plants and these herbivorous animals are consumed by carnivorous animals. Again both plants and animals are attacked by microorganisms. Although a biome contains many components, it is usually named after its vegetal characteristics or dominant plant groups. This is because these plants tend to reflect the general climatic features of the area more precisely than the animals or decomposers. As we know that a consumer depends on the producer for its food and energy. Therefore, vegetations determine the type of animals to be present in a particular biome. Some biomes (e.g. grassland biome) can be influenced by the activities of animals such as grazing, trampling and burrowing.

In the light of the above facts, we can say that biome is the largest geographical biotic unit that deals with the major community of plants and animals with similar life forms and environmental conditions. Several similar biomes comprise a biome type. Many types of biomes have been proposed by geographers and ecologists. The main biomes of the world are the tundra biome, grassland biome, mountain biome, desert biome, cave biome and forest biome. The tropical forest biome has the richest faunal resources. It is estimated that more than five million reported species occur in this biome alone. Avifaunas found in the tropical region have many ecologically significant characteristics as compared to temperate species. The tropical birds, for example, have distinct breeding seasons.

Ecosystem:

The term "ecosystem" was first described by the British ecologist Sir Arthur Tansley in 1935. According to him, an ecosystem is "The system resulting from the integration of all living and non-living factors of the environment". However, in the Rio Convention on Biodiversity, the ecosystem has defined as "A dynamic complex of plant, animal and microorganism communities and their non-living environment interacting as a functional unit". From the above definitions, it is evident that an ecosystem is a functional unit of ecology. It has two components-physical or non-living (abiotic) and biological or living (biotic) and it works through these components (Fig.2.1). Moreover, both the components interact in such a way that any major changes may cause disruption of the whole system. An ecosystem may be a forest, a pond, a cropland or a desert.

An ecosystem functions through various steps. These include:
i)   Reception of energy (sun is the source of energy) by the plants;
ii) Preparation of food by the producers (plants);
iii) Consumption of organic matter by consumers (e.g. deer, cow, sheep etc.);
iv) Decomposition of organic compounds by saprophytes; and
v)   Release of inorganic nutrients in the environment.

Fig. 2.1: An ecosystem works through different components.

In general, there are two types of ecosystem. They are natural ecosystem and artificial (man-engineered) ecosystem. The natural ecosystems are operated by themselves under natural conditions without any major hindrance by human beings. They include terrestrial (e.g. forest, desert, grassland etc.) and aquatic ecosystems. The aquatic ecosystem may be of two types– marine (e.g. estuarine) and freshwater (e.g. pond) ecosystems. Artificial ecosystem is an ecosystem which is modified by man. Cultivated lands and fish farms are the examples of this kind. In this type of ecosystem, man tries to manipulate the living community and physio-chemical environment as well.

As mentioned earlier that the abiotic and the biotic components are the two major structural components of an ecosystem. Abiotic components constitute inorganic substances (e.g. water, carbondioxide, nitrogen, calcium etc.), inorganic chemicals (e.g. chlorophylls), organic materials (e.g. protein, carbohydrate, lipid etc.) and the climate of the given area. The biotic components, on the other hand, are autotrophic components (producers) and heterotrophic components (consumers and decomposers). Producers of the autotrophic components are mainly the green plants. They trap solar energy and manufacture food by utilizing the non-living simple inorganic substances. Photosynthetic bacteria are also the autotrophic components. Consumers along with decomposers are the heterotrophic components of an ecosystem. The consumers or phagotrophs are of three types- primary (herbivores), secondary and tertiary consumers (carnivores or omnivores). Consumers subsist on the matter built up by the producers. Decomposers are saprotrophs (e.g. bacteria, fungi etc.) that break down complex compounds of plant and animal matters into simpler compounds. These simpler compounds are again used by producers.

In ecosystem, the transfer of energy occurs through food chains (Fig.2.2) and successive links of the chains are known as trophic levels. As stated earlier that the green plants are the producers of an ecosystem. They incorporate the energy (from sun), carbondioxide (from air) and mineral nutrients (from soil) into their tissues to form organic compounds. These organic compounds serve as resources

Fig. 2.2: Transfer of energy occurs through food chains.

of foods to their communities in the ecosystem. The producers represent the first trophic level. When animals consume the plant materials, these compounds are broken down and there is release of energy that is used for many physiological processes. Herbivores are the primary consumers and represent in the second trophic level. Some carnivores are known as secondary consumers and they subsist on herbivores or other animals. Predator animals and small insects are examples of secondary consumers. Secondary consumers represent the third trophic level. Some carnivores are eaten by other carnivores and they are called tertiary consumers. Tertiary consumers constitute the fourth trophic level. However, some organisms that eat producers as well as consumers at their lower level in the food chain are called omnivores. The human species eats both plant and animal materials. Some animals are herbivorous at one season and become carnivorous at other time. The great tits and blue tits of Europe, for example, feed chiefly on moth caterpillars in summer and beechmast seeds during winter months. This is because of the fact that caterpillars are not available during winter months. Therefore, such organisms may occupy more than one trophic level in the food chain. Another important feature of the trophic level is that nearer a level is to the source of energy, greater is the diversity of species involves.

As mentioned earlier that the food energy is transformed through food chains. Various food chains are again joined at different trophic levels forming food webs. In this regard, it is to be mentioned that the food web explains the exact meaning of trophic relations in the ecosystem. This is due to the fact that food chain simply describes a kind of dependency where one kind of animal or organism consumes only one other kind, whereas food web explains a more complex overlapping relationship in which many different animals eat a wide variety of other animals or plants or be eaten by them . Therefore, diversity of the organisms in the ecosystem is more important for understanding the complexity of any food web. As an example, the food web in the grassland ecosystem is given below:

a)   grass-grasshopper-hawk.

b)   grass-grasshopper-lizard-hawk.

c)   grass-hare-hawk.

d)   grass-rat-hawk.

e)   grass-rat-snake-hawk.

The diagrammatic representation of food web in a grassland ecosystem is given in Fig. 2.3.

The food web explains how all chains are interlinked with each other at different points. So the basic law of energy flow is that the solar energy is captured by green plants, then converted into chemical energy and is stored as food materials. Later the food energy is passed from one level to the next through food chains and food webs of an ecosystem. Therefore, the destruction of any link in the food chains and food webs may lead to disaster to the whole ecosystem.

Fig. 2.3: The food web in a grassland ecosystem.

In the light of the above discussion, we can precisely explain the general characteristic of an ecosystem. Firstly, it is a biological system which supports life. Secondly, it maintains stability through dynamic equilibrium. The ecosystems function through a constant flow of energy and cycling of minerals. Finally, it has a functional unit with a trophic structure and biotic diversity.

Ecological niche:

The term "niche" was first described by Joseph Grinnel in the year 1917. The ecological niche is described by him as "The ultimate distributional unit, within which each species is held by its structural and instinctive limitations...., no two species in the same general territory can occupy for long identically the same ecological niche". Therefore, an ecological niche of an organism deals with not only the place where it lives but also what it does i.e. its functional role in the community. Many species may be present in one habitat at any point in time, but their niches vary. Carnivores like tiger, wild dog and leopard live in a particular habitat but their niches are different.

Ecological equivalents:

There may be similar ecological niches in different geographical regions. When organisms exist in similar ecological niches of different geographical regions they are called ecological equivalents. Species found in same niches of different regions usually differ from each other taxonomically. The pronghorn antelope of North America is the ecological equivalent to the kangaroo of Australia. The African

15

lion and the North American mountain lion (puma) are said to be the ecological equivalents.

Habitat:

The "habitat" may be defined as a place or type of site where an organism or population naturally occurs. In other words, it is the type of environment in which an organism exists and to which it is therefore adapted. However, according to the Wildlife (Protection) Amendment Act 2002, the habitat includes land, water, or vegetation which is the natural home of any wild animal. Plants and animals are the living elements of a habitat, whereas, terrain, slope, water etc. are the non-living elements. Nevertheless, the habitat plays an important role for wild animals. As we know that every species has its own physical and biological needs that must be satisfied in habitat for survival. Therefore, a habitat fulfills all the requirements (e.g. food, water, cover etc.) of animals and their populations in the habitat are regulated by the availability of these essential factors. In other words, we can say that all their needs are met in the habitat. Another important fact is that the more mobile species can use more than one type of habitat in order to meet their needs. Each species or community has a unique role in its habitat. In this regard, it is to be mentioned that both habitat and niche conceptually differ from each other. A habitat describes geography and a set of conditions or resources that are used by the organisms. Niche, on the other hand, accounts more as an expression of function i.e. interaction of the species and its environment.

There are four main types of habitat in the biosphere. Each habitat differs from other by its characteristic features. A summary of each habitat is described below:

- **Terrestrial habitat**: The terrestrial habitat is concerned with the study of organisms that grow on land. Temperature, light, moisture, soil etc. are the most important factors in the terrestrial habitat. The different ecosystems found in the terrestrial habitat are the results of an interaction between living and non-living communities. Trees, grasses and shrubs are the producers of this habitat. Insects, mammals and birds are called the macro consumers while protozoa, fungi etc are the micro consumers of the terrestrial habitat.

- **Marine habitat**: It is the largest among all the habitats found in the biosphere. Marine habitat covers about 70 per cent of the earth's surface. The communities found in this habitat are chiefly the pelagic (live in the open sea) and the benthic (bottom dwellers) types.

- **Freshwater habitat**: There are generally two types of freshwater habitats-lentic and lotic. Pond, swamp, lake, bog etc. are the examples of lentic or standing water type, whereas spring, stream, river etc are the lotic or running water type. Bacteria, fungi, algae, aquatic insects, fish etc. are the common flora and fauna of freshwater habitat. Freshwater habitats are small as compared to terrestrial and marine habitats.

16

- **Estuarine habitat**: It is a semi-enclosed coastal body of water that freely connects with the open sea. Coastal bays, tidal marshes, river mouths and water bodies behind barrier beaches are some important estuarine habitats. The nature of an estuarine habitat is highly affected by tidal action. The salinity is recorded maximum during high tides. Communities found in this habitat include endemic species and those come in from the sea.

Habitat components:

A wildlife habitat is influenced by both physical and biological factors. As mentioned earlier that the physical factors in a habitat are temperature, water, light, soil, day length, air current and atmospheric pressure. The biological factors, on the other hand, are composed of food, community etc. All these components play an important role in regulating the distribution of wild animals in a particular habitat. When these factors are present in limited amounts in the habitats, it will cause constriction of population growth or distribution. Therefore, wildlife population in a habitat can be limited by various factors. Some salient components such as water, food, cover and space are important for animals. The details of these components are given below:

- **Water:** It is an essential factor for survival of animals. Nothing can live without water or proper kind of water. Therefore, water should be available at every point in the protected area where wild animals exist. The water requirement of desert animals is normally met by succulent plants or metabolic water. The regular supply of water is an important part of the habitat management programme.

- **Food:** Food is required for energy and is a pre-requisite for sustenance of any organism. Plants obtain energy directly from sun while animals get their energy by eating plant or animal matters. Wild animals consume a wide variety of food materials. Herbivores, for example, feed on selected plants and their preferences towards selection of particular plants are related to the palatability. Many wild animals thrive on restricted diet. The availability of food in the habitat is generally matched with the seasonal changes. Therefore, movement of wild animals from one place to other depends on the availability of food in the habitat where they live. Wild sheep and goat of higher mountain regions, for example, climb down into the plains during extreme winter months in order to obtain food. Seasonal movement of wild elephants for food and water in India is very well known.

- **Cover:** It is described as the kind of materials that make up vegetative or other shelter for wildlife. The main function of a cover is to provide shelter to various animals. Cover also protects animals from inclement weather as well as from enemies. A cover may be 'vegetal' or 'non-vegetal' in nature or may be 'natural' or 'artificial'. Culvert, for example, is non- vegetal type and it may serve as a cover for leopard during summer months. A broad classification of cover includes ambush cover (a predatory animal uses for abusing its prey), loafing cover (spread for aimless movement at some specific points), breeding cover (for building nest, parental care etc.) and roosting cover (a place for resting). For proper management of wildlife enough cover is vital.

- **Space**: The space is required for various day-to-day activities. The space required for wild animals depends on the size of the population and other essential factors. Inadequate space may cause certain behavioural changes in wild animals. Therefore, adequate space for proper growth and normal behaviour of wild animals is important.

The habitats of some selected wild fauna are given below:

Animal	Habitat
Lion	Open grassland and woodland.
Tiger	Tropical rain forest, grassland, forest and mangrove swamp.
Cheetah	Open plain, grassland and open woodland.
Leopard	Cold mountains, rain forest and arid savannah.
Hyena	Open forest, open plain and brush land.
Rhinoceros	Open grassland interspersed with swamp, lake and marsh.
Blue bull	Hilly grassland and lightly wooded region.
Brow-antlered deer	Swampy land.
Sambar	Swampy land, coastal forest, mountain and agricultural field.
Swamp deer	Grassland, river bank, near swamp and forest.
Bear	From arctic coast to tropical forest.
Giant panda	Cool and dense bamboo forest.
Red panda	High altitude bamboo forest.
Pangolin	From tropical rain forest to arid and wooded savannah.
Gaur	Hilly forest.
Chimpanzee	Grassland, woodland and rain forest.
Elephant	Forest and savannah grassland.
Jungle fowl	Dry scrub, low altitude forest etc.
Pea fowl	Open dry forest.
Stork	Shallow water, marsh and dry land.
Crane	Marsh land, wet plain, prairie etc.
Ostrich	Open and arid country.

Home range:

It is defined as an area where normal daily activities of an animal or population take place. Animals usually restrict their activities to a particular area. All the physical and biological needs of an individual or groups are met within the home range. The home range of an animal may vary according to its food habit, size, age and sex. The other factors such as season and intraspecific competition are also important. Elephant bulls and family herds normally restrict their daily movements to a few kilometers to 10-20 km. However, their movements may be more than 1000 sq km. The home range of a tiger is normally 40-50 sq km. Animal maintains its home range for shelter, food, water and reproduction. Home range is marked by defaecation, urine and other various markings. Some animals may overlap the home ranges of other animals. The size of a home range also varies according to the type of animals found in a particular habitat. Carnivores usually have larger home ranges than those of herbivorous of same body size. Even males have larger home ranges than females of same species. Home range is not defended by the animal.

Territory:

A particular area with in the home range that is partly or wholly defended by the animal against intruder is termed as territory. A territory, however, may not always be defended by the animal. Moreover, all animals do not have territories. Territorialism may also be seasonal. Sometimes territories are made during breeding periods. Territory of a particular animal is associated with food, reproduction, maternal behaviour, care of young etc. Territories of animals are maintained by sounds, songs, urination, defaecation etc. Blue bull, for example, marks its territory by defecating at specific spot. Tigers are highly territorial animals. A few species usually maintain very large territories.

Core area:

A certain area within the home range that is more frequently used by the animal than other part is known as core area. For example, each tiger reserve has a core area and all human activities are banned in that area.

Critical area:

It is defined as a part of home range where limiting habitat resources are found. This is also called key area. The management of critical area within the home range of a wildlife population is an important part for wildlife managers. During summer months, some animals depend on resources that are located near water bodies.

Wildlife corridor:

It is defined as the habitat that facilitates animals for free movement between ecological isolates. Wildlife corridors support long term viable population of wild animals. It increases the rate of local migration of animals and provides additional feeding and breeding grounds. In this regard, the 'Project Elephant' has given importance on corridors linking different parts of the composite elephant ranges.

Dispersal:

It is defined as the outward movement of an animal from one home range to another home range for permanent residence. It is also called emigration. Wild animals seek a new home range in search of food and other purposes like avoiding predators and social intolerance. The juveniles of fox and beaver, for example, leave their parental home ranges due to social intolerance.

Migration:

Migration is a two way movement in which the animal returns again to the area from where it had moved. Migration usually takes place during unfavourable conditions in the original home range where animals are inhabited. Therefore, they move to other home ranges where conditions for living are favourable. It is usually a periodical or seasonal phenomenon. Migration is very common to many birds, mammals and reptiles. Factors like weather, photoperiod, availability of food etc. play crucial roles in migration. Elephants from the Dalma wildlife sanctuary (Jharkhand state), for example, migrate to the South West Bengal forests annually in order to search food and other reasons. Some young birds migrate to other areas after hatching and then return to their natal places for breeding.

Introduction:

It is defined as the release of animals into a habitat in which they have never occurred naturally. So, it is a process of establishing a population of animals in an area that is remotely located from their original wild habitats. Introduction is normally practised in both wild-caught and captive born animals. It is done for a variety of reasons such as economic development. The introduction may be accidental or deliberate. The British, for example, introduced European red foxes (*Vulpes vulpes*) to Australia and North America for sport hunting. Introductions normally disrupt natural populations. For instance, the introduction of grey squirrels (*Sciurus carolinensis*) into England has resulted in a decline in the native red squirrel (*Sciurus vulgaris*) population due to interspecific competition.

Reintroduction:

The term "reintroduction" may be defined as the release of animals into an area in which they have either declined or disappeared due to natural causes, human pressures or some other factors. In other words, it is the process of re-establishing a population of animals within the area of its original wild habitats. For example, Rhinoceros were caught from the Kaziranga national park (Assam) and the Royal Chittwan national park (Nepal) and subsequently they were reintroduced in Dudhwa national park (Uttar Pradesh). Reintroduction involves a precisely defined population being returned to a precisely defined location. The important point regarding reintroduction is that animals to be reintroduced, wherever possible, should be originated from as near as possible to the selected habitat. Animals to be introduced may be either wild-caught or captive born. When animals are ready for reintroduction, they are required to be taken to the site of release, acclimatized and trained. There should be a follow-up period for scientific observation. The planning of reintroduction includes various stages which are as follows.

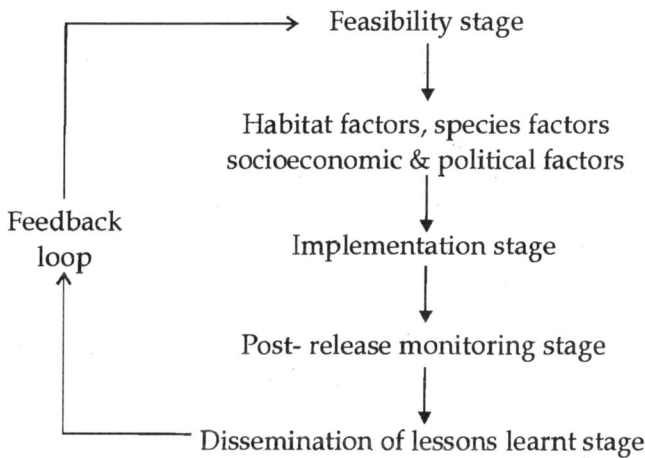

The main objectives of reintroduction of wild animals are (i) to enhance the long term survival of the species, (ii) to provide long term benefits to local people, and (iii) to maintain biodiversity.

Translocation:

Translocation can be defined as the capture and transfer of feral or wild animals from one part of their natural ranges to another with minimum time spent in captivity. It encompasses introduction, reintroduction and restocking. Translocation is done to reduce overpopulation in existing protected areas or to remove animals from areas of doomed habitats to more secured sites. It is one of the important techniques for conservation of rare and endangered species. However, each translocation project should take into account of all biological, ecological, geographical and epidemiological factors. In India, the translocation of lions from Gir forest to Kuno sanctuary in Madhya Pradesh is under process.

Carrying capacity:

Carrying capacity is the property of a habitat. It is defined as the number of animals of a specified quality that a habitat can support while sustaining a specified level of habitat resources. In other words, we can say that carrying capacity denotes the maximum number of animals the environment can support at a given time in a given area. It is broadly classified into two types: ecological carrying capacity and economic carrying capacity. The ecological carrying capacity is determined only by limited habitat resources. The economic carrying capacity, on the other hand, describes the maximum number of animals which can be economically sustained in a habitat. There are many factors such as availability of forages, quality and quantity of food etc. affecting carrying capacity. Populations below the level of carrying capacity normally have access to greater amounts of food and other welfare factors.

Pinch period:

The pinch period is defined as a period when the availability of food or water or both are minimum in their quality, quantity and distribution. For example, the water and food become scare during summer months. The animals remain stressful condition during this period.

Flagship species:

Flagship species are those species which are having large home ranges and diverse habitats. Examples are tiger, elephant, giraffe, rhinoceros etc.

Indicator species:

It indicates the welfare of other species or it represents a specific habitat condition called indicator species. The presence of para grass (*Brachiara mutica*) in a particular area indicates the marshy habitat.

Keystone species:

There are animals and plants which by virtue of their presence or absence alter the structure and community. The example is *Ficus* species.

Hot spot:

It is a specific small area within a landscape exhibiting significantly higher levels of plant and animal diversity or unique composition of plants and animals. In other words, the hot spots are heavily threatened areas of great ecological diversity providing shelter to large numbers of endangered species, many of which are endemic. The hot spot is also known as environmental emergency room. There are 34 hotspots all over the world covering 2.3 per cent of the earth's surface and are home to 75 per cent of the world's most threatened species.

IUCN Red List Categories:

Species that are on the verge of extinction are known as endangered species. There are not any fixed rules for identification of endangered species. Each country develops its own parameters to identify species that face the threat of extinction. However, the IUCN (International Union for Conservation of Nature and Natural Resources) and WCMC (World Conservation Monitoring Centre) do maintain a global list of endangered and vulnerable animal species called Red List. The aim of the Red List is to provide an objective framework for the classification of species according to their extinction risk. It assesses the status of, and threats to, animal species worldwide. The Red List places species in several categories which are as follows:

(a) **Extinct (EX):** A taxon is thought to be extinct when there is no reasonable doubt that its last individual has died, after exhaustive surveys in known or expected habitat, at appropriate times (diurnal, seasonal, annual) throughout its historic range have failed to record an individual. Surveys should be over a time frame appropriate to the taxon's life cycle and life form. In India, the cheetah, the pink-headed duck and the Himalayan quail are extinct.

(b) **Extinct in the Wild (EW):** A taxon is extinct in the wild when it is known only to survive in cultivation, in captivity, or as a naturalized population well outside the past range. A taxon is presumed extinct in the wild when exhaustive survey in known or expected habitat, at appropriate times (diurnal, seasonal, annual) throughout its historic range have failed to record an individual. Surveys should be over a time frame, appropriate to the taxon's life cycle and life form.

(c) **Critically Endangered (CR):** A taxon is critically endangered when it is facing an extremely high risk of extinction in the wild in the immediate future. A species can be determined as critically endangered based on one of several criteria. For instance, if the population has decreased by over 80 per cent in the last 10 years or 3 generations, whichever is longer, irrespective of whether the cause for the drastic reduction in numbers has been identified and rectified or not. The other criteria include if the extent of occurrence of a species is less than 100 sq km or if the area of occurrence is less than 10 sq km or if there are less than 50 mature individuals of species left. The pink pigeon *(Columba mayeri)* in Mauritius is listed as critically endangered species.

(d) **Endangered (EN):** A taxon is endangered when it is not critically endangered but is facing a very high risk of extinction in the wild in the near future. A species can be determined as endangered species which have the criteria such as either a 50 per cent reduction in population in the last 10 years or extent of occurrence less than 5000 sq km or an area of occurrence is less than 500 sq km or less than 250 mature individuals left. The hawksbill turtle *(Eretmochelys imbricate)*, tiger, elephant and snow leopard, for example, are listed as endangered species.

(e) **Vulnerable (VU):** A taxon is vulnerable when it is not critically endangered or endangered but is facing a high risk of extinction in the wild in the medium term future. A species can be determined as vulnerable species such as a 50 per cent reduction in the last 20 years, extent of occurrence less than 20,000 sq km or an area of occurrence less than 2000 sq km, or less than 1000 mature individuals. The frog *Dyscophus antongilii* is listed as vulnerable species. Dhole and Nilgiri leaf monkey are also listed in this category.

(f) **Low Risk (LR):** A taxon is in low risk when it has been evaluated and does not qualify for any of the categories critically endangered, endangered and vulnerable. Taxa in this category are further classified into three sub-categories. They include conservation dependent, near threatened and least concern.

- **Conservation Dependent (CD):** Taxa that do not currently qualify as critically endangered, endangered or vulnerable may be classified as conservation dependent. To be considered conservation dependent, a taxon must be the focus of a continuing taxon-specific or habitat-specific conservation programme which directly affects the taxon in question. The cessation of this conservation programmes would result in the taxon qualifying for one of the threatened categories above.

- **Near Threatened (NT):** A taxon is near threatened when it has been evaluated against the criteria but does not qualify for critically endangered, endangered or vulnerable now, but is close to qualifying for or is likely to qualify for a threatened category in the near future.

- **Least Concern (LC):** A taxon is least concern when it has been evaluated against the criteria and does not qualify for critically endangered, endangered, vulnerable or near threatened. Widespread and abundant taxa are included in this category.

(g) **Data Deficient (DD):** A taxon is data deficient when there is inadequate information to make a direct or indirect assessment of its risk of extinction based on its distribution or population status. Bengal fox is one of the taxa in this category.

(h) **Not Evaluated (NE):** A taxon is not evaluated when it is has not yet been evaluated against the criteria.

Critically endangered species are included in the endangered list and vulnerable list whereas endangered species are included in the vulnerable list. These three categories are together called as 'threatened'. There are 8,322 species which have been identified as threatened throughout the globe. Of these, 4,328 species are classified as endangered and 2,853 are critically endangered. It is reported that as many as 60 species have been identified as extinct in the wild. In India, 272 species have been identified as threatened. Of these, 206 are endangered and 84 have been classified as critically endangered.

Wildlife Census:

In wildlife, the word 'census' is used for estimation of abundance of animal populations. The adjectives absent, rare, occasional, common, abundant denote measures of abundance of a species or population.

A census technique is an important aspect of wildlife management. Through census techniques, we get information like population estimates per area, relative abundance, population trend etc. The followings are the objectives of wildlife census.

- To determine how well the introduced stock is doing.
- To compare population densities in core areas and buffer areas.
- To compare population densities in different areas before and after management intervention.
- To make a future strategy for further improvement of wildlife.

Wildlife census techniques are based on direct observations or indirect evidences. A direct census technique is done through observation on actual individual, whereas, an indirect census technique is done on the basis of indirect evidences. The water whole census technique or animal counts on transects is the example of direct census technique, whereas, dung counts and pug mark are the examples of indirect census techniques. However, feeding signs, calls, nesting etc are also the indirect evidences of animals present in an area. The direct technique is important to those species in the habitat where their frequency are relatively high densities and animals visible when searched for. Rhinoceros, elephants or medium to large sized ungulates, for example, can be counted by the direct method. Species which occurs in low densities or difficult to observe due to poor habitat visibility or cryptic behaviour, the census should be done by indirect methods. Nocturnal mammals and most carnivores, for example, the census can be done by this technique. In this regard, it is to be mentioned that most indirect methods are suitable for obtaining relative indices of population size.

Pug mark: The pug mark or foot print is a census technique used for identifying and counting tigers and other carnivores like leopard. This was first developed by S.R. Chaudhary for counting tigers in Orissa in 1969. This technique of census is based on recording several distinct morphological characteristics of tiger's foot prints. As we know that the tigers generally walk along a well defined network of forest roads, paths, stream banks etc. During normal gait the front feet of a tiger are completely or partially superimposed by the hind feet. However, during faster movement, the hind foot overshoots the front one and hence, it is possible to distinguish both the hind and front feet. As the hind foot mark is invariable smaller than the front one (Fig.2.5), the inner outline is the outline of the hind foot in superimposed foot print and it remains intact in both the cases.

Therefore, the prints of hind feet are obtained for census purpose. Moreover, pug marks are also important for distinguishing the sex of the animal. The male pug mark is nearly square in outline while it is rectangular shape in case of female pug mark (Fig.2.4). Another important fact is that the shape of toe pad is also the indicator for identifying of sex. The female toe pad is more pointed in the form while the male leaves round impression.

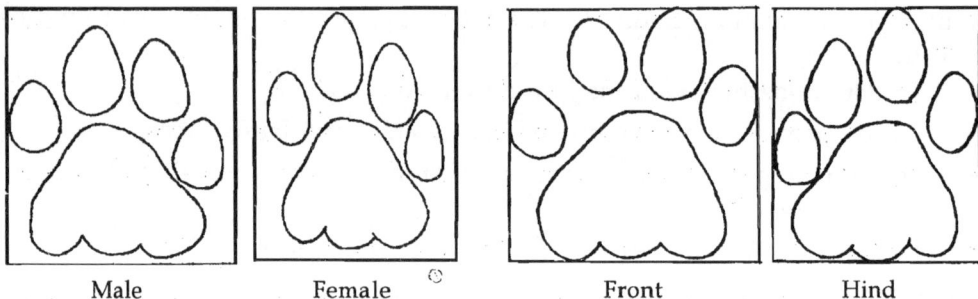

Male	Female	Front	Hind

Fig. 2.4: The male pugmark is square shaped while it is rectangle in female

Fig. 2.5: Hind pug is smaller than the front.

The best prints are obtained on hard surface. A film of dry or moist fine soil of about 5 mm thick can be overlaid on open surface in order to get foot prints. Various equipments like tracing paper, field note book, ball pen, forest map, tape, plaster of paris etc are necessary for taking foot prints.

Ecotourism :

The concept of ecotourism has gained importance only recently. The ecotourism is defined by the World Tourism Organization (WTO) as, "Tourism that involves traveling to relatively undisturbed natural areas with the specified object of studying, admiring and enjoying the scenery and its wild plants and animals, as well as any existing cultural aspects (both of the past or the present) found in these areas". The key elements of ecotourism are (a) a natural environment as the prime attraction, (b) an optimum number of environment friendly visitors, (c) activities which do not have any serious impact on the ecosystem, and (d) the positive involvement of the local community in maintaining the ecological balance. Through ecotourism the travellers come to a better understanding of unique natural and cultural environments around the globe.

Ecotourism can be of different forms such as losing oneself in a beautiful natural forest, watching animals, birds and trees in a forest, corals and marine life in sea, engaging in boating or wandering amongst sand dunes. However, the followings are the principles of ecotourism:

- Minimize the negative impacts on nature and culture that can damage a destination.

- Educate the traveller on the importance of conservation.

- Stress the importance of responsible business which works cooperatively with local authorities and people to meet local needs and deliver conservation benefits.

- Direct revenues to the conservation and management of natural and protected areas.

- Seek to ensure that tourism development does not exceed the social and environmental limits of acceptable change as determined by researchers in cooperation with local residents.

- Minimizing the use of fossil fuels, conserving local plants and wildlife and blending with the natural and cultural environment.

- Maximize economic benefit for the local business and communities particularly peoples living in and adjacent to natural and protected areas.

Chemical Restraint:

Chemical restraint is defined as any method that mainly uses a chemical agent or drug from restraining animal freedom i.e. walking, running and aggressiveness. The state of restraint can vary from immobilization (arresting movement), tranquilization (calming) to anesthetization (complete loss of consciousness). Chemical restraint is used for various purposes such as capture of problematic animal in distress, veterinary care of captive and wild animals, experiment of drug and for the purpose of translocation. There are many advantages of chemical restraint. It can be used in various situations. Chemical restraint enables selection of time of capture. Moreover, the equipments used for this purpose are easy to carry and shift from one place to another in the field. However, there are many disadvantages also. Sometimes the undesirable side effects of drugs are observed because of unknown physiological status of animal. The animal may be injured due to delivery of dart on the wrong side. There may be equipment failure. However, the followings are some basic considerations for chemical restraint.

- The aggressive, potentially dangerous to restrainer or health status of the animal should be known.

- Type of terrain is important. The terrains like dense cover, steep slopes and large water bodies nearby should be avoided.

- The high or low environmental temperatures are not suitable for delivering dart.

- The delivery of the drugs should be done on the correct sides of the animals (Fig.2.6).

27

- The emaciated, sick or old animals should be avoided for chemical restraint unless it is urgently required.
- The safety drugs should always be kept to meet emergency situation.
- The other factors like sufficient day light, keeping a human antidote (narcan) against morphine drugs etc are important.

Deer   Elephant   Small cat

Lion   Rhinoceros

Fig. 2.6: Darting sites for different wild animals.

Chapter 3

Zoo-geographical Regions and Wildlife Distribution in India

Introduction :

Zoo-geographical regions are also called as faunal regions. Each zoo-geographical region differs from the other from its fauna and flora. Not all animals are found in every environment. A few animals occur in their identical ranges. Few animals inhabit whole of any one geographical region. Some animals extend their ranges from one geographical region to another. This is due to the fact that presence or absence of an animal in an area depends on ecological and historical factors. In a particular environmental condition temperature, light, humidity, oxygen and carbondioxide have a direct role in controlling the distribution of animals. Indirectly, vegetation growth of a particular zoo-geographical region also plays an important role in maintaining the status of fauna. Therefore, the presence or absence of fauna in a particular zoo-geographical region is the result of the combining effect of geography, climate and evolutionary process. Two same faunas found in two different regions, however, are not absolutely different from one another.

The zoo-geography is the study of past and present animal distributions, including the evolution, spread, recession and extinction of species. It deals with both ecological and historical aspects. Both are intimately related and each helps to elucidate the other. The land masses of the world were first divided by a British ornithologist, Philip Lutley Sclater into six zoo-geographical regions in 1858. These include Neotropical region, Nearctic region, Palaearctic region, Ethiopian region, Oriental region and Australian region. The Palaearctic and Nearctic regions are sometimes considered as a single zoogeographical region called Holarctic region. The Antarctica now is treated as a separate zoogeographical region. Scientists believe that most animal groups had originated in the warm central landmasses of Palaearctic, Nearctic, Ethiopian and Oriental regions and from these warm landmasses they spread towards the peripheral regions of Southern Africa, Australia and Americas. The following text describes the salient features of different zoo-geographical regions:

- **Neotropical region:** The Neotropical region covers the whole of South and Central Americas, West Indies and Galapagos islands. This region is classified into Chilian, Mexican, Brazilian and Artillian sub-regions. Most parts of this region are covered by tropical forests. However, a small portion of its land mass covered by desert. The Amazon valley is reputed all over the world for its swamp and evergreen tropical rain forests. The pampas is also found in this region. The region is enriched with great faunal resources. They include New World monkeys (e.g. tamarin, marmoset etc.), anteaters, tree sloths, armadillos, capybaras and guinea pigs. Llamas and vampire bats are the unique mammals of the Neotropical region. This region has also very rich avifauna resources. Some unique birds inhabiting in this region include sun bittern, oilbird, rhea and trumpeter.

- **Nearctic region:** The Nearctic region comprises Greenland and North America up to the centre of Mexico. In the west, there are many large lakes and island seas. The northern fringe of this region is covered by tundra. The southern part is having coniferous and deciduous forests. This region has also other natural vegetation zones such as mixed forests, short grass savannah, tall grass etc. The Nearctic region is enriched by a large number of unique faunas. Pronghorn antelope, gila monster (a poisonous lizard) and a primitive type of frog *(Ascaphus* sp) are solely found in this region. The other notable faunas found in this region include lynx, timber wolf, bear, flying squirrel, mule deer, raccoon, coyote, wild turkey, Canada goose and American alligator.

- **Palaearctic region:** The Palaearctic region covers Europe, most of Asia and a small northern part of Africa. It is divided into European, Mediterranean, Siberian and Manchurian sub regions. Some salient features observed in this region are common with the Nearctic region. The main habitats of this region are the tundra taiga, mixed and deciduous forests, savannah and desert. Many predator mammals and birds such as wild cat, lammergeier vulture and golden eagle are found in rocky cliffs of Europe. The brown bear, elk, musk deer, panda, red squirrel and Siberian weasel inhabit in the Asian part of this region. The large-eared pika, alpine marmot, Siberian ibex, snow leopard and wild yak occur in the highlands of Central Asia. Many birds of Palaearctic region are summer migrants.

- **Ethiopian region:** This region covers the entire Africa except a narrow coastal stripe to the north west which falls in the Palaearctic region. The Ethiopian region also covers the very important faunal region of Madagascar (now Malagasy Republic). Much of the areas of this zoo-geographic region are covered by grassland or thorn-scrub. The equatorial region of West and Central Africa, however, is covered with rain forest. The Ethiopian region is blessed with some typical faunas such as giraffes, gorillas, chimpanzees and zebras. African predators like cheetah, leopard, lion, hyena, civet and bat-eared fox are found in this region. Africa is home to many antelopes. Some well known antelopes such as greater kudu, oryx, gerenuk and dikdik occur in northern and southern bushlands. All great apes except orangutan are exclusively found in this region. Other African primates found in this region include bush baby, potto, mandrill, baboon and colubus monkey. Most African primates live in the forests and savannah regions. Ungulates like okapi, bongo, bushbuck and bush pig occur in the rain forests of Ethiopian region. Other notable faunas such as black rhinoceros, elephants, pangolins, chevrotains and many snakes are also found in this region. The distinctive mammal, the lemur is well known in Madagascar.

- **Oriental region:** The Oriental region is represented by India, southeast Asia and the islands west of Wallace's line (Wallace's line passes between the islands of Bali and Lombok). Dr. Salim Ali, the noted ornithologist of India, however, classified this region into three sub-regional divisions: the Indo-Chinese, Indo-Malayan and Indian sub-regions. Both the Indo-Chinese and Indo-Malayan regions are predominantly covered by humid tropical and sub-tropical forests. The Indo-Chinese sub-region that inclines to the Palaearctic region is having the faunas like flying lemurs, tapirs and moles. Indo-Malayan sub region has unique faunas like proboscis monkey, tree shrew, orangutan, gibbon etc. The Indian sub region roughly covers the same area as the geographical area of Indian sub-continent. Faunas, in general, found in this sub region include leopard, clouded leopard, elephant, tiger, lion, bear, deer, snow leopard, langur, gibbon, chevrotain etc. Many of these faunas are, however, originally extended from the Ethiopian region. The Oriental region boasts of having three species of rhinoceros out of five recognized species of the world, twenty one out of the 36 wild cat species, 18 of the 41 deer species and five of the 8 bear species. This region as a whole has the strong representation of the pheasant family (jungle fowl, peacock, argus pheasant etc.). The Oriental region is the home of many migratory birds. Many reptiles such as Chinese alligator and gharial occur in the Oriental region.

- **Australian region:** The Australian region covers Australia, New Zealand and nearby islands. This region is separated from the Oriental region by an imaginary line called Wallace's line which is supposed to run between the islands of Bali and Lombok. The climate in the Australian region is temperate and tropical types. Some parts of the continental Australia are covered with grassland and evergreen eucalyptus forests. This region has less fresh water bodies as compared to other land masses. Tropical rain forests are found in New Guinea. The natural vegetations such as alpine, scrub, moorland and tussock grasslands occur in this region. The Australian region is blessed with kangaroos, koalas, wombats, wallabies, dingos (wild dogs), platypuses, spiny anteaters and many Australian bats. The notable avifaunas found in this region include emu, cassowary, megapode, frogmouth, lyrebird and honey eater. New Guinea is reputed for being a home to birds of paradise. Kiwi, takahe (flightless rail), owl parrot, tuatara and tailed frog *(Leiopelma* sp) are the unique faunas found in New Zealand. Marsupials are the key animals in Australian region.

However, the world's most desolate area, the Antarctica has now been classified into another zoo-geographical region. Life in this region is mainly restricted by the lack of warmth and moisture. More than two-thirds of its surface is covered with ice. The lowest temperature in the world was recorded in the Antarctic region. Lichens, moss and only higher plants *(Colobanthus quitensis* and *Deschampsia antarctica)* are found here. Faunas occur in this region include seals, whales, penguins, gulls etc.

Wildlife distribution in India:

India has a rich heritage of flora and fauna. They are found in different habitats. Each habitat is characterized by having distinctive faunal resources. On the basis of present physiography and climate, the country is divided into three broad ecological sub divisions- the Himalayan mountain system, the Peninsular Indian sub-region and the tropical evergreen forests or Indo-Malayan sub-region.

- **The Himalayan mountain system:** The Himalayan mountain system has three distinct sub-zones. These include the Himalayan foothills, the high altitude regions of the Western Himalayas and the Eastern Himalayan sub-region. The Himalayan foothills that cover forest slopes of the Himalayas range from the eastern frontiers of Kashmir to Bhutan. It comprises the terai, bhabar and siwalik regions of the Western, Central and Eastern Himalayas upto an elevation of about 2000 m. The terai region is characterized by tall grassy meadows and savannah vegetation. The typical faunas occur in this region include elephant, leopard, tiger, wild dog, hyena, jackal, wild boar, sambar, barking deer, hog deer, swamp deer, cheetal, Himalayan black and sloth bears and porcupine. Rhinoceros, gaur and wild buffalo occur in the foothills of eastern Nepal, Bhutan and Arunachal Pradesh. The Himalayan foothills are also represented by pygmy hog, golden langur, hispid hare and gangetic gharial. The thamin (brow-antlered deer) is now only confined to Manipur.

The high altitude region of Western Himalayas ranges from Kashmir and Laddakh to Kumaon. Oaks, birches, mangolins, laurels are found at an altitude of 1500 to 1800 m. The area with an elevation of about 2700 to 3700 m is covered by fir, pine, rhododendron, birch and dwarf bamboo with alpine pasture. The typical mammals found in this region include wild goat, Tibetan wild ass, wild sheep, goat antelope, musk deer, hangul (Kashmir stag), Tibetan gazelle, yak, mouse hare, marmot, weasel, woolly flying squirrel, snow leopard, black and brown bears and Pallas's cat. The Himalayan golden eagles and snow partridges are well documented avifauna found in this region.

The Eastern Himalayan sub-zone is covered with semi-evergreen and evergreen forests. The coniferous forests are found at higher elevations. This area receives much higher rainfall. Snowfall is less common. Red pandas, hog badgers and crestless porcupines are the unique faunas of this region.

- **The Peninsular Indian sub-region:** This region represents the 'true home' of our faunal resources. The Peninsular Indian sub-region is broadly divided into two zones: a) the tropical deciduous woodland peninsular area and its extension into the drainage basin of the Ganges, and b) the desert region of Rajasthan. The former zone is characterized in having the tropical moist and dry deciduous forests and scrub vegetation, while the latter zone has a distinct variety of vegetations such as khair, kheira and rohira. The Indian or Thar desert is lying to the east of Indus valley. The woodland Peninsular India and the drainage basin of the Ganges have some common animals such as tiger, leopard, elephant, wild boar, nilgai, cheetal, sambar, four-horned antelope, blackbuck, wild dog, lion, gaur, chinkara and barking deer. The Indian desert is represented by some typical faunas like Asiatic wild ass and the great Indian bustard. The chinkara, blackbuck and four-horned antelope are very common. During winter months desert also attracts some migratory birds e.g. flamingos, pelicans, teals, cranes etc.

- **Tropical evergreen forests**: The tropical evergreen forest or Indo-Malayan sub-region covers the north east India and the Western Ghats of the south including the Malabar coast. The area receives heavy rainfall. The Western Ghats are characterized by thick rain forest, grassy downs and densely forested gorges of evergreen vegetation known as 'sholas'. The Nilgiri tahr and pine marten are found in the Nilgiri. Elephant and barking deer are found in the tropical rain forest. Hoolock gibbon (only ape found in India), golden langur, Assam macaque, pig-tailed macaque, capped langur and slow loris inhabit in the tropical rain forest of the north-eastern part of India. The lion-tailed macaque, the Nilgiri langur and the slender loris occur in the tropical rain forest of the south.

Apart from the above three well-defined ecological zones, there are three important zones in India that attract many naturalists and wildlife biologists all over the world. They include the Andaman and Nicobar Island, the mangrove swamps of Sunderban and the Keoladeo Ghana National Park, Bharatpur (Rajasthan).

Biogeography of India:

As we know that some animals and plants are exclusively confined to one particular place while others are almost cosmopolitan in distribution. Therefore, it is very important to have a comprehensive biogeographic classification that will represent all the biological regions and communities. The classification is also important for proper understanding of plants and animals. According to the Wildlife Institute of India (WWI), the country is classified into 10 biogeographic zones (Fig. 3.1). The biogeographic zones deals with a specific ecological pattern, biome representation, community and species. Each biogeographic zone has again been classified into different biotic provinces (further level of detail within zone contains some distinctive species elements- e.g.Western and Eastern Himalayas). The biotic province that forms a secondary unit within biogeographic zone helps in studying a particular community separated by dispersal barriers or gradual change in environmental factors. The biotic provinces have been further classified into biogeographic regions (distinctive geographic subdivision- e.g. Garhwal and Kumaon in Western Himalayas) and biomes (major ecosystem groupings found within each province and region- e.g. alpine, sub-alpine and temperate conifer forest within Western Himalayas). The biogeographic classification of India, as documented by WII, is given below:

Biogeographic zones	Biotic provinces	
1. Trans-Himalayas	A)	Laddakh
2. Himalayas	A)	North-West Himalaya
	B)	West Himalaya
	C)	Central Himalaya
	D)	Eastern Himalaya
3. Indian Desert	A)	Kutch Desert
	B)	Thar Desert
4. Semi-arid Zone	A)	Punjab Plain
	B)	Gujarat-Rajwara
5. Western Ghats	A)	Malabar Plain
	B)	Western Ghats
6. Deccan Peninsula	A)	Southern Deccan
	B)	Central Deccan
	C)	Eastern Highlands
	D)	Chota Nagpur
	E)	Central Highlands
7. Gangetic plain	A)	Upper Gangetic plain
	B)	Lower Gangetic plain
8. North-east India	A)	Brahmaputra Valley
	B)	Assam Hills
9. Islands	A)	Andamans
	B)	Nicobars
	C)	Lakshadweep
10. Coasts	A)	West Coast
	B)	East Coast

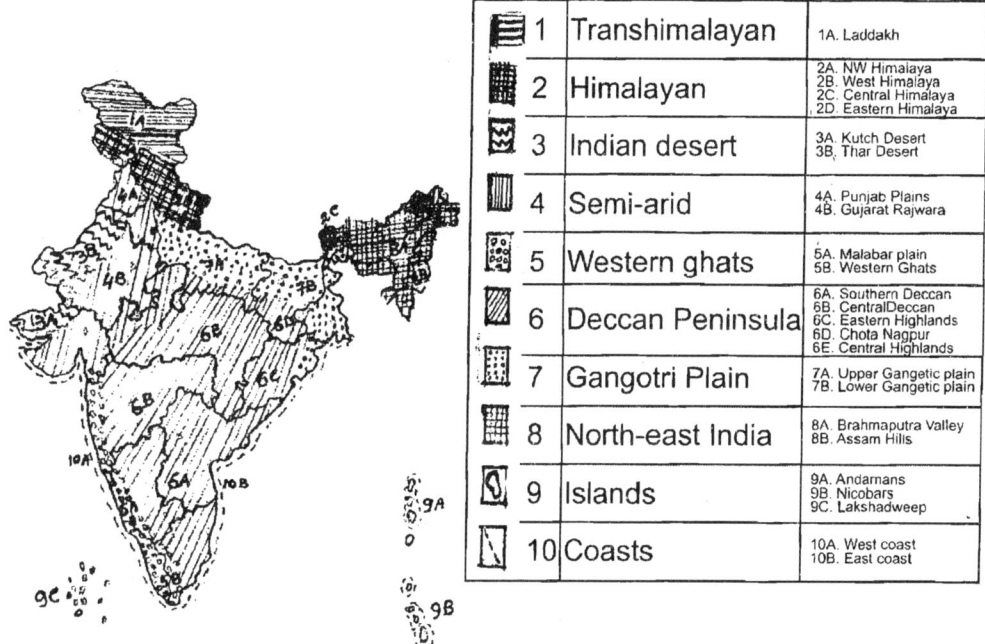

	1	Transhimalayan	1A. Laddakh
	2	Himalayan	2A. NW Himalaya 2B. West Himalaya 2C. Central Himalaya 2D. Eastern Himalaya
	3	Indian desert	3A. Kutch Desert 3B. Thar Desert
	4	Semi-arid	4A. Punjab Plains 4B. Gujarat Rajwara
	5	Western ghats	5A. Malabar plain 5B. Western Ghats
	6	Deccan Peninsula	6A. Southern Deccan 6B. CentralDeccan 6C. Eastern Highlands 6D. Chota Nagpur 6E. Central Highlands
	7	Gangotri Plain	7A. Upper Gangetic plain 7B. Lower Gangetic plain
	8	North-east India	8A. Brahmaputra Valley 8B. Assam Hills
	9	Islands	9A. Andamans 9B. Nicobars 9C. Lakshadweep
	10	Coasts	10A. West coast 10B. East coast

Fig. 3.1: Biogeographic zones and biotic provinces of India as documented by WII, Dehradun.

As mentioned earlier that India has a rich source of wild flora and fauna. Here the estimated population of some faunas in their wild states in India is given below.

Species	Population	Species	Population
Tiger	3660	Asiatic lion	359
Elephant	19,000-24,000	Rhinoceros	1100
Wild buffalo	1000-1500	Gaur	5000
Pygmy hog	100-150	Malabar civet	less than 500
Rhesus monkey	3, 00,000-3, 50,000	Pig-tailed macaque	1500-2000
Stumped-tailed macaque	1000-1500		
Assam macaque	5000-6000	Lion-tailed macaque	3000-4000
Langur	3,00,000	Indian wolf	1000-2000
Fishing cat	Rare	Tibetan ass	About 2000
Thamin deer	106	Urial	400-500
Blackbuck	29,000-38,000	Adjutant stork	About 500 pairs
Black-necked crane	About 300	Great Indian bustard	1200
Sarus crane	8,000-10,000		
Marsh crocodile	3000-5000	Saltwater crocodile	2000
Gharial	2200	Olive Ridley turtle	6,50,000 or less

Chapter 4

Biodiversity and Sustainable Development

Definition of biodiversity:

The term 'biodiversity' or 'biological diversity' is simply described as the variety of life in all its manifestations. However, according to the International Convention on Biological Diversity, it is defined as "Variability among the living organisms from all sources including *inter alia* terrestrial, marine and other aquatic ecosystems and the ecological complexes of which they are the part; this includes diversity within species, between species and of ecosystems." The above definition has been mentioned in the Article 2 of the International Convention on Biological Diversity. The convention was held in the Rio de Janeiro, Brazil in 1992 during "Earth Summit" of the United Nations Conference of Environment and Development (UNCED). There are 42 articles in the Convention covering a wide variety of issues like general measures for conservation and sustainable use, research and training, public education and awareness, exchange of information and impact assessment, and minimizing adverse impacts. The idea of the Convention on Biological Diversity (CBD) was, however, framed at the Third World Congress on National Parks which was held in 1982 in Indonesia. The convention came into force at the end of 1993. It has been ratified by more than 165 countries. India became the member in 1994. As it has been mentioned that the Convention addresses many issues, it however covers three main aspects: (a) conservation of biodiversity, (b) the sustainable utilization of biological resources and (c) the equitable sharing of benefits arising from such utilization. It is also important that the convention covers not only the ecological aspect of biodiversity but also addresses economic and social issues. The CBD is the first treaty that deals with the entire range of life forms seen on the earth. It is a historic commitment in the sense that the convention has been ratified by all the leading developed and developing nations.

As we know that diversity is a basic property of any living system. Therefore, biological diversity can be explained as the property of classes or groups of living entities to be varied. Each class of entity such as gene, cell, species or ecosystem has more than one kind. Diversity expresses itself at every level from molecules to ecosystems. The loss of species diversity and the reduction in the genetic diversity of plants and animals may lead to loss of ecosystem stability and function. Mutation is one of the important biological events that generates diversity. In this context, it is to be noted that any process that creates genetic diversity, however, does not create new genetic information. The existing information is rather divided into discrete entities. Although, species have independent existence, most species are functionally linked with other species thus forming communities. So the key elements of our biological diversity are the ecological diversity (e.g. biomes, ecosystems, bioregions, habitats and niches), genetic diversity (e.g. populations, individuals and genes) and organismal diversity (e.g. kingdom, phyla, family, genera and species).

Earth is home to many animals, plants and other living things. The present day biodiversity is estimated at about 13.5 million species, though exact numbers remain unknown. Out of the total informed species, about 1.75 million species have been scientifically described so far. The present day biodiversity is represented by a few groups of organisms and much of the population accounts insects.

The biodiversity wealth of India comprises 81,250 known species of animal kingdom, over 46,286 species of plants including 15,000 medicinal herbs and 32,000 wild relatives of agricultural crops, and 131 wild relatives of domesticated animals. It is reported that about 5,150 species of plant kingdom and 1,837 species of animal kingdom are endemic species that are found only in small pockets in the country. It is also the fact that India provides 8 per cent biodiversity and ranks 6th amongst 12 mega-biodiversity countries of the globe. The report of the "World Development Indicators 1998" covers the whole gamut of biodiversity such as the amount of naturally protected land area, the number of species of mammals, birds and higher plants etc. (Table 4.1). All countries have a number of threatened species. Another important fact highlighted in that report is that the low income countries have 5.2 per cent naturally protected areas, while the high income countries have 11.9 per cent.

Importance of biodiversity:

The importance of biodiversity in the development of agriculture, animal science, medicine and industry is immense. The values of biodiversity are well known for human lives. It has both direct and indirect use values. Direct use values are concerned with the consumption or production of marketable commodities. These include food, medicine, industrial material, recreational harvesting and ecotourism. We get food in the form of fruits, nuts, vegetables and meat. Out of the 25,00,00 or so species of flowering plants, about 3000 plants have been regarded as a food source and 200 have been domesticated for food purpose. Only 20 kinds of plants, however, supply more than 80 per cent of world's food base. Another important fact is that a significant portion of drugs is derived directly or indirectly from biological resources. It is reported that around 119 chemical substances have

Table 4.1: Biodiversity of selected countries:

| Country | Naturally protected areas | | Mammals | | Birds | | Higher plants | |
	Thousand sq. km., 1994	%of total land area ,1994	Species, 1994	Threatened species, 1994	Species 1994	Threatened species 1994	Species, 1994	Threatened species, 1994
Argentina	43.7	1.6	320	20	976	40	9000	170
Australia	940.8	12.2	252	43	751	51	15000	1597
Bangladesh	1.0	0.7	109	16	684	28	5000	24
Brazil	321.9	3.8	394	45	1635	103	55000	463
China	580.8	6.2	499	94	1186	83	30000	1009
Cuba	11.5	8.1	31	10	342	13	6004	811
Egypt	7.9	0.8	98	7	439	10	2066	84
Germany	91.9	26.3	-	-	-	-	-	-
India	143.4	4.8	316	40	1219	71	15000	1256
Indonesia	185.6	10.2	463	57	1531	104	275000	281
Iran	83	5.1	140	9	502	12	-	1
Japan	27.6	7.3	132	17	583	31	4700	704
Kenya	35	6.2	359	16	1068	22	6000	158
South Korea	6.9	7.0	49	6	372	19	2898	69
Malaysia	14.8	4.5	286	20	736	31	15000	510
Mexico	98.5	5.1	450	24	1026	34	25000	1048
Nigeria	29.7	3.3	274	22	862	8	4614	9
Norway	55.4	18	54	3	453	3	1650	20
Pakistan	37.2	4.8	151	10	671	22	4929	12
Russia	705.4	3.9	-	17	-	35	-	127
South Africa	69.7	5.7	247	25	790	16	23000	953
Sri Lanka	8.0	12.3	88	4	428	11	3000	436
United Kingdom	51.1	21.2	50	1	219	2	1550	28
United States	1302.1	11.4	428	22	768	46	16302	1845
Vietnam	13.3	4.1	213	25	761	45	-	350

Source: The Hindu Survey of the Environment, 1998.

extracted from some 90 species of higher plants. These chemical substances are used in medicine worldwide. Fibres, dyes, resins, gums, rubber, oils, waxes, agricultural chemicals, perfumes etc. are the important industrial materials obtained from both plant and animal resources. Report indicates that more than 3.8 million cubic metres wood are harvested annually all over the globe for the purpose of fuel, timber and pulp. Another important direct use value of biodiversity is the recreational harvesting like fishing, hunting and gardening from harvested plants.

Indirect use values of biodiversity, on the other hand, are related with atmospheric regulation, climatic regulation, hydrological regulation, photosynthesis, pollination, pest control and soil formation. The intrinsic values obtained from our vast variety of plants and animals can not be quantified in terms of money. Creative works like paintings, literature, music etc. are inspired by nature.

As mentioned earlier that out of the total edible plant species, only 20 or so supply more than 80 per cent of the globe's food base. Some pillar crops like maize, barley, rice, wheat, millet, sorghum, soyabean, bean, peanut, banana, coconut, sugarcane, potato and cassava form the major dietary ingredients of human beings. Therefore, there is enough scope to utilize thousands of wild strains for further improvement of many crops. The massive production of tomatoes, for example, has been possible today due to the introduction of genes of a few wild species of tomatoes from Peru and Ecuador. The sugarcane production has been commercially profitable worldwide due to introduction of genes from wild sugarcane species of Indonesia. The productivity of many cereals such as rice, barley, wheat and maize has substantially been improved with the genes obtained from closely related wild species. Moreover, the wild varieties of many crops have certain known qualities. Wild rice in India, for example, has disease resistance genes. Chinese large-leafed spinach containing resistance genes has saved the spinach crop of United States from blight and wilt diseases. All the high yielding varieties for wheat in India are derived from the five lines obtained from CIMMYT (Mexico) or their related crosses. There are many more examples where wild varieties have been introduced in the field of agricultural sciences.

Likewise animals have also been improved by taking the advantage of the biological diversity. Ninety eight per cent of world's livestock production comes from cattle, buffalo, sheep, goat, pig and poultry. Wild animals are being used to improve many traits of domesticated livestock. The American wild buffaloes, for example, have been crossed with cattle for higher meat production. Almost all broiler chickens that we see today are the results of crossing between the white Plymouth Rock and White Cornish.

In the light of the above facts, therefore, it is quite obvious that there is enough scope to utilize wild species of plants as well as animals from the explored and unexplored natural environments. In this way, we can also maintain and preserve the genetic diversity of the earth's valuable living resources. The maintenance of genetic diversity is very important for sustaining life support system. This will ultimately fulfill three main objectives: (i) preservation of genes for future use, (ii) the protection from genetic vulnerability, and (iii) the protection of endangered species.

Threat to biodiversity:

Today our biodiversity is gravely threatened. Report says nearly 228 thousand tones of timber were lost during First World War. About 30 per cent of the natural world has been destroyed by human activities in the last 25 years. According to the World Wide Fund for Nature (WWF-Nature), it is the period of greatest destruction in the planet since the extinction of the dinosaurs. WWF-Nature has unveiled the Living Planet Index highlighting the fact that there has been a loss in abundance of species (not extinction), the deterioration in oceans, lakes and rivers and the loss of forest cover. The index reveals a fall of 10 per cent for forests, about 30 per cent for oceans and a staggering 50 per cent for freshwater ecosystems. The loss of 45-50 per cent of the world's tropical forest is a matter of great concern to the naturalists. According to the Red Data Book of IUCN (1996), 86 mammalian species have been extinct. Many birds, reptiles and amphibians have already been extinct and many more are threatened with extinction.

The first comprehensive world assessment report on biodiversity prepared by UNEP (United Nations Environment Programme) reveals that between five and 20 per cent of some groups of animal and plant species is to be threatened with extinction in the near future. Many factors that have major detrimental impacts on our biological diversity include loss and fragmentation of habitat; exploitation of wild living resources, expansion of agriculture, forestry and aquaculture; pollution of soil, water and atmosphere; global climate change; introductions of non-indigenous species; hunting; deliberate extermination and economic and political policies. There is a growing pressure on forest lands for subsistence needs, industries, agriculture, power and irrigation projects. So, the environmental and ecological stability have been threatened by all these factors. Therefore, the time has come to save our biological diversity from possible threat to human survival.

Measures to protect biodiversity:

There is a widespread concern for the present status of biological diversity among various sections of people. Naturalists are not only concerned for wild flora and fauna but also for the future of human beings. It is a well known fact that land, water, air, plant and animal are the main components for man's life-support system. Increased human population demands for more and more natural resources. We are persistently altering the environment to fulfill our ever increasing desires. The negligence towards protection and conservation of biological diversity ultimately leads to threaten our life-support system. Therefore, it is the paramount importance to save our biological diversity at this juncture and suitable strategies should be evolved so that all are to be benefited not at the cost of others. The followings are some important measures to protect biodiversity.

• The separation of ecosystems, species and genetic resources from human activities is indeed an important step for protection of our biological diversity. So, the traditional approach to protecting biodiversity through establishment of more protected areas is a right direction.

- The preservation of germplasm is an important step for conservation of our biological diversity. Therefore, the establishments of more seed banks, semen banks etc are vital.

- The harvesting of endangered species should be prohibited.

- Government should provide greater support for conservation projects. The participation of local people in designing and implementation of ecological projects has to be given paramount importance. At the same time the involvement of private sectors in conservation of biodiversity should be highly encouraged. In this regard, it is to be noted that the first Global Environment Facility (GEF) was started in India with the aim at giving more emphasis on effective conservation measures and participation of local people.

- Finally any development project for economic reasons should be eco-friendly as far as possible.

Sustainable development:

The term 'sustainable development' is more commonly used nowadays in describing any developmental activity. According to the Brundtland Commission (1987), "Sustainable development is development that needs of the present without compromising the ability of future generations to meet their own needs". In other words, it is actually a process in which the exploitation of resources, the orientation of technical development and the institutional changes are matched in context with future as well as present needs. Ecological harmony, economic efficiency, equity, social justice and endogenous choices are the preconditions for sustainable development. It is a well known fact that the earth's natural resources are limited. The reckless use of planetary resources may cause disruption of our ecosystem. Therefore, the judicial use of limited resources is the prime concern to any development activity. The resources should be used in such a way so that it will not adversely affect the sustenance of biological diversity on long term basis. The rational use of natural resources should, therefore, be given more priority to meet the needs and aspirations of the present and future generations. Nevertheless, the concept of sustainable development is based on the following two important facts:

a) The symbiotic relationship between the human race and the natural (producers) system.

b) The compatibility between ecology and economics.

Sustainable development is closely related with population, poverty and pollution (3 Ps). The population explosion may lead to a drastic reduction in the earth's limited resources. The stabilization of human population at some point should be the prime agenda for the whole world in general and the developing countries in particular. The present day available technologies should be blended in the development process in such a way that it does not lead to the long term decline of biological diversity and at the same time it suits environment friendly. Ultimately, the sustainable development will help in maintenance of life support systems, preservation of genetic diversity and judicious utilization of species and ecosystems.

Chapter 5

Wildlife Conservation and Conflict

Conservation of wildlife:

The term 'conservation' is derived from two Latin words *con* means together and *servare* means to keep or guard. So the meaning of conservation is to keep together. The Oxford dictionary meaning of conservation is the preservation from destructive influence, decay or waste. However, conservation may be defined as the judicious use of the planetary resources through planned and effective management. The resources may be biotic (e.g. forest, wildlife and fish) and abiotic (e.g. land, water and minerals) resources. As the natural resources are limited, hence, the conservation of these resources is very important for the existence of human beings. Any irrational use of resources may lead to the disruption of living communities and their environments resulting in extinction of many plant and animal species. Although, it is said that extinction is a 'biological reality', man's direct or indirect involvement causing extinction of some wild flora and fauna is a matter of great concern. The majority of recent extinction of plants and animals have been due to man's interference.

There are three principal objectives of conservation of living resources:
- To maintain essential ecological process and life-support systems.
- To preserve genetic diversity.
- To ensure the sustainable utilization of species and ecosystems.

Nevertheless, conservation is generally influenced by society's needs, ideas etc. The conservation of wildlife, however, highlights three important aspects: a) maintaining the habitat, b) maintaining the breeding stock, and c) prohibition of killing of any animal unless situation demands for. Conservation may be achieved through *in situ* conservation (in their natural habitats) or *ex situ* conservation (the multiplication and conservation of germplasm through natural or *in vitro* means, outside their natural habitats). The conservation of wildlife resources is intimately related to the conservation of other natural resources.

Wild animals are recklessly hunted for their valuable products. Elephants are poisoned for tusks; rhinoceros are being killed for horns which are mistakenly believed to have medicinal values; musk deer are hunted for musk glands; bears are being killed for their gall bladders (bile); a large number of snakes are being killed to procure their skins for preparation of many valuable goods. Man's greed for bones, meat, oils, horns, antlers, fur and feathers is causing a great concern to wildlife conservationists. There are innumerable examples where human intervention is a major cause towards the loss of wild animals. The white-rumped vultures (*Gyps bengalensis*), for example, have declined 97 per cent within a span of ten years. Many precious wild animals such as tiger, rhino, wild buffalo, thamin deer and whale are threatened with extinction. It is said that at least 10 per cent of India's recorded wild flora and 20 per cent of its wild mammal species are threatened with extinction. Therefore, the concern for wildlife is a matter of great concern for man himself. This is due to the fact that nature is delicately balanced and a time will come when the whole biotic community would be under great threat if proper measures are not taken.

Reasons for dwindling of wildlife resources:

The reasons for which the wildlife resources are in great danger are not difficult to search. The following factors highlight, how wildlife in recent times is under threat.

• Habitat destruction:

The habitats are destroyed by demographic pressure, advancement of human technology, deforestation, unrestricted grazing, fire, wetland drainage and natural calamities. All these factors are responsible for causing a great threat to wildlife. The ever increasing human population causes more pressure on forest lands. Poverty and demographic pressure are responsible in many ways for the extinction of many wild flora and fauna. The demand for the exploitation of mineral resources for modern industries is one of the factors responsible for wildlife habitat destruction. Once abundantly found pelican and painted storks in Kukrebellur village in Karnataka, the population has decreased rapidly following severe deforestation and drying up of lakes. The non-human primates which are said to the home to the country's largest number of population are facing a threat to their survival in the northeastern states. Habitat destruction is also causing a great problem for survival of many wild animals such as elephant, tiger, leopard, wild cat, pygmy hog, Asiatic wild buffalo, rhinoceros and swamp deer.

Uncontrolled and unregulated grazing by domestic livestock is one of the important factors causing dwindling of wildlife population. They are not only detrimental to the grazing area but also play an important role in spreading the dreadful diseases. It is reported that the gaur population has been dwindling as a result of death due to cattle borne diseases in many places of northeast India. There are many cases of wild animals succumbing to epidemics spread by domestic livestock have been reported. Except remotely located forests, almost all forest land within the wildlife reserves are degraded by overgrazing. According to the National

Commission on Agriculture (NCA, 1976), 88 per cent of the forest area is open to grazing and rest of the forest area remains closed to grazing for regulation purpose. Reports indicated that 67 per cent of national parks and 83 per cent of sanctuaries have illegal grazing. Overgrazing by livestock also causes destruction of seedlings, trampling of plants and destruction of perennial grass species. It also reduces soil cover. Another important adverse effect of overgrazing is that it reduces the species diversity of flora and fauna which is essential for preservation of the environment. Grasslands in the Himalayan region are so overgrazed by domestic livestock that very little wildlife can be sustained in them. However, stringent measures taken by the authorities have reduced the total number of livestock population to a great extent within the protected areas. The cattle population in Ranthambore national park, for example, has drastically been reduced from 40000 to about 4000 during 1997-98.

According to some people, the constructions of big projects (e.g., Sardar Sarovar dam) play a catastrophic role in destruction of natural habitats of wildlife. Natural calamities like flood and drought are also equally responsible for habitat destruction. The recent devastating flood in Assam, for example, has caused a havoc loss of wild faunas (e.g., deer) in the Kaziranga National Park. This has resulted in the change of feeding behaviour of some animals like tigers. This has been evidenced by their frequent attacks on rhino calves.

• Poaching, hunting and killing of wild animals:

Poaching is one of the important factors causing loss of wildlife. Over the years rampant poaching has drastically reduced wildlife population. Many criminals and smugglers have been involved in the destruction of forests as well as killing of a large number of wild animals. Reports indicate that about 30 tiger skins, 13 tiger skeletons, 184 leopard skins, 28 elephant teeth, 20 rhino horns, 2 lion skins and 443 fox skins have been confiscated from illegal traders and dealers during the last 23 years in the eastern part of the country. Poaching for elephant meat is one of the causes for dwindling of elephant population in north-eastern part of India. About 30 per cent of birds in the Nilgiri biosphere reserve are trapped for sale as pets, meat and oil. Many species of Indian birds are being pushed on the verge of extinction because of excessive capture. It is reported that no less than 1,00,000 birds are illegally traded from the main markets of northern India annually. Parakeets and munias account about 80 per cent of the above trade. Birds are caught with traps or hanging nets. Their feathers are shaved or dyed. Their wings are clipped and often they are debeaked. A falcon can be sold anywhere between Rs.500 to 80,000 in India and the Middle East depending on the bird and the consumer. There are about 33 per cent of reptiles in the Nilgiri biosphere reserve reported to be illegally traded for skins, flesh, traditional medicines and live displays. The crocodile population in India is reducing due to high demand for skins in the international market. The largest killing of tigers ever took place in

India is in Ranthambore tiger reserve in August,1993. Today there is not a single tiger found in Sariska tiger reserve in Rajasthan. There are endless examples of poaching taking place in India. Sometimes, domestic livestock are lifted by wild animals. Therefore, people living in the vicinity of protected areas kill and poison the wild animal through bait.

Hunting caused by man is one of the contributing factors to the disappearance of many wild species. There are three major purposes of hunting: commercial hunting, subsistence hunting and sport hunting. Among these the commercial hunting is widely spread and animals are killed with the object of selling their body parts. The illegal hunting for wild animals is, however, not a national phenomenon. It is now stated to be a global business. In Columbia, for instance, 25 million exotic small animals and birds are smuggled each year. The illegal hunting is a problem of many African countries. The main reason for illegal hunting of wild animals is due to their lucrative demands in the international market. A tiger skin can fetch upto Rs. 70,000 in the world market. A leopard skin can be sold at Rs. 50,000 in the market. About Rs. 20,000 can be earned by selling a crocodile skin. A single rhino horn is worth as much as Rs. 96,000 in the black market. Another reason for poaching is that forest officials are sometimes ill equipped to counter poachers. Poachers use sophisticated weapons , camouflaged traps and blinding spotlights to hunt many wild animals.

• Pollution:

Major contributors to the environmental pollution are pesticides, herbicides, industrial and human waste and gaseous pollutants. The indiscriminate use of persistent insecticides like DDT, aldrin, dieldrin, lindane and chlordane to protect agricultural crops has caused a great concern to the wildlife. Birds get killed after feeding on grains sprayed with pesticides. The study in USA indicates that only one percent of the applied pesticides reaches to the target pests and rest of the amount goes for killing non-target organisms. Excessive use of DDT causes less production of viable eggs in birds. Moreover, it inhibits the secretion of carbonic anhydrase enzyme resulting in the thinning of the egg shell. Water pollution caused by DDT impairs the fish production. Many birds (e.g., eagles and hawks) become the victims of pesticides. Chemical pesticides are responsible for serious health hazards to wild animals. The excessive use of chemical fertilizers and pesticides are detrimental to vegetation and aquatic flora and fauna. The aquatic faunas are being threatened due to water pollution caused by excreta, industrial effluents and other chemicals. The industrial effluents containing lead, mercury, chlorine and other heavy toxic metals are very detrimental to aquatic and marine life. Lead poisoning is the common metal poisoning in water fowl throughout the world. The literature is flooded with the information about the toxic effects of lead to the wild animals. The other factors that cause water pollution include accidental spills, mining and agricultural operations. Many species of sea turtles, fish, birds and mammals are still striving hard for their existence because of water pollution caused by oil slick in the Gulf during Gulf War.

- ## Introduction of exotic predators:

The predators introduced to a particular habitat sometimes causes extermination of wild animals. For example, rat which was introduced in Big Cape Island during 1964 was responsible for exterminations of many bird species. Introduction of non-native animals into a wildlife habitat also plays a role in reducing wildlife population.

- ## Introduction of competitors:

Competition in the same habitat is sometimes responsible for extermination of a particular animal. One striking example is the Australian burrowing kangaroo *(Bellongea lesueur)*. This animal once abundantly found in Australia was exterminated by rabbits. This was happened due to the fact that they were the competitors for each other for same food and burrows.

- ## Introduced pathogens:

Many diseases are responsible for death of various wild animals in their natural states. In African countries, many species of wild animals like antelopes, giraffes and wild buffaloes die due to rinderpest. Considerable loss in the population of gaur, wild pig, elephant and deer has taken place due to another fatal bacterial disease called anthrax. Pasteurellosis, foot and mouth disease and many other infectious diseases causing death of wild animals have been reported. Mortality due to outbreak of myxomatosis in rabbit is reported to be as high as 90 per cent.

- ## Other factors:

Many wild animals are killed due to road accidents, agricultural practices and other unavoidable causes.

Wildlife conservation measures:

We have already lost many precious wild animals. Cheetah, two horned rhinoceros of Sunderbans and pink-headed duck are extinct from our country. Naturalists also foresee that wild animals such as tiger, swamp deer and vulture will disappear from the soil of India sooner or later if proper protection measures are not taken at this juncture. It is a known fact that the establishment of wildlife population in a particular region depends on many factors like reproductive capabilities of the animals, predation, competition, prevalence of diseases and physical factors (e.g. water, land, climate etc.). Nevertheless, the following measures are important in order to save precious wildlife from imminent danger of threat of disappearing.

1) The areas where wild population exist beyond carrying capacity, the excessive animal population should be culled below the optimum level. This is

more important where the rate of reproduction is high, and the environmental effect is low.

2) The habitats of the animals must be preserved. Manipulation of habitats should be carried out in such a way that they provide an ideal place for survival of wild animals. The protection or destruction of forest determines the relative abundance of many wild animals. Some factors like feeding habits of the animals, proper plantation, development of water spots etc should be taken into consideration while manipulating the habitats. Deforestation should be checked as far as possible. Indiscreet felling of trees and lush green forest leads to soil erosion, flash floods etc. The hill region should have at least 60 per cent area under forest for ecological stability. Therefore, the habitats of the animals must be preserved.

3) Some management practices like controlled burning of grasslands, protection of trees for shelter, regeneration of trees and fertilization of fishponds are important.

4) Large scale land development projects have a long term adverse effect on the environment. Therefore, the land development programmes should be planned in such a way that it will have minimum adverse effects on environment.

5) Introduction of wild animals to a new habitat should be encouraged only after detailed studies on the possible effects on habitat and animals or plants.

6) As stated earlier that the pollution is one of the important factors that influences the depletion of wildlife. The pollutants generally come from three sources: sewage discharged into the river; surface run off from agricultural land where pesticides, insecticides, chemical fertilizers and manures are used; and industrial effluents discharged into the rivers without any pretreatment. Pollution has an adverse effect on the growth of fauna and flora. The following measures should be taken in order to prevent pollution.

- The industrial waste water should be treated to the maximum possible extent;
- There should be a sewage treatment plant where the sewage from the city drains and the partially treated industrial effluents could be processed;
- Heavy metals and pesticides have harmful effects on reproduction, hatching, feeding and growth of birds. Therefore, the use of more harmful pesticides should be restricted.
- The biological control of pest is important; and
- The existing laws for the prevention and control of water pollution need to be implemented forcefully.

7) The establishment of more sanctuaries, national parks and bioreserves for

preservation of representative samples of biotic communities should be given more importance. At least five per cent of the country's geographical land should be reserved under protected areas.

8) Sometimes heavy traffic in protected areas causes soil erosion and vegetation damage. This also causes disturbance to wild animals. Restriction on traffic control, wherever necessary, should be taken into consideration. So, there should be a sound tourism policy that will preserve its natural assets. Sound tourism also brings sufficient revenue as we find in many African countries like Kenya.

9) Some developmental activities like construction of biogas plants, upgrading of local cattle pupation, fodder development and adequate veterinary services should be provided to improve the economic condition of the people living in the protected areas. Organized dairying is important for better returns.

10) The existing wildlife protection laws should be firmly implemented in order to control illegal hunting of wild animals. Smugglers, poachers and hunters should be dealt with a firm hand. More funds should be allocated for purchasing of vehicles, weapons and other sophisticated equipment so as to combat the poachers. There should be a coordination among different bodies like National Board for Wildlife, National Committee on Environmental Planning and Coordination and Steering Committee on Project Tiger.

11) Finally the emphasis should be given more on research, education and extension activities related to nature and wildlife. In this regard, it is to be mentioned that the "wildlife week" is observed to create a general awareness of nature and wildlife conservation in India every year in the first week of October. The establishment of captive breeding centres is an important measure for conservation of wild animals in general and endangered wild animals in particular. A sizeable animal population can be produced in these breeding centres and these animals subsequently will be released in the wild. However, it is important to note that captive breeding programme is an emergency measure and is no substitute for maintenance of a species in its natural habitat.

Wildlife conservation projects :

India accounts about 8 per cent wildlife in the world. Though the country boasts of a rich wildlife heritage, we are unable to protect many of our wild flora and fauna from imminent danger of extinction. Unabated poaching, land encroachment, grazing, loss of habitat and diseases are some of the much highlighted factors causing threatening to the wildlife. With the fast disappearance of forests, wild animals too have fallen an easy prey to poachers. The present status of wildlife has already reached at an alarming situation and some rare wild species like musk deer, Kashmir stag, thamin deer, snow leopard and tiger are threatened in their respective habitats. It has already revealed that 10 per cent of the flowering plants and 20 per cent of the animals in India are threatened with extinction. About 20,000 tigers were killed for sport between 1860 and 1970 by the Rajas, Maharajas and ruling elites. Once roamed 3000 blackbuck in the Karera wildlife sanctuary in Madhya Pradesh, the population has now reduced to merely 500. The swamp deer population along the Satthiana belt near Dudhwa in Uttar Pradesh has been reduced from 1600 in the seventies to just 50 in recent times. As a result, there has been a rapid depletion of wildlife resources from the natural states. A large number of endangered or rare species will be extinct sooner or later if corrective measures are not taken immediately. Realizing its importance, various wildlife conservation projects were taken up by both Central and State Governments. The State Governments sponsored wildlife projects taken up for conservation of animals include captive breeding of pheasant (Himachal Pradesh), project on snow leopard (Jammu and Kashmir), captive breeding of Blythe's tragopan pheasant (Nagaland), special project for rehabilitation of the lion-tailed macaque (Tamil Nadu) and rehabilitation project on blackbuck and wild ass (Gujarat). The Central Government sponsored wildlife projects like project tiger, project elephant and musk deer project were also taken up. These wildlife projects have been instrumental in conservation and protection of wild animals. The programmes on improvement of habitat have been taken up simultaneously. They include development of water points, protection of large trunk trees, protection of fruit trees for the welfare of the fruit eating mammals, and allowing more light on the ground in order to encourage growth of fodder grasses, herbs and shrubs. The most important part of all these projects is the active participation of the local tribal people. They play an important role in the wildlife management. However, some special wildlife projects were started by the Government of India with collaboration of other organizations. They include:

1) Project tiger.
2) Gir lion sanctuary project.
3) Himalayan musk deer project.
4) Manipur brow-antlered deer project.
5) Project hangul.
6) Crocodile breeding project.
7) Project elephant.
8) Eco-development project.

The details of each project are described below.

Project tiger:

The tiger has been a symbol of power, freedom and beauty all over the world. Tigers, the apex animals of the complex food chain, are the indicators of the stability of the ecosystem as well as an index of environmental quality. The conservation of tiger is essential to maintain an ecological balance in nature. The IUCN first listed this animal in its Red Data Book in 1969. It is said that there were 1, 00,000 tigers at the beginning of the 20th Century. Today there are only about 5000-7500 tigers throughout the world. Ninety five per cent of our tiger population declined in the last 100 years. Once roared with 40,000 tigers, India now only has 3660 tigers in the natural state. However, many national and international organizations are very septic about the officially claimed statistics on tiger population. According to an estimate, about 200-300 tigers have been reducing annually in India since 1989. Some authorities, however, are of the opinion that at least one or two tigers are vanishing every day in India. So, the time is not far off when the total tiger population will be wiped out from the Indian soil. The reasons for declining tiger population are not a difficult task to seek off. Unrelenting human population pressure, heavy poaching, depletion of the prey base, constant shrinking of the tiger habitats and finally lack of professionalism in park management are some real causes that lead to our tiger population to such a sorrowful state. A tiger can be sold at much higher price for its skin, bones, claws, teeth etc. in an international market. China is reported to be the primary destination for Indian tigers. It is the main supplier of medicines preparing from tiger parts in the international market. Despite comprehensive efforts for conservation of tiger population, we are unable to prevent poaching and hunting of tigers in many tiger reserves. The biggest ever seizure of tiger bones in Ranthambore tiger reserve in 1993 where 40 tigers were killed by smugglers, is perhaps an indication of our inability to save tigers from the hands of poachers. Today, there is not a single tiger found in Sariska tiger reserve. A marked fall in the population of cheetal, sambar, swamp deer and wild boar has also been reported in the tiger reserves and the prey base has an effect on the tiger population. Smugglers more often use strong poisons to kill tigers. Sometimes local villagers are also responsible for keeping illegally hunted tiger's hides and bones in their houses.

Realizing its imminent danger of vanishing of tiger population from our country, the Project Tiger was launched on 1 April, 1973 with the following objectives.

- To ensure the maintenance of viable tiger population in its natural environment in the country; and

- To preserve for all times areas of biological importance as a national heritage for the benefit, education and enjoyment of the people.

Management plans, however, were formulated for each tiger reserve on the following principles:

- Elimination of all forms of human exploitation and biotic disturbance from the core area and rationalization of such activities in the buffer zone;

- Recovery of damaged ecosystem to its natural state; and

- Monitoring the floral and faunal changes over time and conducting research about wildlife.

Corbett national park was the venue for launching this popular project. Initially nine tiger reserves were established covering an area of 16,339 sq km with a population of 268 tigers in different states. Today there are 27 tiger reserves in the country (Table 5.1) covering about 37,761 sq km with a population of 1498 tigers. Tiger reserves constitute about 1.14 per cent of the total geographical area of the country. Over 300 small and large rivers originate in tiger reserves. Each tiger reserve has a core area which is kept free of biotic disturbances and forestry operations in terms of collection of minor forest produce, grazing, human disturbances etc. However, in buffer zone some activities are allowed so long as it will not have any adverse effect on wildlife. The buffer zone is used as "multiple use area" with the objectives of providing habitat supplement to the spill over population of wild animals from the core conservation unit. It also provides site specific ecodevelopmental inputs to surrounding villages for relieving the impact on the core area. In each reserve, a core zone of a minimum of 300 sq km with a sizeable extent of buffer zone was recommended. The overall administration is looked after by the "Steering committee", an apex body under the Ministry of Environment and Forests. The execution of the project tiger is done by the respective State Government. A full-fledged director at the centre level is present to coordinate the work. A field director is appointed for each tiger reserve. Initially the project was a central sector scheme. Now all the expenditures are shared by both Central and State governments. Since its inception some other activities like eradication of weeds, soil conservation, fire protection and intensive anti-poaching activities are being carried out as part of the project networks. A new proposal regarding issuing identity cards specifying number of family members and cattle head to those villagers who bring their cattle from outside for grazing in the tiger reserve has been proposed.

Many non-government organizations (NGOs) have also been playing an instrumental role in protecting tigers in India. A popular network programme called "Tiger link" was started by NGOs in order to save this majestic animal. The Tiger Conservation Programme (TCP) was started by WWF-International in January,1997 to support tiger conservation programme in India and other Asian tiger range countries. Under the tiger conservation programme, the compensation to villagers for their cattle killed by tigers has been launched. This will definitely reduce man-animal conflict. The TCP also aims at improving the entire chain of food supply in nature.

However, many conservationists are very much worried about the current activities that are going around the tiger reserves. According to them the current strategies are seriously flawed. Examples are many. Here some are mentioned to justify the above statement made by the conservationists. Illegal mining activities

Table 5.1 : Tiger reserves in India:

Name of Tiger Reserve	Core Area	Buffer Area	Total Area(km²)	Tiger Population 1995
Corbett TR(Uttaranchal; 1973-74)	338.0	796.0	1134.0	128
Similipal TR (Orissa; 1973-74)	845.70	1924.30	2770.0	97
Ranthambore TR (Rajasthan; 1973-74)	392.0	433.0	825.0	38
Kanha TR (Madhya Pradesh; 1973-74)	940.0	1005.0	1945.0	97
Manas TR(Assam; 1973-74)	470.0	2370.0	2840.0	94
Bandipur TR(Karnataka; 1973-74) ; extended upto Nagarhole (1999-2000) with an area of 643 sq.km.	523.0	351.0	874.0	74
Palamau TR(Jharkhand; 1973-74)	213.0	715.0	928.0	47
Melghat TR(Maharashtra; 1973-74)	448.0	1170.0	1618.0	71
Sunderban TR(West Bengal; 1973-74)	1330.0	1255.10	2585.10	242
PeriyarTR (Kerala; 1978-79)	350.0	427.0	777.0	39
Sariska TR(Rajasthan; 1978-79); at present there seems to be no tiger here.	498.0	302.0	800.0	-
Indravati TR(Chhattisgarh; 1982-83)	1258.37	1545.0	2803.37	15
Buxa TR(West Bengal; 1982-83)	315.0	443.82	758.82	31
Nagarjunasagar-Shri SailamTR (now Rajiv Gandhi Tiger reserve, Andhra Pradesh; 1982-83)	1200.0	2368.0	3568.0	34
NamdaphaTR (Arunachal Pradesh; 1982-83)	177.0	1808.0	1985.0	47
Kalakad-Mundanthurai TR(Tamil Nadu; 1988-89)	571.0	229.0	800.0	17
Dudhwa TR(Uttar Pradesh ; 1987-88); extended upto Katerniaghat(1999-2000) with an area of 551 sq.km.	648.0	163.0	811.0	98
Valmiki TR(Jharkhand; 1989-90)	336.0	504.0	840.0	49
Pench TR(Madhya Pradesh; 1992-93)	292.85	465.0	757.85	27
Bandhavgarh TR(Madhya Pradesh 1993-94);	624.75	537.72	1161.47	46
Tadoba-Andheri TR(Maharashtra; 1993-94)	-	-	620.0	36
Panna TR(Madhya Pradesh; 1994-95)	340.0	202.66	542.66	22-26
DampaTR (Mizoram; 1995-96)	340.0	160.0	500.0	4
Bhadra TR(Karnataka; 1998-99)	-		492.0	-
Pench TR(Maharashtra; 1998-99)	-	-	257.0	-
Bori-Satpura-Panchmarhi TR, M.P.(1999-2000)	-	-	1486.0	-
Pakhui –Nameri TR, Arunachal Pradesh-Assam(1999-2000)	-	-	1206	-

are being operated in the Panna tiger reserve; the Melghat tiger reserve is threatened by the Upper Tapti stage II hydroelectric project; the Palamau tiger reserve is threatened by the world bank-financed Kotku dam; and an international steamer route is to be cut through the Sunderbans tiger reserve. Nevertheless, the project tiger has been instrumental in conservation of our tiger population. Now the administration is geared up with full facilities such as fast moving jeeps, flash lights, motorcycles, jet boats, camels, ward staff and modern instrument. Moreover, an immune zone has been established around the tiger reserves so that wild animals can be protected from many dreadful diseases that occur in domestic livestock.

Gir lion Sanctuary project:

The Asiatic lions are found only in the dry deciduous and thorn forest of Gir national park in Gujarat. But it is a fact that the Asiatic lions once were distributed in the northern and central areas of the Indian sub-continent. Nevertheless, in 1965, about 1,265 sq km area was notified as Gir wildlife sanctuary and a core area of 140.4 sq km was declared as national park in 1975. The Gir national park is also the habitat of other magnificent wild animals such as sambar, Indian gazelle, nilgai, spotted deer and four-horned antelope. In 1969, the participants at the technical session of the IUCN in New Delhi had mentioned the following points about the Gir forest:

1) The whole ecosystem of the sanctuary is under threat owning to over grazing caused by domestic cattle. As a result, the natural vegetation of the forest is disturbed, and;

2) The population of cheetal, sambar, nilgai, wild boar and four-horned antelope which form the natural prey base of the lion is drastically reduced in the habitat. As a consequence lions in the forest started preying on domestic livestock. Later on it became the point of conflict with the livestock owners. Recognizing the imminent danger to the Asiatic lions in the Gir forest, as evident from the above technical session, the State Government launched the Gir Lion Project in 1972 for the conservation of lions as well as for social upliftment of the pastoral communities. The five year plan scheme was laid down on certain guidelines which include conservation and improvement of the habitat, exclusion of all exotic forms of animals and plants, minimizing human interference and proper rehabilitation of the people living within the sanctuary. The scheme also highlighted the need for the construction of barricades along the periphery of the sanctuary. The Central Government had also assisted in the above programme.

Owning to the protection measures taken under this project the population of lions has been increased in the Gir forest. The Asiatic lion population has steadily been grown from 177 in 1968 to 359 in 2005. Today it is reported that the Gir forest is practically overflowing with lion population.

However, the Rabbari, Maldhari and Gujjar herdsmen are still living in the Gir forest. It is estimated that there are about 15,000-20,000 heads of cattle moving freely inside the sanctuary area. These animals compete with the wild ungulates for food. Wildlife biologists are of the opinion that it is very difficult to increase the natural prey population of lions unless these cattle are removed from the forest. It is also reported that a sizeable cattle population is being preyed upon by the lions. Poaching, increased number of pastoralists settled in the forest, and disappearance of open grass and scrub forest land are the important issues that still needed to be properly addressed.

As stated earlier that the Gir forest is the only home to the Asiatic lion. The International Union for Conservation of Nature and Natural Resources (IUCN) has declared the Asiatic lion as endangered species. Wildlife experts have an opinion that it is always considered unwise to manage endangered species as a single population at one site. In wildlife management, the maintaining of endangered species at one pocket always runs the risk of any man-made or natural disaster. The outbreak of any disease may wipe out the entire population. Therefore, a second home is always important for maintaining an endangered animal population and this will save the animals from extinction. Accordingly, in 1957, the first attempt to find a second home for the Asiatic lions was made in Chandraprabha sanctuary near Varanasi in Uttar Pradesh. However, the effort to find a second home to lions was not success on account of lacking effective protection against poaching and paucity of natural prey population. In 1979, the second attempt to find a suitable home to lions was initiated in Barda hills near Porbandar in Gujarat. The plan was also failed. Finally, the Wildlife Institute of India (WII) explored a year long survey programme in 1993-94 to study the suitability of potential sites for lion translocation. Three protected areas namely the Kuno-Palpur wildlife sanctuary in Madhya Pradesh and both Sitamata wildlife sanctuary and Darrah Jawahar Sagar wildlife sanctuary in Rajasthan were intensively and systematically surveyed. Among the three sites studied, the Kuno-Palpur wildlife sanctuary has been adjudged as best for lion translocation. An area of 350 sq km has been earmarked for territory where lions are to be released. However, the area would be increased up to 3700 sq km by 2015. The work would be done in three phases. In the first phase the work to be completed includes preparation of infrastructure, development of habitat, removal of biotic pressure, building the prey base, and scientific assessment of the ecological dynamics. In the second phase the work to be done includes the actual translocation of lions and close monitoring. In the third phase, it is planned for follow-up and consolidation of the achievements. The first phase of work has already been carried out. The second phase of work is yet to be implemented.

Lions require an open space, deciduous forest, an adequate supply of perennial water, sufficient viable prey population and limited human disturbances. Another important fact is that each male lion eats about 50-55 deer in a year. Therefore, it is essential to establish a good number of ungulates base. The sufficient amount of vegetation growth is vital for survival of the ungulate population. If everything goes well, then this "centrally sponsored lion translocation project" will be the first major exercise of such kind in India. Nevertheless, this particular wildlife translocation programme may face with greater challenges because of limited financial and professional resources, differences in public attitudes and political realities.

Himalayan musk deer project:

Musk deer inhabit in the Himalayan region. They are hunted for their musk glands. Musk is used in the preparation of perfumes and medicines. Deforestation, unrestricted grazing by livestock and poaching have brought about a sharp decline in musk deer population. In 1974, this graceful animal was listed in the Red Data Book by IUCN. The project on the "Ecology and conservation of the Himalayan musk deer" was launched in January 1979 by the Government of India with the active cooperation of the IUCN and WWF. The basic objective was to maintain a viable population of musk deer in their natural habitat by undertaking an ecological study with emphasis on conservation. The emphasis was also given to protect musk deer population in the national parks and sanctuaries. The project was launched at Kedarnath sanctuary in Uttaranchal. Three farms namely the Kanchula Khark musk deer farm in Kedarnath (Uttaranchal), the Kufri musk deer farm in Shimla (Himachal Pradesh), and the musk deer research centre in Almora (Uttaranchal) have been carrying out breeding programmes. For effective conservation of this vulnerable species, a concerted effort is required. Protective measures like strict action against illegal hunting, controlling the musk trade, and education and research should be given more priority.

Manipur brow-antlered deer project:

The thamin or Manipur brow- antlered deer is the most endangered deer in Asia. In 1950, it was regarded as extinct until a small population of it was discovered in Keibul Lamjao on the Loktak lake area in Manipur. Therefore, it was a topmost priority to conserve this deer in its natural habitat. Accordingly, the area was declared as a sanctuary and subsequently it was declared as national park in 1977. Construction of Loktak dam, cutting of grasses, grazing, fishing and poaching were prevalent until the area was declared a national park. However, villagers surrounding the national park still depend largely on cultivation or fishing for their livelihood. The grassy vegetation with rich nutrients and tropical climate is an ideal habitat for thamin deer. According to a report, the present population of thamin deer has estimated to be 106. Despite all the measures taken by the authority, the thamin is threatened to extinction.

Project hangul:

The hangul or Kashmir stag, a cousin of the European red deer, is an endangered mammal. They are distributed mainly in Dachigam sanctuary. However, they are also found in some other small pockets of Jammu & Kashmir and Himachal Pradesh. Now the sanctuary has been given the status of a national park. The term "Dachigam" means "ten villages" This national park has the ideal habitats such as grassland, coniferous forest, alpine pasture and scrub and broad leaved woodland forest for hangul deer. Indiscriminate destruction of forest for fuel, timber and fodder, overgrazing by domestic livestock and poaching are some factors leading to the present sorrowful status of hangul population. It is said that the country had 3000 hangul population in 1940. Their number was drastically reduced to a mere 140 to 170 by the year 1970. With the realization of its imminent danger, the "Project hangul" was initiated in 1970 by Jammu & Kashmir Government with the help of IUCN and WWF. A comprehensive ecological cum management plan was taken for conservation of this endangered species. These include protection measures against indiscriminate habitat destruction, protection against grazing of sheep and cattle, complete banning on cutting of woods, improvement of fodder, controlling of fire, strict actions against poaching, and vaccination programmes of domestic cattle against dreadful diseases. As a result, there has been a steady increase in hangul population numbering 810 in 1987. The four large predators namely the Himalayan brown bear, the Himalayan black bear, the leopard and the snow leopard have also some roles in reducing the hangul population in the wild state.

Crocodile breeding project:

Three species of crocodile namely the mugger or freshwater crocodile, the salt water or estuarine crocodile and the gharial are found in India. The crocodilian population has declined considerably due to unabated poaching for their valuable skins, construction of dams and canals, loss of natural banks, and fishing. With the assistance of the United Nations Development Programme (UNDP) and Food and Agricultural Organization (FAO), the Government of India had launched a crocodile breeding and management project on 1st April, 1975. The objectives laid down in the project include location of best crocodile habitats, collection of eggs for hatching, rearing of the young and subsequently release of young into their natural habitats. The project was initially started in Orissa. Under this project, gharial eggs were hatched in captive condition for the first time in the world. The scheme was subsequently spread to many other states. The Nandan Kanan Zoological Park Captive Breeding Project in Orissa (captive breeding and husbandry of 3 species), Nehru Zoological Park Rehabilitation and Captive Breeding Project in Andhra Pradesh (captive breeding and husbandry of 2 species), Sunderban Salt Water Crocodile Scheme in West Bengal (husbandry and rehabilitation of saltwater crocodile), Parambikulam Mugger Scheme in Kerala (husbandry and rehabilitation of saltwater crocodile), Kukrail National Park in Uttar Pradesh (husbandry band rehabilitation of mugger and gharial) are some notable crocodile breeding projects.

Project elephant:

Elephants have been playing an important role in our socioeconomic and political lives since time immemorial. Today, there are about 37,000 to 48,000 wild elephants found in Asian countries, while 6,00,000 elephants have been recorded in their wild state in Africa. In India, the total elephant population is between 19000 and 24000 in the wild and there are about 3000 elephants found in the captivity. Elephants all over Asia are, however, disappearing rapidly due to various reasons which include rapid and violent denudation of natural habitats, growing human encroachment into the protected areas, paucity of food, fodder and water, ruthless felling of forest cover for agriculture and developmental projects, and indiscriminate poaching. Elephants are also killed for meat which is traditionally consumed by many people of north-east India. Due to massive deforestation in the protected areas, elephants more often enter into human habitats resulting loss of properties, agricultural crops and lives.

After realizing the above facts, the Government of India formulated a task force in the year 1990 with the aim at preparing a guideline for improving the plight of elephants. Based on the recommendations given by the task force, the Government of India had launched a popular programme almost on the lines of the Project Tiger, called "Project Elephant" or "Gajatame" in the year 1992. The objectives laid down in the project elephant are as follows:

- To ensure the full protection to elephants for maintaining a viable elephant population in our country.
- To give emphasis more on reducing the elephant-human conflicts.
- To take effective measures against ivory trade and poaching.
- To provide a package of practices for improvement of the plight of elephants including health care, breeding, capturing, management and training of elephants.
- To give emphasis on maintaining corridors for the free movement of elephants in their natural habitat without impediments.
- To compensate the monetary loss to farmers for their crops damaged by the elephants.

The project will cover an area of 57,994 sq km with 19,040 elephant population. The task force has, however, identified eleven "elephant ranges" that will cover reserved forests, national parks, sanctuaries and other land categories. The part of each range falling within a state has been designated as 'elephant reserves'. There are 25 designated elephant reserves in eleven elephant ranges and out of which 14 elephant reserves (e.g. Mayurjharna, Singhbhum, Wayanad, Nilambur,

Kameng, Shiwalik and Garo hills) have been set up so far. The proposed elephant ranges (ER) are given below.

1) Northern India ER (Uttaranchal: Shivalik elephant reserve covering 3000 sq km with 750 elephant population);

2) North Brahmaputra ER (Arunachal Pradesh-Assam: Kameng-Sonitpur elephant reserves covering 4300 sq km with 1580 population; Namdapha tiger reserve also included);

3) Eastern Dooars ER (Assam-West Bengal : Manas-Buxa-Jaldapara elephant reserves covering 3800 sq km with 800 elephants);

4) Kaziranga ER (Assam-Nagaland: Kaziranga — Karbi- Anglong-Intanki elephant reserves covering 4900 sq km with 1800 population);

5) South Brahmaputra ER (Assam-Arunachal Pradesh : Dibru-Deomali elephant reserves covering 4400 sq km with 500 elephant population);

6) Eastern India ER (South West Bengal–Jharkhand-Orissa: Mayurjhana-Singhbhum-Mayurbhanj-Mahanadi-Khalasuni elephant reserves covering 9694 sq km with 1560 elephants; Simlipal tiger reserve (Orissa) and Dalma sanctuary (Jharkhand) also included;

7) Garo hills ER (Meghalaya: Garo hills elephant reserves covering 3500 sq km with 1700 population);

8) Periyar ER (Kerala- Tamil Nadu: Periyar-Madurai elephant reserves covering 3300 sq km with 1500 elephants) ;

9) Western Ghats ER (Tamil Nadu- Kerala: Anamalai-Perambikulam elephant reserves covering 5700 sq km with 1600 population);

10) South Nilgiri ER (Kerala- Tamil Nadu: Nilambur-Silent Valley-Coimbatore elephant reserves covering 2400 sq km with 950 elephants);

11) Nilgiri-Eastern Ghats ER (Karnataka-Kerala-Tamil Nadu-Andhra Pradesh: Mysore- Wyanad –Mudumalai- Kaundinya elephant reserves covering 13000 sq km with 6300 population).

Eco-development project:

An important conservation project called "Eco-development project" was started in 1996 to save and restore a viable population of our vast flora and fauna. This World Bank aided project was initiated with the following objectives:

*   To conserve biological diversity in some protected areas of the country;

*   To enable local communities in order to meet their needs in an environmentally sustainable manner and use traditional knowledge in resource management;

*   To control poaching and fire in the selected areas to be brought under this project; and

- To promote conservation through environmental education, people's participation and increasing understanding of some socially perceptible issues relevant to protected area management.

The seven protected areas (Gir national park, Periyar tiger reserve, Palamau tiger reserve, Buxa tiger reserve, Pench tiger reserve, Ranthambore tiger reserve and Nagarhole elephant reserve) have been selected for implementation of the project. The socioeconomic welfare activities like silvipastoral development, soil water conservation measures in cultivated fields, safe drinking water facilities, establishment of veterinary centres, supply of improved chullahs, construction of bio gas plants and training of village level workers in ecodevelopment and related activities have been taken in the project area. The eco-development project also highlights many important issues like alternate non-forest based employment opportunity to the people. Some areas would be created as a core area where human activities are not to be allowed. As India is one of the twelve mega biodiversity countries in the world, this project will have a great impact on conservation of wild flora and fauna.

National wildlife action plan (2002-2016):

Based upon the decision taken in the meeting of the Indian Board for Wildlife held in 1982, the first National Wildlife Action Plan (NWAP) was adopted in 1983. The strategies and action points for conservation of wildlife outlined in that meeting are still relevant. However, some problems have become more acute and new concerns have become apparent. Therefore, it is required a change in priorities. Increased commercial use of natural resources, continued growth of human and livestock populations and changes in consumption patterns have an impact on flora and fauna. So, the new National Wildlife Action Plan (2002-2016) has given emphasis on the following parameters.

- Strengthening and enhancing the protected area network.
- Effective management of protected areas.
- Conservation of wild and endangered species and their habitats.
- Restoration of degraded habitats outside protected areas.
- Control of poaching, taxidermy and illegal trade in wild animal and plant species.
- Monitoring and research.
- Human resource development and personnel planning.
- Ensuring peoples' participation in wildlife conservation.
- Conservation awareness and education.
- Wildlife tourism.
- Domestic legislation and international conventions.

- Enhancing financial allocation for ensuring sustained fund flow to the wild-life sector.
- Integration of National Wildlife Action Plan with other related programmes.

Conservation and conflict:

The Stockholm Declaration on Human Environment (1972) has rightly pointed out that "The natural resources of the earth, including air, water, land, flora and fauna and especially representative samples of natural ecosystems must be safeguarded for the benefit of present and future generations through careful planning or management….". The above declaration has also given emphasis by saying that "Man is both creature and moulder of his environment, which gives him physical sustenance and affords him the opportunity for intellectual, moral, social and spiritual growth; both aspects of man's environment - natural and man-made are essential to his well-being and to the enjoyment of this basic human rights even the right to life itself." However, we often come across some practical findings that need to be properly resolved in the light of the present context.

Many a time it is argued that whether tea estate owners should shoot elephants to protect their cash crops; whether the protection of Kashmir stag is more important than the nomadic graziers in Kashmir whose livelihood depend on meat and wool bearing animals; and whether supply of water to wild animals is more useful than those people who live in the vicinity of the protected areas. It is a matter of fact that many poor people have lost their livelihood as a result of ban on circus animals and trapping and trade in birds. Hundreds of human settlements are being displaced in the name of development and conservation projects. Wild animals cause a great damage to the agricultural crops. Leopards and other many wild animals sometimes grab domestic livestock. We also come across with many unfortunate remarks given by the villagers inhabiting in the protected areas. One such remark that is frequently narrated is that "We are dying of hunger and you are more bothered about raising tigers." Therefore, the issues raised in connection with conservation of wildlife are in fact is the subject of "conservation and social conflict". Our planners, administrators and social activists are more often faced with these issues.

Many noted environmentalists are of the opinion that protection of forest and wildlife should not be viewed as people versus forest or animal rights. But it is a fact that conservation is essential and important. The most important understanding at this juncture, therefore, should be the proper utilization of the limited resources in such a way that the long term needs of human beings are given equal priorities with the immediate needs of today's population. Report indicates that about 3 million tribal people depend on forest produce. Some measures like stall feeding of domestic livestock, fodder development, upgradation of local cattle, supply of smokeless chulhas, land development, supply of drinking water, establishment of primary education and health centres and many such socio-economic development programmes need to be urgently implemented. The most important point is that conservation programmes will be effective only if local people have the greater access to their environment.

Chapter 6

Protected Areas, Biosphere Reserves and Wetlands

Protected area:

The need for protection and conservation of wild flora and fauna is well documented all over the world. The present-day depletion of wildlife population can be attributed to the uncontrolled exploitation of natural resources to fulfill the needs of human beings. Much of our natural resources have been depleted due to habitat destruction, commercial exploitation, and lack of awareness of the need and importance of conservation of natural resources. Therefore, more emphasis has been laid to protect and conserve our natural resources through establishment of protected areas. As a consequence, more than 9000 protected areas have been created worldwide to conserve and protect wild flora and fauna. These protected areas are designated by more than 140 names. A protected area can be described as a geographically defined area, which is designated or regulated and managed in such a way so that the specific conservation objectives can be achieved. However, according to IUCN, a protected area is "An area of land and/or sea especially dedicated to the protection and maintenance of biological diversity, and of natural and associated cultural resources and managed through legal or other effective means."

The fundamental objectives of creating such large numbers of protected areas are to safeguard our varied ecosystem as natural as possible, and also to provide a base from which a sustained yield can be obtained from the natural resources. However, most of our protected areas are facing a serious threat in achieving the desired goals. The reasons for degrading the ecosystem in the protected areas are manifold. Demographic pressure, demand for more natural resources in the protected areas (e.g. fuel wood, fodder, small timber etc.), and establishments of mining development projects for exploitation of mineral deposits in the protected zones are some of the much highlighted causes for which our national parks and wildlife sanctuaries are facing a great crisis today. It is reported that about 3 million people inhabit inside the protected areas. However, the conflict over the use of resources within the protected areas has to be viewed in the context of present socioeconomic condition of the tribal people.

The Commission of National Parks and Protected Areas (CNPPA) of IUCN earlier classified the protected areas into ten categories. However, the IUCN has slightly modified the earlier classification of protected areas. This includes the addition of wilderness areas to category 1, definition of a new category VI, and the recognition of those areas of interest in the former categories VI to X into the new categories I to VI. So, the present classification of protected areas with broad management objectives is given below:

Category I: Strict Nature Reserves/Wilderness Areas: Protected areas are managed mainly for science or wilderness protection.

Category II: National Parks: Protected areas are managed mainly for ecosvstem conservation and recreation.

Category III: Natural Monuments: Protected areas are managed primarily for conservation of specific natural features.

Category IV: Habitats/Species Management Areas: Protected areas are managed mainly for conservation through management intervention.

Category V: Protected Landscapes/ Seascapes: Protected areas are managed mainly for conservation and recreation.

Category VI: Managed Resource Protected Areas: Protected areas are managed primarily for the sustainable use of natural ecosystems.

In India, the protected areas belong to the categories of II, IV, V and VI. National Parks and Sanctuaries in India fall into the category of II and IV respectively, while the Biosphere Reserves fall into the category VI. Most India's forest areas belong to the category VI.

National park and sanctuary:

It is a well known fact that national parks and sanctuaries have been created for providing maximum protection and conservation of wildlife in their natural environment. The definition of national park, according to the Wildlife (Protection) Act 1972 is, " Whenever it appears to the State Government that an area, whether within a sanctuary or not is by reason of its ecological, faunal, floral, geomorphological, or zoological association or importance, needed to be constituted as a national park for the purpose of protecting, propagating, or developing wildlife therein or its environment, it may, by notification, declare its intention to constitute such area as a national park".

Sanctuary, on the other hand, has been defined in the Wildlife (Protection) Act 1972 as "Any area comprised within any reserve forest or any part of the territorial waters, which is considered by the State Government to be of adequate ecological, faunal, floral, geomorphological, natural or zoological significance for the purpose of protecting, propagating, or developing wildlife or its environment, is to be included in a sanctuary".

So any area included for the purpose of protecting, propagating or developing wildlife can be declared as a national park or a sanctuary by the State Government. Nevertheless, there are minor differences between sanctuary and national park. In sanctuary, the process of determination of people's right and their elimination/ acquisition is followed after an area to be declared as sanctuary, whereas this process is gone through first prior to an area is declared as national park. Another difference is that in national park, no human activities are allowed i.e. it is completely free from all forms of exploitation. In sanctuary, restricted human activities such as regulated grazing, timber harvesting and fishing are, however, allowed so long as they are consistent with the object of wildlife management. A sanctuary can always be given to the status of a national park. Moreover, the drastic manipulation of the habitat in the national park is not favoured.

In India, there is a long and proud history of national parks and sanctuaries for conservation of wildlife. We find the concept of sanctuary in Kalidas' much celebrated book *Abhigyan Shakuntalam*. During third Century B.C. Kautilya recorded some ideas regarding abhayaranyas in his widely published book *Arthashastra* and the *abhayaranyas* were created for protection of elephants and other wild animals. The ancient concept of wildlife sanctuary can also be traced in Vana Bhatt's *Kadambari* and other Indian classics. The Vedanthangal Bird Sanctuary, the first sanctuary in India, was established in Tamil Nadu as early as 1878. Since then, the country has made an impressive progress in the establishment of national parks and sanctuaries. A major impetus towards conservation of wildlife, however, was given during the seventies. The decades of the seventies and eighties had witnessed a phenomenal growth in the creation of protected areas.

In the year 1970, India had only 10 national parks and 127 sanctuaries that covered an area of 25,000 sq km The number of protected areas has steadily been increased. There were 24 national parks and 372 sanctuaries that accounted for 3.3 per cent of the total geographical area in 1989. Presently the country boasts of having 89 national parks and 500 wildlife sanctuaries covering an area of 1.56 lakhs sq km. This constitutes 4.6 per cent of the country's total geographical area. However, many conservationists are of the opinion that the country should have at least five per cent protected areas of total landmass in order to conserve the full range of biological resources effectively. The status of protected areas in India is given in Table 6.1.

Nevertheless, all the national parks and sanctuaries were not created for same reasons. Some were created for specific purposes. The Kaziranga national park, for example, was created for the protection of one-horned rhinoceros. The Keoladeo Ghana national park at Bharatpur in Rajasthan was established for protection and conservation of avifauna. Likewise, the Gir national park for Asiatic lion, the Dachigam national park for Kashmir stag and the Keibul Lamjo national park for brow-antlered deer are some widely known wildlife reserves that were created for conservation and protection of specific wild animals. In India, the Corbett national park which was founded in 1936 is the oldest national park. The Yellowstone national park, the first national park in the world was established in 1872 in USA.

Table 6.1: Status of wildlife protected areas (national parks and sanctuaries) in India:

State/Union Territory	Status	Fauna	Well known national parks (NP)/ wildlife sanctuary(WLS)	Remarks
Andhra Pradesh	Total geographical area-2,75,045.0 sq km; national park-4; sanctuary-21; wildlife reserves-4.69% of total land area and 20.2% of total forest area.	Tiger, leopard, sambar, blackbuck, gaur, jungle cat, sloth bear, nilgai, langur, wild pig, cheetal, wild dog, four-horned antelope, marsh crocodile, pelican, flamingos, painted stork, teal etc.	Sri Venkateswara NP,Mahaveer Harina Vanasthali NP, K.B.Reddy NP, Coringa WLS, Nagarjunsagar Srisailam(Rajiv Gandhi) WLS, Kolleru WLS, Krishna WLS,Nellapattu WLS,Pulicate lake WLS, and Manjira WLS.	Pulicate lake is the largest lagoon in South Asia. The Nagarjunsagar-Srisailam WLS is the largest protected area comprising 3568.09 sq km
Arunachal Pradesh	Total geographical area-83,743.0 sq km; national park-2; sanctuary-10; biosphere reserve-1; protected area-11.44 % of total land area and 18.5% of forest area.	Snow leopard, clouded leopard, elephant, tiger, panther, binturong, gaur, Himalayan black bear, Assam macaque, sambar, golden cat, musk deer, red panda, hog deer, barking deer, slow loris, stumped-tailed macaque, hoolock gibbon, wild dog, python, king cobra, monitor lizard and many birds.	Mouling NP, Namdapha NP, Dibang WLS, Kamlang WLS, Kane WLS, D'Ering memorial (Lali) WLS, Itanagar WLS, Sessa orchid WLS and Tale valley WS.	Blessed with 4 popular cats-tiger,leopard, snow leopard and clouded leopard. Recently two deer species the leaf deer(*Muntiacus putaoensis*) and the black muntjac(*Muntiacus crinifrons*) have been discovered. The Dibang WLS has the largest area comprising 4,149.0 sq km

State/Union Territory	Status	Fauna	Well known national park (NP)/ wildlife sanctuary(WLS)	Remarks
Assam	Total geographical area-78,438.0 sq km; national park-5; sanctuary-12; biosphere reserve-2; protected area comprises 2.7 % of total geographical area and 6.8% of total forest area.	Rhinoceros, elephant, wild buffalo, pygmy hog, hoolock gibbon, capped langur, golden langur, tiger, gaur, civet, otter, swamp deer, clouded leopard, hog deer, hispid hare, barking deer, geese, whistling teal, pelican, white winged woodduck, hornbill python, cobra etc.	Kaziranga NP, Manas NP, Nameri NP, Bardoibum beelmukh WLS, Gibbon WLS, Pobitora WLS, Garampani WLS and Sonai-Rupai WLS.	The state boasts of having hoolock gibbon.The pygmy hogs, the highly endangered animals are found. Maximum populations of Indian rhinoceros are found in this state. Manas NP is the largest protected area comprising 500.0 sq km Both the Kaziranga NP and Manas NP are inscribed as World Heritage Site.
Bihar	Total geographical area- 94164.0 sq km; National park:1; sanctuary: 11; wildlife reserves comprise 3.48% of the total geographical area.	Leopard, sambar, cheetal, wild boar, gaur, wolf, barking deer, wildcat, blue bull, peafowl, crocodile, florican, teal etc.	Valmiki NP, Gautam Budha WLS, Nakti dam WLS, Rajgir WLS, Valmiki WLS, Kaimur WLS and Vikramshila Gangetic Dolphin WLS.	The Kaimur WLS is the largest protected area comprising 1342.0 sq km
Chhattis-garh	Total geographical area-1,35,191.0 sq km;National park: 3; sanctuary-10; wildlife reserves account for 4.69% of total land area.	Tiger, wild buffalo, , cheetal, gaur, sloth bear, leopard, wild boar, hyena, wolf, python, hill mayna etc.	Indravati NP, Kangerghati NP, Sanjay NP, Achankmar WLS, Pomed WLS, Sitanadi WLS and Udanti WLS	Sanjay NP is the largest protected area comprising 1,471.13 sq km
Delhi	Total geographical area-1,483.0 sq km; wildlife sanctuary-1 protected area comprises 0.89 % of land area and 31.4% of total forest area.	Faunas are found in captivity.	Indira Priyadarshini (Asola)WLS.	The total protected area is 13.20 sq km

State/Union Territory	Status	Fauna	Well known national parks(NP)/ wildlife sanctuary(WLS)	Remarks
Goa	Total geographical area- 3,702.0 sq km; national park-1; sanctuary-6; wild life reserves constitute 20.37% of the land area and 23.5% of total forest area	Gaur, sambar, mouse deer, flying squirrel, porcupine, panther, civet, grey jungle fowl etc.	Mollem NP, Mollem WLS, Bondla WLS, Chorao Island(Dr.Salim Ali) WLS, Netravali WLS and Cotigaon WLS.	Netravali sanctuary is the largest protected area comprising 211.0 sq km
Gujarat	Total geographical area-1,96,024.0 sq km; national park-4; wildlife sanctuary-21; biosphere reserve-1;wildlife reserves constitute 8.62% of total land area and 87.1% of total forest area(includes desert wildlife sanctuary).	Lion, Indian wild ass, chinkara, wild boar, four-horned antelope, wolf, hyena, blackbuck, cheetal, leopard, nilgai, sambar, dolphin, sea turtle, dugong, crocodile etc.	Gir NP, Bansda NP, Marine NP, Blackbuck(Velavador) NP,Lala Great Indian bustard WLS, Kachchh desert WLS, Khijadiya WLS, Nal sarovar WLS, Hingolgarh nature reserve WLS and Wild ass WLS.	Asiatic lion is found only in Gir forest. The state is also blessed with Indian wild ass. Kachchh desert WS is the second largest wildlife reserve in India comprising 7506.22 sq km
Haryana	Total geographical area-44,212.0 sq km; national park-1; sanctuary-9; wildlife reserves constitutes 0.63% of geographical area and 16.7% of total forest area.	Painted stork, coot, spoonbill, sarus crane etc.	Sultanpur NP, Abubshehar WLS, Bir bara ban WLS, Nahar WLS, Chhilchila WLS and Khaparwas WLS.	Abubshehar WLS is the largest protected area comprising 115.30 sq km
Himachal Pradesh	Total geographical area-55,673.0 sq km; national park-2; sanctuary-32; wildlife reserves constitutes 12.93% of total geographical area and 17.6% of total forest area.	Musk deer, panther. snow leopard, black bear, brown bear, flying fox, ibex, goral, blue sheep, great Tibetan sheep, yak, civet, marten, Himalayan mouse hare, crane, pheasant, tragopan etc.	Great Himalayan NP, Pin valley NP, Bandli WLS, Kanawar SLW, Gobindsagar WLS, Manali WLS, Naina Devi WLS, Sainj WLS, Kibber WLS and Tundah WLS.	Kibber wildlife sanctuary is the largest protected area comprising 1400.5 sq km The state is blessed with many high altitude wildlife fauna.

State/Union Territory	Status	Fauna	Well known national parks(NP)/ wildlife sanctuary(WLS)	Remarks
Jammu & Kashmir	Total geographical area-2,22,236.0 sq km; national park-4; sanctuary-15; wildlife reserves account for 6.69% of total land area and 73.4% of total forest area.	Musk deer, Kashmir stag(hangul), Himalayan black and brown bears, snow leopard, goat antelope, Tibetan ass, wild yak, wild sheep, wild goat etc.	Dachigam NP, City forest(Salim Ali) NP, Hemis NP, Kistwar NP, Gulmarg WLS, Nandini WLS, Karakoram WLS and Ramnagar rakha WLS.	The Karakoram WLS is the largest wild life reserve (5,000.0 sq km). Many high altitude wildlife are found in this state.
Karnataka	Total geographical area-1,91,791.0 sq km; national park-5; sanctuary-20; wildlife reserves account for 3.49% of geographical area and 16.5% of forest area.	Elephant, gaur, tiger, sloth bear, sambar, wild dog, leopard, cheetal, wild boar, barking deer, four-horned antelope, Malabar squirrel, flying squirrel, civet cat, hare, porcupine, wolf, lion-tailed macaque, peacock, jungle fowl, python, crocodile, cobra, krait etc.	Bandipur NP, Anshi NP, Bannerghatta NP, Kudremukh NP,Nagarahole NP, Bhadra WLS, Cauvery WLS, Pushpagiri WLS, Sharavathi valley WLS, and Arabithittu WLS.	The Bandipur NP is the largest wildlife reserve accounting 874.0 sq km
Kerala	Total geographical area-38,863.0 sq km; national park-3; sanctuary-12; wildlife reserves account for 6.91% of geographical area and 23.8.% of forest area.	Elephant, Nilgiri tahr, lion-tailed macaque, gaur, tiger, sambar, leopard, cheetal, civet, nilgai, langur, wild boar, Malabar squirrel, wild dog, barking deer, grey jungle fowl, crocodile etc.	Periyar NP, Eravikulam NP, Silent valley NP, Aralam WLS, Chinnar WLS, Iddukki WLS, Parambikulam WLS, Periyar WLS and Wayanad WLS.	Periyar wildlife sanctuary is the largest protected area accounting 777.0 sq km
Madhya Pradesh	Total geographical area-308144.0; National park-9; sanctuary-25; wildlife reserves constitute 3.52% of total land area.	Tiger, leopard, nilgai, sambar, cheetal, swamp deer, blackbuck, gaur, barking deer, chinkara, sloth bear, wild boar, wild dog, hyena, wolf, mouse deer, gharial, peafowl, etc.	Bandhavgarh NP, Kanha NP, Panna NP, Pench(Priyadarshini) NP, Sanjay NP, Satpura NP, Pachmarhi WLS, National Chambal WLS, Ken gharian WLS and Palpur-kuno WLS.	Nauradehi WLS is the largest wildlife reserve comprising 1,194.67 sq km . Palpur-Kuno wild life sanctuary has been selected for second home to Asiatic lion.

State/Union Territory	Status	Fauna	Well known national park (NP)/ wildlife sanctuary(WLS)	Remarks
Maha-rashtra	Total geographical area-3,07,690.0 sq km; national park-5; wildlife sanctuary-34; wildlife reserves constitute 5.09% of geographical area and 23.5% of total forest area.	Tiger, panther, sloth bear, wild boar, cheetal, sambar, nilgai, gaur, chinkara, barking deer, four-horned antelope, mouse deer, jungle cat, common langur, hyaena, jungle fowl, peafowl, partridge, crocodile etc.	Gugamal NP, Nawegaon NP, Pench NP, Sanjay Gandhi(Borivilli NP, Todaba NP, Bhimashankar WLS, Chandoli WLS, Great Indian bustard WLS, Koyana WLS, Melghat WLS and Nagzira WLS.	The great Indian bustard wildlife sanctuary is the largest protected area of India covering 8496.44 sq km
Manipur	Total geographical area-22,327.0 sq km; national park-1; wildlife sanctuary-1; wildlife reserves accounts for 1.06 % of geographical area and 1.4% of total forest area.	Brow-antlered deer, hog deer, fishing cat, leopard, leopard cat, wild boar, water birds etc.	Keibul-Lamjao NP and Yangoupokpi-Lokchao WLS.	Brow-antlered deer , the most endangered deer in south Asia is found only in Manipur.
Meghalaya	Total geographical area-22,429.0 sq km; national park-2; sanctuary-3; wildlife reserves cover 1.34% of geographical area and 3.1% of total forest area.	Elephant, bear, tiger, leopard, barking deer, gaur, wild boar, pangolin, slow loris, capped langur, clouded leopard, Himalayan black bear, sun bear, red panda, red jungle fowl, grey peacock pheasant etc.	Balphakram NP, Nokrek ridge NP, Baghmara Pitcher plant WLS, Nongkhyllem WLS and Siju WLS.	Balphakram is the largest protected area covering 220.0 sq km
Mizoram	Total geographical area-20,987.0 sq km; national park-2; sanctuary-4; wildlife reserves account for 4.86% of geographical area and 5.5% of total forest area of state.	Tiger, gaur, leopard, clouded leopard, hoolock gibbon, capped langur, slow loris, sloth bear, Malayan giant squirrel, elephant, Himalayan bear, barking deer, sambar, wild dog, wild boar, leopard cat, python, king cobra, hornbill , pheasant etc.	Murlen NP, Phawngpui blue mountain NP, Dampa WLS, Khawnglung WLS, Lengteng WLS and Ngengpui WLS.	Dampa wildlife sanctuary is the largest protected area in the state covering 500.0 sq km

State/Union Territory	Status	Fauna	Well known national park (NP)/ wildlife sanctuary(WLS)	Remarks
Nagaland	Total geographical area-16,579.0 sq km; national park-1; sanctuary-3; wildlife reserves account for 1.34% of geographical area and 2.6% of total forest area.	Elephant, tiger, bear, panther, barking deer, gaur, wild boar, clouded leopard, pangolin, khaleej pheasant , traggopan and reptiles.	Intanki NP, Fakim WLS, Puliebadze WLS and Rangapahar WLS.	Intanki national park is the largest protected area accounting 202.02 sq km
Orissa	Total geographical area-1,55,707.0 sq km; national park-2; sanctuary-18; wildlife reserves cover 5.2% of total land area and 13.9% of total forest area.	Tiger, elephant, leopard, sambar, hyena, cheetal, wild boar, giant squirrel, civet, jungle cat, sloth bear, four-horned antelope, leopard cat, gaur, mouse deer, flying squirrel, crocodile, pelican, crane, painted stork, sea turtle, jungle fowl, python, king cobra etc.	Bhitarkanika NP, Simlipal NP, Badrama WS, Chandaka dampara WLS, Chilka (Nalban) WLS, Gahirmatha (Marine) WLS, Kuldiha WLS, Satkosia George WLS and Sunabeda WLS.	Chilka lake is the home to many migratory birds. Gahirmatha(Marine) WLS is the largest protected area(1435.0 sq km) in the state.
Punjab	Total geographical area-50,362.0 sq km; sanctuary-10; wildlife reserves account for 0.62% of total land area and 10.9% of total forest area.	Nilgai, blackbuck, hog deer, wild boar, hare, pigeon, partridge, peafowl, parakeet etc.	Abohar WLS, Bir Aishvan WLS, Bir Bhadson WLS, Harike lake WLS and Takhni Rehampur WLS.	Abohar WLS (186.50 sq km) is the largest protected area in the state.
Rajasthan	Total geographical area-3,42,239.0 sq km; national park-4; sanctuary-24; biosphere reserve-1; wildlife reserves cover 2.79 % of geographical area and 30.0% of total forest area.	Blackbuck, caracal, chinkara, desert fox gazelle, hare, nilgai, sambar, cheetal, mongoose, fishing cat, sloth bear, leopard, wild boar, tiger, jackal, hyena, civet, four-horned antelope, great Indian bustard, jungle fowl, crane, stork, crocodile etc.	Desert NP, Keoladeo Ghana NP, Ranthambore NP, Sariska NP, Jawahar sagar WLS, Mount Abu WLS, National Chambal WLS and Sitamata WLS.	Desert NP is the largest protected area consisting 3162.0 sq km The Keoladeo NP is inscribed as World Heritage Site.

State/Union Territory	Status	Fauna	Well known national park (NP)/ wildlife sanctuary(WLS)	Remarks
Sikkim	Total geographical area-7,096.0 sq km; national park-1; sanctuary-5; wildlife reserves account for 28.87% of geographical area and 77.3% of total forest area.	Snow leopard, wolf, Tibetan ass, blue sheep, clouded leopard, marbled cat, civet, binturong, serow, goral, takin, musk deer, red panda, Himalayan black bear, green pigeon, pheasant etc.	Khangchendzonga NP, Barsey Rhododendron WLS, Maenam WLS and Shingba (Rhododendron) WLS.	Khangchendzonga is the largest protected area accounting 1784.0 sq km Many high altitude wildlife are found in this state.
Tamil Nadu	Total geographical area-1,30,058.0 sq km; national park-5; sanctuary-19; biosphere reserve-2; wildlife reserves account for 2.18% of geographical area and 12.8% of total forest area.	Elephant, tiger, panther, sloth bear, sambar, cheetal, gaur, lion-tailed macaque, Nilgiri langur, Nilgiri tahr, wild dog, black buck, crocodile, waterfowl etc.	Mudumalai NP, Guindy NP, Indira Gandhi(Annamalai) NP, Gulf of Mannar marine NP, Kalakad WLS, Point Calimere WLS, Indira Gandhi (Annamalai) WLS, Vedar thangal WLS and Vaduvoor WLS.	The Indira Gandhi WLS is the largest protected area accounting 841.49 sq km
Tripura	Total geographical area-10,491.69 sq km; sanctuary-4; wildlife reserves account for 5.75% of geographical area and 9.5% of total forest area.	Elephant, leopard,clouded leopard, barking deer, civet,capped langur, pig-tailed macaque, slow loris, spectacled langur, wild boar, wolf, tiger(recently found), water fowl, wader etc.	Gumti WLS, Rowa WLS, Sepahijala WLS and Trishna WLS.	The largest protected area is the Gumti wildlife sanctuary (389.54 sq km).
Uttar Pradesh	Total geographical area-2,38,566.0 national park-1; sanctuary-23 ; wildlife reserves constitute 2.39% of total geographical area.	Tiger, elephant, leopard, wild boar, cheetal, sambar, nilgai, sloth bear, four-horned antelope, porcupine, , barking deer, peafowl, hyena, red jungle fowl, partridge etc.	Dudhwa NP, Chandraprabha WLS, National Chambal WLS, Okhala WLS, Turtle WLS and Vijai Sagar WLS.	The Hastinapur WLS is the largest protected area comprising 2,073.0 sq km Swamp deer are found in Dudhwa NP.

State/Union Territory	Status	Fauna	Well known national park (NP)/ wildlife sanctuary(WLS)	Remarks
West Bengal	Total geographical area-88,752.0 sq km; national park-5; sanctuary-15; biosphere reserve-1; wildlife reserves account for 3.09% of geographical area and 23.5% of total forest area.	Tiger, elephant, rhinoceros, leopard, gaur, black bear, wild boar, hog deer, barking deer, sambar, cheetal, crocodile, turtle, pelican, gangetic dolphin etc.	Buxa NP, Gorumara NP, Neora valley NP, Singhalila NP, Sunderban NP, Jaldapara WLS, Sajnekhali WLS and Chapramari WLS.	Sundarban national park is the largest protected area comprising 1330.10 sq km It is also inscribed as World Heritage Site.
Andaman & Nicobar Islands	Total geographical area-8,249.0 sq km; national park-9; sanctuary-96; biosphere reserve-1; wildlife reserves constitute 18.54% of geographical area and 17.7% of total forest area.	Andaman wild pig, crab-eating monkey, palm civet, dugong, dolphin, white-breasted swiftlet, megapode, teal, crocodile, monitor lizard, turtle, king cobra etc.	Campbell Bay NP, Galathea NP, Mahatma Gandhi Marine (Wandoor)NP, Middle button Island NP, Mount Harriett NP, Bamboo island WLS, Chanel island WLS,Mangrove island WLS, Megapode island WLS and Turtle island WLS.	Campbell Bey national park is the largest protected area accounting 429.00 sq km
Chandigarh	Total geographical area-114.0 sq km; sanctuary-1; wildlife reserves constitute 22.29% of total geographical area and 82.0% of total forest area.	Many birds.	Sukhna lake WLS.	The total protected area is 25.42 sq km
Jharkhand	Total geographical area- 79,714.0 sq km; National park-1; sanctuary-10; wildlife reserves account for 2.63% of total geographical area.	Elephant, tiger, leopard, hyena, wolf, pangolin, cheetal, sambar, bear, wild boar, python etc.	Betla NP, Dalma WLS, Hazaribagh WLS and Palamau WLS.	The Palamau WLS is the largest protected area(794.33 sq km).

State/Union Territory	Status	Fauna	Well known national park (NP)/ wildlife sanctuary(WLS)	Remarks
Uttaranchal	Total geographical area- 53,483.0 sq km; National park- 6; sanctuary-6; wildlife reserves accounts for 12.10%.	Tiger, elephant, musk deer, cheetal, sambar, nilgai, wolf, leopard, hyena, bear, python etc.	Corbett NP, Gangotri NP,Nanda Devi NP,Rajaji NP, Valley of flowers NP, Kedarnath WLS and Sonanadi WLS.	Gangotri NP is the largest protected area(1,52.0 sq km) in the state. The Nanda Devi NP is inscribed as World Heritage Site.
Daman & Diu	Total geographical area- 112.0 sq km.; wildlife sanctuary-1; wildlife reserves accounts for 1.94% of total geographical area.	Different species of mammals, birds and reptiles are found.	Fudam WS	The total protected area is 2.18 sq km

Biosphere reserve:

The biosphere reserve is a coordinated global network of national parks, biological reserves and other protected areas serving conservation as well as educational and research needs. Biosphere reserves also address a bio-regional approach to land-use and development planning for conservation of flora and fauna in the existing representative ecosystems. It was started by UNESCO in 1973-74 under the Man and Biosphere (MAB) programme. The following are the major objectives of the biosphere reserves.

- To conserve the diversity and integrity of biotic communities of plants and animals within natural and semi natural ecosystem and to protect the genetic biodiversity of species;
- To promote ecological and environmental research;
- To provide facilities for education, awareness and training; and
- To prepare a management plan for biosphere reserve and to cooperate between the national and the international network.

A biosphere reserve has five essential components. They include (i) a strictly protected core area which is normally a national park or nature reserve; (ii) a comparatively large buffer zone surrounding the core area with allowing of limited activities and a transition area; (iii) development of a relationship between man and environment through applied and basic research projects; (vi) trained scientists and managers of protected areas; and (v) contribution of data to the Global

Environmental Monitoring System of UNEP through various monitoring projects.

For proper management the biosphere reserve has different schematic zones.

1. Core area (strictly protected).
2. Buffer zone.
3. Transition area.

There are 300 biosphere reserves covering more than 70 countries world wide. In India, 13 biosphere reserves have been established .They cover all the major bio-geographical regions of the country. The following are the biosphere reserves (BR) in India (Table 6.2).

Table 6.2: The biosphere reserves of India:

S. No.	Name of site	Area (sq.km)	Location (state)	Biogeographic zone
01	Nilgiri BR	5520.0	Part of Waynad, Nagarhole, Bandipur and Mudumalai, Nilambur, Silent Valley and Siruvani hills of Tamil Nadu, Karnataka and Kerala.	Western Ghats
02	Great Nicobar BR	885.0	Southern most islands of Andaman & Nicobar Islands.	Islands
03	Manas BR	2837.0	Part of Kokrajhar,Barpeta, Nalbari, Kamprup, Bongaigaon and Darang districts of Assam.	East Himalayas
04	Nanda Debi BR	5860.69	Part of Pithoragarh, Chamoli and Almora districts of Uttaranchal.	West Himalayas
05	Dehang-Debang BR	5112.00	Part of Siang and Debang valleyofArunachal Pradesh.	East Himalayas
06	Nokrek BR	820.0	Part of Garo hills in Meghalaya.	East Himalayas
07	Sunderbans BR	9630.0	Part of delta of Ganges and Brahamaputra river system in West Bengal.	Gangetic delta
08	Gulf of Mannar BR	10500.0	Indian part of Gulf of Mannar (Tamil Nadu) between India and Sri Lanka	Coasts
09	Pachmarhi BR	4926.28	Parts of Betul, Chindwara and Hoshangabad of Madhya Pradesh.	Semi arid
10	Similipal BR	4374.0	Part of Mayurbhanj district in Orissa	Decan Peninsula
11	Khangchendzonga BR	2619.92	Sikkim	East Himalayas
12	Dibru-Saikhowa BR	765.0	Part of Dibrugarh and Tinsukia districts of Assam.	East Himalayas
13	Agasthyamalai BR	1701.0	Neyyar, Peppara and Shenduruny wildlife sanctuaries and their adjoining areas in Kerala.	Western Ghats

For effective conservation, certain factors such as public awareness, people's participation, and use of modern technologies have to be given more emphasis in order to achieve the desired results. The biosphere reserves would ultimately help in protecting a large number of rare and endangered animals. Nevertheless, some basic conflicts like local needs vs. national demand, marketed benefits vs. non-marketed values, resource use for basic needs vs. production of non-basics, and social priorities vs. individual freedom need to be properly addressed.

Wetland:

Wetland forms an important part of our ecosystem. They cover a broad spectrum of specialized animals and plants. There are a variety of formal definitions of wetland. Wetland is generally soggy and low lying place on earth where the land is perpetually saturated or even partially submerged. In other words, it is the land where the soil is saturated with water for permanent basis or for some period during the year. Specific plants and animals are adapted to the wetland for their sustenance. The two important characteristics of wetlands are: (a) the soils remain waterlogged or are submerged under water for whole or part of the year; and (b) the wetland biota are adapted to this water logging or submergence during at least a part of their life cycle. So, by the term 'wetland' we simply mean ponds, shallow water bodies, shallow peripheral areas of large lakes etc. However, as per definition adopted in Ramsar Convention on Wetlands (an important inter-governmental treaty on wetlands was adopted in the Iranian city of Ramsar in 1971), wetlands are, "Areas of marsh, fen, peat land, or water, whether natural or artificial, permanent or temporary, with water that is static or flowing, fresh, brackish or salt, including areas of marine water the depth of which at low tide does not exceed six metres".

Wetland serves as spawning ground for many aquatic mammals and birds. They filter out many water borne pollutants. We are immensely benefited from wetlands. There may be three types of benefits:

- Benefits directly cognizable in monetary terms (e.g. fishing, agriculture, tourism etc);
- Indirect hydrological and ecological benefits of great economic values (e.g. ground water discharge); and
- Aesthetic values (e.g. creative stimulation).

There are 726 wetlands worldwide covering an area of 43.84 million hectares. They are estimated to cover approximately six per cent of earth surface and are found in all climates and regions.

In India, there are 19 wetlands designating as Ramsar sites. Bhoj wetland (MP), Keoladeo national park (Rajasthan), East Calcutta wetland (WB), Chilika wetland (Orissa) and Loktak lake (Manipur) are important wetlands in India.

Classification of wetlands:

There are many classifications of wetlands. However, here a simple classification system of wetland types is given below:

- **Marine and coastal wetlands:** There are various types of marine and coastal wetlands. The examples are i) Marine waters- permanent shallow waters less than six meters deep at low tide. It includes sea bay straits; ii) Inter-tidal marshes including salt marshes, salt meadows, tidal brackish and freshwater marshes; iii) Inter-tidal forested wetlands (e.g. mangrove swamps and tidal freshwater swamp forests); iv) Brackish to saline lagoons with one or more relatively narrow connections with the sea; and v) Freshwater lagoons and marshes in the coastal zone (e.g. delta lagoon and marsh systems).

- **Inland wetlands:** There are many types of inland wetlands. The examples of inland wetlands are i) Permanent rivers and streams; ii) Seasonal and irregular rivers and streams; iii) Riverine flood plains; iv) Permanent freshwater lakes (over 8 ha); v) Permanent and seasonal brackish, saline or alkaline lakes, flats and marshes; and vi) Freshwater swamp forest.

- **Man-made wetlands:** The examples of man-made wetlands are i) Water storage areas (e.g. reservoirs, barrages and hydroelectric dams); ii) Ponds including farm ponds, stock ponds and small tanks; iii) Aqua culture ponds; iv) Irrigated land and irrigation channels, rice fields and canals; and v) Seasonally flooded arable land and farm land.

Management of wetlands:

As stated earlier, wetlands are the refuges of many plants and animals. They serve as breeding grounds for fish, birds and many other animals. Therefore, proper management of wetlands is an important step towards improvement of wildlife. Followings are some measures to improve wetlands:

- Regulation of optimum water level and maintenance of water quality;
- Adequate protection to the habitat;
- Proper management of grass and peat resources;
- Controlled and sustainable exploitation of fish;
- Delayed exploitation of wetland resources;
- Establishing artificial nesting and loafing habitats; and
- Produce and support wildlife.

Ramsar Convention on Wetlands:

It is an important inter-governmental treaty on wetlands adopted on 2nd February 1971 in the Iranian city of Ramsar. In this regard, it is to be mentioned that it was the first of the modern global treaties on conservation and wise use of natural resources. Initially there was an emphasis on providing habitat for water birds.

However, over the years, the scope of the Convention has expanded to cover all aspects of wetland conservation and judicial use. The Convention came into effect in 1975 and has 133 contracting parties with 1201 wetland sites totalling 105.8 million hectares designated for inclusion in the Ramsar list of wetlands of international importance.

India became the party of the Convention in 1982. It has been mentioned earlier that there are 19 wetlands as Ramsar sites in India.

Chapter 7

Forest and Wildlife

Definition of forest and other related terminologies:

The term "forest" is derived from the Latin word *foris* meaning outside the village boundary or away from inhabited land. According to the Indian Forest Record (1936), a forest is defined as "An area set aside for the production of timber and other forest produce or maintained under woody vegetation for certain indirect benefits which it provides e.g. climatic or protective". However, according to the Food and Agriculture Organization (FAO) a forest is defined as land having a tree canopy cover of more than ten per cent over an area of more than 0.5 hectare with forestry as the principal land use. Forest is generally called an area occupied by different kinds of trees, shrubs, herbs and grasses and is maintained as such. It is a complex ecological system in which trees are the main life form. A forest has five important components. Firstly, it should have a defined land area. Secondly, the land area should be occupied different vegetation types e.g. trees, shrubs etc. Thirdly, the trees found in a particular area should form a closed or partially closed canopy. Fourthly, the trees and other forms of vegetation should be managed in order to get forest produce and finally, it should provide shelter to other flora and fauna.

As mentioned earlier that a forest is essentially composed of trees, shrubs, herbs and grasses. In this regard, therefore, it is important to familiarize with the actual meaning of some common terminologies which are described below.

Tree: A tree is defined as a large woody perennial plant, usually with single stem from which branches sprout at some distance above the ground to carry a spreading crown of leaves. Trees form the upper strata in forest. A tree is usually more than 6 metres in height and may be recorded up to 127 metres height.

Shrub: It may be defined as a woody perennial plant which is larger than herb but smaller than tree and it measures usually below 6 metres in height. It has several woody stems growing from the same root. The canopy density of shrub is less than 10 per cent.

Herb: Herbs are the annual, biennial or perennial plants whose heights are usually not more than a metre. Its stem is green and tender.

Grass: Grasses are small plants that cover the ground.

Deciduous tree: Deciduous trees are those plants that remained leafless for some time during the year (e.g.*Acacia catechu*). The period for which they remain leafless, however, varies according to species. The sal tree *(Shorea robusta)*, for instance, remain leafless for a period of one week.

Evergreen tree: Evergreen trees are the perennial plants that are never entirely without green foliage and the old leaves persist until a new set has appeared (e.g. *Pinus wallichiana*).

Savannah: It is defined as tropical or subtropical grassland containing scattered trees or shrubs.

Importance of forest:

The importance of forests to our day to day life is enormous. Forests have been a vital source of subsistence employment, revenue earning and raw materials to many industries. It has been reported that the gross value of goods and services provided by the forestry sector is estimated at Rs. 26,329.8 crores. In terms of GDP, the forestry sector now contributes 2.37 per cent at market price. Its role in maintaining the environmental stability has been widely recognized. That is why, forests are often called as a regulator of ecosystem. Forests also play a crucial role in conservation of biodiversity, food security and sustainable development. Some benefits that we get from the forests are as follows:

- Raw materials such as industrial wood, fodder and fuel wood are provided for industries, domestic uses and other purposes.
- A large number of non-wood products are obtained from the forests. These include oils, fibres, resins, tans, dyes, spices, drugs, insecticides and many other edible products.
- Forests maintain soil fertility by adding large amounts of organic matter as well as recycling of the nutrients and regulate the water cycle in the ecosystem.
- Forest has an important role in conservation of both soil and water. It also checks flood.
- Forests maintain carbondioxide balance in the atmosphere. It protects us from various types of pollution. Forests are reported to increase five to 15 per cent annual precipitations.
- Forests are known as a store house of genetic diversity. Many birds, mammals reptiles, amphibians and insects harbour in the forests. India's forests are reported to be the home of over 75,000 species of animals.
- Forests play an important role in alleviating poverty through gainful employment. Their role in the socioeconomic development and cultural life of the tribal people has been widely recognized.

Status of forest:

It is said that about 8000 years ago there were about 8080 million hectares of forests worldwide. But today, the total area under forests is reported to be only 3040 million hectares. However, the existing forests are being destroyed at an alarming rate. Tropical rain forest that contains 30 to 40 per cent of the world's biological diversity covers only 7 per cent of the total geographical land area. The tropical forest is, however, depleting at an alarming rate of 17 million hectares per year and it may lead to elimination of approximately five to 15 per cent of the world's species in the tropical forests. Environmentalists have already forecasted that there will be virtually no natural forests in some countries like Pakistan, Thailand, Malaysia and Costa Rica in less than fifty years if the present trend of forest degradation is continued. Nevertheless, deforestation continues to be the single most important factor for losing forest in most parts of the globe.

The present status of forest in India has been reported in the State of Forest Report (2003), Ministry of Environment and Forests, Government of India (Table 7.1). The report indicates that the actual forest cover in the country is 6,78,333 sq km, accounting for 20.64 per cent of its geographical area. This comprises about two per cent of the total forest area in the world. Out of the total covered forest in India, very dense forest (land with a forest cover of trees with canopy density over 70 per cent) constitutes 51,285 sq km (1.56 per cent), moderately dense forest (land with a forest cover of trees with canopy density between 40 and 70 per cent) constitutes 3,39,279 sq km (10.32 per cent) and open forest (canopy density between 10 per cent and 40 per cent) constitute 2,87,769 sq km (8.76 per cent).

There is an overall increase of 2,795 sq km (0.41 per cent) forest cover in the country as compared to 2001 report. The mangrove forest (salt tolerant forest ecosystem) covers to the extent of 4,461 sq km (0.14 per cent of geographic area) accounting for 1,162 sq km very dense (26.05 per cent),1657 sq km (37.14 per cent) moderately dense and 1,642 sq km (36.81 per cent) open mangrove cover. Mangrove forests which account 5 per cent of the globe are spread along the coastal areas of the country. In India, the largest area under mangrove forests is found in Sunderbans in West Bengal. The scrub forest (land having bushes and/or poor tree growth with canopy density less than 10 per cent) covers 40,269 sq km (1.23 per cent).

The largest area under forest among the states and Union Territories is found in Madhya Pradesh (76,429 sq km) followed by Arunachal Pradesh (68,019 sq km) and Chhattisgarh (55,998 sq km). At the time of independence the recorded forest area in India was said to be about 40 million hectares in which the total forests owned by the government was about 26 million hectares, while private and community owned forests were about 14 million hectares. At present, the per capita availability of the forest area is only 671 sq mts. However, the country needs a forest cover of 33 per cent in the plains and 66 per cent in hilly regions for maintaining the ecological balance. It is learnt that India had lost about 43,420 sq km forest area between 1951 and 1983. The total forest area was diverted at the rate of 0.1 million hectares between 1950 and 1980 to non forestry purposes such as river valley projects, agriculture, industries, townships, roads, transmission lines etc. In India, the forest area is also damaged by repeated annual forest fire. Many tribal communities still practice shifting cultivation. For this purpose, they clear a forest, burn the slash and raise one or two crops for a period of two to four years and then shift to clear another forest. This practice is primarily confined to north eastern hill states and Orissa. The report given by the Task Force of Shifting Cultivation (1983) indicates that the forest area affected by shifting cultivation is recorded to the extent of 4.35 million hectares. It has been estimated that 4,07,292 sq km (36.91 per cent of the total geographic area) of India's forest cover is found in 187 tribal districts. The forest cover in these tribal districts accounts for 60.04 per cent of the total forest cover of India. As the socioeconomic conditions of these people much depend on the forest produces, studies indicated that over 80 per cent of the forest dwellers collect twenty five to 50 per cent of their food from the forests. It is also an established fact that most of the fuel woods are extracted from forests by the villagers living in the vicinity of forest. There is, however, a wide gap between the requirement and supply of forest biomass and more and more forest biomass are being utilized by rural communities in order to meet their requirements.

Another important fact regarding the status of Indian forest is that more than 90 million domestic livestock graze freely in forest areas. The large number of livestock population which is not proportionate to the carrying capacity of forests is causing a great damage to regeneration and productivity of forests. In this context, it is to be mentioned that only 2.5 million livestock were grazed in the forest during 1950-51. The availability of fodder at present is accounted for 435 million tons. These come from permanent pastures, agricultural and forest lands and other grazing lands. The uncontrolled and excessive grazing has led to severe soil erosion. As mentioned earlier that the forest cover is affected by jhum cultivation in North eastern states of India and the maximum effect has been noticed in Nagaland followed by Arunachal Pradesh, Manipur, Mizoram and Meghalaya.

Forest policies and programmes:

Though, it is a matter of great concern that our forest resources are depleting at an alarming rate, the country, however, has a long tradition for professional management of forest resources. As stated earlier that the reasons for depletion of forest resources in India are manifold. The intense demographic pressure on forests for fodder, fuel, and agricultural land, the low priority for preservation of tree cover, inadequate allocation of financial and managerial resources for aforestation, failure to involve local people in the ecological planning and legal

Table 7.1: The present status of forest in India (2003).

States/UTs	Forest cover (sq km)	Of State's geographical area (%)	Countries forest cover (%)	Tribal population (%)	Livestock population (million)
Andhra Pradesh	44419	16.15	6.55	6.30	32.91
Arunachal Pradesh	68019	81.22	10.03	63.70	0.8
Assam	27826	35.48	4.10	12.80	16.06
Bihar	5558	5.90	0.82	0.80	47.93 (including Jharkhand population)
Chhattisgarh	55998	41.42	8.26	32.50	13.49
Delhi	170	11.46	0.03	-	0.32
Goa	2156	58.24	0.32	-	0.24
Gujarat	14946	7.63	2.20	14.90	18.6
Haryana	1517	3.43	0.22	-	9.14
Himachal Pradesh	14353	25.78	2.21	4.20	5.11
Jammu & Kashmir	21267	9.57	3.14	-	8.7
Jharkhand	22716	28.50	3.35	22.50	-
Karnataka	36449	19.0	5.37	4.30	29.57
Kerala	15577	40.08	2.30	1.10	5.8
Madhya Pradesh	76429	24.79	11.27	19.90	33.26
Maharashtra	46865	15.23	6.91	9.30	36.4
Manipur	17219	77.12	2.54	34.40	1.29
Meghalaya	16839	75.08	2.48	85.50	1.20
Mizoram	18430	87.42	2.72	94.70	0.2
Nagaland	13609	82.09	2.01	87.7	1.1
Orissa	48366	31.06	7.13	22.20	22.7
Punjab	1580	3.14	0.23	-	10.2
Rajasthan	15826	4.62	2.33	12.40	48.4
Sikkim	3262	45.97	0.48	22.40	0.39
Tamil Nadu	22643	17.41	3.34	1.0	25.0
Tripura	8093	77.18	1.19	31.0	1.6
Uttar Pradesh	14118	5.86	2.08	-	64.8(including Uttaranchal population)
Uttaranchal	24465	45.74	3.61	3.0	-
West Bengal	12343	13.91	1.82	5.6	35.31
Andaman & Nicobar Islands	6964	84.42	1.03	9.50	0.15
Chandigarh	15	13.16	0.002	-	0.03
Dadra & Nagar Haveli	225	45.82	0.03	79.0	0.07
Daman & Diu	8	7.14	0.001	11.50	0.01
Lakhadweep	23	71.88	0.003	93.20	0.02
Pondichery	40	8.33	0.006	-	0.14

Source: State of Forest Report, 2003, Government of India.

constrains are some highlighted facts causing deterioration of forests. Therefore, there is need for proper implementation of existing forest policies and programmes so that our present forest resources can be saved from further losses.

Way back in 1864, the contemporary forest legislation and policy were formulated by the British Government. As a result, the forest became almost exclusively the state property. However, a great thrust towards development of forest was given in 1952 with the declaration of new National Forest Policy. The major objectives laid down in the above forest policy are as follows:

- To maintain one third of the country's land under forest cover for ecological reason;
- To provide adequate facilities for management of forest;
- To protect wildlife through proper management and scientific studies; and
- Weaning of tribal people away by persuasion from the baneful practice of shifting cultivation.

The above national forest policy also highlighted the need for protection of rare animals such as lions, tigers and rhinoceros. National level planning was also done in forestry by establishment of the National Commission on Agriculture (NCA) in 1976. The NCA focused on the importance of conservation of forest resources. The Commission classified forest lands into protection forest, production forest and social forest. The concept of social forestry was highlighted by the NCA. The NCA also recommended for the establishment of Forest Development Corporation (FDC).

The implementation of the Forest (Conservation) Act, 1980 had resulted in significant reduction of forest land for non-forest purpose. However, the new National Forest Policy formulated in the year 1988, has been instrumental in conservation and protection of our forest resources. The main objectives laid down in the new forest policy are listed below:

- To maintain of environmental stability through preservation;
- To conserve the natural heritage with the vast variety of flora and fauna by preserving the remaining natural forest;
- To check soil erosion and denudation in the catchment areas of rivers, lakes and reservoirs in the interest of soil and water conservation and to mitigate floods and droughts and for the retardation of siltation of reservoirs;
- To increase substantially the forest/tree cover through massive aforestation and social forestry programmes, especially on all denuded, degraded and unproductive lands;
- To meet the requirements of fuel wood, fodder, and small timber;
- To increase the productivity of forests to meet essential national need.
- To encourage efficient utilization of forest produce and to maximize substi-

tution of the wood; and

- To create a massive peoples movement with the involvement of women and minimizing pressure on existing forest.

The importance of forests was also highlighted in different Five-Year Plans. In the First and Second Five-Year Plans, the emphasis was given to the rehabilitation of degraded forest, introduction of economic species and survey and demarcation of forest. The objectives of the Third and Fourth Five-Year Plans were to increase productivity of forest through fast growing special plantation, scientific assessments and modern logging. However, during the Fifth Five-Year Plan, the policy makers gave importance on social forestry and fuel wood reserves in order to save natural forests. The emphasis during the Seventh Five-Year Plan was on forest conservation, massive afforestation and wasteland development. The Eighth Five-Year Plan was very instrumental in conservation of our forest resources. The main emphasis in that plan was given on preservation of biodiversity, protection of forest against biotic interference, utilization of wasteland, and promotion of people's participation through Joint Forest Management Schemes.

The concept of National Forestry Action Programme (NFAP) came into existence in 1993. The main objectives of the new strategic planning process are to enhance the contribution of forestry and tree resources to ecological stability and people's oriented development through qualitative and quantitative measures. The document was prepared by the Ministry of Environment and Forests and was finalized in February 1998. The following specific objectives have been laid down in the draft documents:

- To effective conserve, rehabilitate, replenish, enhance, develop and manage the forest resources as renewable national assets to meet the important requirements of forest goods and services for the benefits of the people.
- To protect the land resource against degradation by deforestation, soil erosion, shifting cultivation, fire, grazing, flood and other natural causes and enhance the protection function of forest and trees.
- To protect wild flora and fauna, preserve biodiversity, conserve ecosystems and improve the environmental services of forests through maintenance, restoration and enhancement of a nation wide system of protected areas.
- To provide increased socioeconomic benefits to the people of the country by contributing income generation and poverty alleviation programmes.
- To develop a network with various institutions.
- To facilitate human resource development for forestry in qualitative and quantitative terms including education, research and training.

The establishment of the Indian Council of Forestry Research and Education (ICFRE) is a milestone in the development of research and education in forestry. There are 8 research institutes/centres under ICFRE. The National Afforestation and Eco-development Board is also playing important roles in rehabilitation of degraded forests and wastelands through state forest departments and non-government organization.

Forest types:

The forest type is a stable unit of forest vegetation. The characteristic feature of a forest type depends on the species composition, type of soil, density of tree cover and the geological history of the forest region. For example, in cool and high-latitude polar region, hardy evergreen forests such as fir, larch, spruce and pine are dominant. However, there are five major forest types found in India. They include tropical forests, montane subtropical forests, montane temperate forests, sub-alpine forests and alpine scrub. The tropical forests are again classified into various types such as wet evergreen forests and semi evergreen forests. Nevertheless, 16 different forest types have been recognized in India. Each forest type is described here with its salient features.

Tropical wet evergreen forests: These occur along the western side of Western Ghats and the coasts of Orissa, Bengal, Assam and Andamans. The area receives annual rainfall over 2500 mm or more and it has hot and humid climate. The mean annual temperature is recorded more than 23^0C. Important species found in this forest type include jamun, agar, mesua, cane and white cedar. Shrubs and forbs grow sparsely in the under storey. The tropical wet evergreen forests are classified into two subgroups - southern tropical wet evergreen forest and northern tropical evergreen forest.

Tropical semi-evergreen forests: The tropical semi-evergreen forests are found in the western coast, lower slopes of eastern Himalayas, Assam, Orissa and Andamans. Some important plant species found in this forest type are mango, semul, kadam, laurel, kanju, haldu and thorny bamboo. Trees become leafless for brief period. The mean annual temperature is about 24^0-26^0C. The mean annual rainfall is recorded at 1500-3000 mm. Like tropical wet evergreen forests, it is classified into two subgroups: southern tropical semi-evergreen forest and northern tropical semi-evergreen forests.

Tropical moist deciduous forests: The tropical moist deciduous forests range in the foothills of Himalayas, eastern part of Western Ghats, Chhota Nagpur, Khasi hills and throughout Andamans. The area receives 1500-2500 mm annual rainfall and the temperature ranges from about 21^0C to 27^0C. Sal, amla, kusum, common bamboo, teak, padauk, irul, semul, chikrasi etc. are the predominant trees in this forest type. The leaves of some tree species are shed for short period, while some other plant species are evergreen and semi evergreen in nature. The tropical moist deciduous forests are classified into 3 subgroups: Andamans moist deciduous forest, north Indian moist deciduous forest and south Indian moist deciduous forest.

Littoral and swamp forests: A forest growing at near the seashore is called littoral forest. They are found all along the coasts. The swamp forest occurs along the deltas of bigger rivers. Plants predominant in these forests include sundry, keora and agar. The mean annual temperature ranges from 26^0C to 29^0C and the rainfall varies from 760 mm to 5000 mm.

Tropical dry deciduous forests: The entire Indian Peninsular region is under tropical dry deciduous forests. The mean annual temperature ranges between 23.5^0C and 29^0C. This area receives about 700- 1500 mm rainfall. Species dominated in this forest types include tendu, teak, anjan, sal, palas, common bamboo, laurel and ghont. Trees are of moderate sizes with sparse canopy. The majority of trees shed their leaves during the dry season and remain bare until the beginning of rain. The tropical dry deciduous forests are classified into two subgroups: southern tropical dry deciduous forest and northern tropical dry deciduous forest.

Tropical thorn forests: The tropical thorn forests are found in Rajasthan and the adjoining areas such as south Punjab, Deccan plateau and upper Gangetic plains. The annual precipitation varies from 250 mm to 750 mm. The mean annual temperature varies form 24^0C to 28^0C. Trees found in this forest type are neem, sandal wood, dhaman, kanju, khejra, and khair. It is classified into two subgroups- southern tropical thorn forest and northern tropical thorn forest.

Tropical dry evergreen forests: They occur only along Carnatic coast from Tirunelveli to Nellore. The mean annual temperature ranges from 28^0C to 29^0C. The average rainfall varies from 870 mm to 1270 mm. The area receives little or no rainfall during summer. Important plant species represented in the tropical dry evergreen forests include neem, palm, cane, jamun, kokko, tamarind, ritha and machkund.

Subtropical broad-leaved hill forests: The subtropical broad-leaved hill forests are found in the lower slopes of the Himalayan region of Assam, West Bengal and other hill ranges such as Khasi, Nilgiri and Mahabaleswar. The mean annual temperature ranges between 16^0C and 22^0C. The average rainfall is recorded from 1500 mm to 11400 mm. The important tree species such as rhododendrons and jamuns are found in this forest type.

Subtropical pine forests: It occurs in central and north-western Himalayas. The mean annual temperature ranges between 5^0C to 20^0C. The area receives 1000 mm to 3000 mm annual rainfall. Oak, rhododendron, jamun and chir are some important trees found in this forest type.

Subtropical dry evergreen forests: The subtropical dry evergreen forests develop in the Siwaliks and the western Himalayas with an elevation up to about 1000 metres. Acacia, olive and pistacia are mostly found in this forest type. The area receives not more than 1000 mm rainfall annually. The salient characteristics of this forest type include long hot and very dry season, and cold winter.

Montane wet temperate forests: They are found in the eastern Himalayas at an elevation between 1800 and 3000 metres and tops of southern hills. Trees dominated in the montane wet temperate forests are birch, Indian chestnut, magnolia, and cinnamon. The mean annual temperature varies from 12^0C to 17 ^{0}C. The annual rainfall is recorded from 1300 to about 6000 mm.

Himalayan dry temperate forests: The Himalayan dry temperate forests occur in the inner ranges of Himalayas such as Ladakh, Chamba, Garhwal and Sikkim. The mean annual temperature is recorded between 6^0C and 17^0C. This region receives annual rainfall ranging from 80 mm to 800 mm. Rainfall mostly occurs during winter. Trees dominated in this forest type are maple, oak, deodar, olive etc.

Himalayan moist temperate forests: The Himalayan moist temperate forests extend along the entire length of the Himalayas at an elevation between 1500 and 3300 metres. Trees represented in this forest type are maple, birch, oak, chestnut, fir etc. The mean annual temperature ranges between 13^0C to 16 ^{0}C. With the exception of the northwest, the annual rainfall varies from 1100 to 2500 mm.

Sub-alpine forests: They are found at the altitudinal zone of 2900 to 3500 metres or more. The mean annual temperature varies form 2^0C to 6^0C and the mean annual rainfall is recorded from 80 mm to 650 mm. Snowfall is a regular feature of this region.

Moist alpine scrub: The moist alpine scrub is found along the entire length of the Himalayas at an elevation of about 3500 metres. Birch, rhododendrons etc are represented in this forest type.

Dry alpine scrub: It occurs in the altitudes over about 3500 metres. Juniper, honey suckle etc develop in dry alpine scrub.

Forest and wildlife:

A forest is usually described as a large area which is densely covered with trees. Naturalists, however, see a forest beyond the trees alone. According to them, it is an intricately interwoven community of living things- plants, animals and microorganisms. Each member plays a role in the continuing cycle of life, death, and rebirth in the forest community. Many living creatures make forests their home. Each forest type has its distinct plant and animal life. Forest is treated as one of the healthy factors that promotes excellent mode of preservation of wildlife. Many animals have adapted their mode of lives accordingly. True forest birds, for instance, are poor fliers.

Animals are normally found in specific layers rather than a random mix in the forest and they are adapted accordingly. Most of the birds that live in trees rarely come down to ground level. Studies reveal that animals living in each level of a forest differ both in their mode of locomotion and feeding habits. Mammals living on the ground are herbivores (e.g. deer and elephant), mixed feeders (e.g. Malayan bear) and carnivores (e.g. leopard and tiger). Herbivores browse on leaves and fallen fruits. Animals occurring in the middle level and canopy of the forests have little contacts with the forest floors. Their modes of locomotion are also different. Some are adapted for climbing, leaping and swinging (e.g. monkey and gibbon), while others glide for long distance (e.g. flying lemur and flying squirrel). Mostly they are fruit and insects eaters, though some are carnivores (e.g. clouded leopard and marten). Birds living in different strata of forests also have their different movements and feeding habits.

Among all the major forest zones found around the world, the tropical rain forests have the richest and most diverse plant and animal lives. It is blessed with various animals ranging from the largest land mammal (elephant of Africa and Asia) to tiny animals (mice and shrews). In tropical Americas, tapirs and jaguars are, however, the largest mammals. Animals found in the rain forest also have remarkably structural adaptations to the mode of life particularly those living in the tree tops. The prehensile tails of the New World monkeys, wing-like skin flaps of the flying squirrels and lemurs, the limbs of sloths, the beaks of parrots (to adapt in cracking nuts and to assist in climbing) and the stiff tails of wood peckers (use for climbing) are some examples of such adaptations. Primates like gorillas and chimpanzees of Africa live nearly on the ground. Limbs of these animals are adapted for walking and climbing. They, however, do not venture to climb high in trees. Some animals in the tropical rain forest live below the soil surface to depth of three or four feet (e.g. armadillos). However, they spend much time above the ground. The above facts highlight that the forest and wildlife are interrelated with each other. Therefore, the studies on wildlife can not be done in isolation without forest.

Chapter 8

Organizations and Institutions Involved in Wildlife Conservation

Introduction:

A large number of organizations have been actively involved in conservation and protection of precious flora and fauna all over the world. These organizations have been a constant source of inspiration to naturalists. There is no denying the fact that many wild flora and fauna are threatened with extinction. Therefore, the role played by many national and international organizations and institutions towards conservation is unique. The conservation of wildlife is achieved through research, education and extension activities. The international organizations like the International Union for Conservation of Nature and Natural Resource (IUCN) and the World Wide Fund for Nature (WWF-Nature) are widely known for their roles in conservation of wildlife throughout the world. Indian organizations such as the Bombay Natural History Society (BNHS), the Wildlife Institute of India (WII) and World Wide Fund for Nature-India have been instrumental in creating awareness among people about the importance of wildlife through their multidimensional activities. The roles played by these organizations in propagation and conservation of the wild flora and fauna are widely known throughout the length and breadth of the country. Therefore, it is important to know the activities of these organizations. The following text highlights the roles played by various institutions and organizations.

National Board for Wildlife (NBWL):

The National Board for Wildlife (earlier called Indian Board for Wildlife) was established in 1952 with the aim at advising the government on matters pertaining to the protection, preservation and improvement of wildlife. This is the highest advisory body of the Government of India regarding wildlife. The board consists of eminent people from different fields. They include ecologists, naturalists, environmentalists, biologists and policy makers. Some well known social activists from non-governmental organizations (NGOs) also represent the member of the National Board for Wildlife. The board was reconstituted in the year 1980 in order to give a fresh impetus to the conservation of wildlife throughout the country. However, the NBWL was instrumental in the enactment of the unified and comprehensive legislation of the Wildlife (Protection) Act, 1972. The main functions of the board are as follows:

- It guides the Central and State Governments on matters relating to wildlife through legislative and practical measures.
- It promotes interest and education in wildlife conservation.
- This body also helps in solving certain policy matters concerning skins, fur, trophies and other wildlife derivatives.
- It has an immense role in the establishment of protected areas such as national parks, sanctuaries and biosphere reserves.
- Finally, the board reviews the progress regarding the activities of wildlife at regular intervals.

Wildlife Advisory Board:

The Wildlife Advisory Board is an apex statutory body of each State Government advising on matters relating to the conservation of wildlife. The Wildlife (Protection) Act, 1972 has made a mandatory provision for the state to set up a State Wildlife Advisory Board. The members representing this board are selected from the various fields. They include the members of the State Legislature, Secretary (Department of Forest), Chief Wildlife Warden, an officer nominated by the Director, Forest officer in-charge of the State Forest Department, officers of the respective State Government and other representatives who according to the State Government are interested in protection of wildlife. The minister in charge of forest or the Chief Secretary (where there is no separate ministry of forest) is the Chairman of the Wildlife Advisory Board. However, the board has the following functions:

- To formulate the policies pertaining to the protection and conservation of wildlife.
- To select areas to be declared as national parks and sanctuaries.
- To formulate some guidelines related to various social issues including people living within the protected areas.

The success to conservation of wildlife in respective state largely depends on the effective roles played by the Wildlife Advisory Board. It also plays a vital role in understanding the local situations and need concerning of Wildlife.

Bombay Natural History Society (BNHS):

The Bombay Natural History Society is a widely known institute in the field of wildlife. This premier institute was started by seven eminent people of Bombay as a private organization in 1883. Since its inception, the society has actively been involving in various scientific activities for the preservation of vast flora and fauna in the Indian sub-continent. This organization has been instrumental in the field of wildlife conservation through research, education and extension activities. The society disseminates knowledge pertaining to wildlife through publications, lectures, field visits and films. The society publishes a reputed journal that covers a wide spread of indispensable information on oriental flora and fauna. It has also the credit of publishing many highly acclaimed books. The handbook of the birds of India and Pakistan by Salim Ali and S. Dillon Repley, for example, is said to be a store house of knowledge in the field of oriental birds.

The society has a rich collection of mammals, birds, reptiles, fishes, amphibians and insects. These enormous collections have been a source of inspiration to students, scientists, naturalists and many other professionals from various fields. The organization has been conducting research in collaboration with many international institutes. Apart from the research and publications, the society also conducts post graduate and doctoral programs in the field of ornithology, herpetology, and mammalogy. The BNHS has two permanent research stations: one at Point Calimere Sanctuary (Tamil Nadu) and the second one at Keoladeo Ghana National Park (Rajasthan). The society represents as the member to various boards such as the National Board for Wildlife and the Wildlife Advisory Board of many states.

Wildlife Institute of India (WII):

The Wildlife Institute of India was established in Dehradun (Uttaranchal) in 1982 with an objective of becoming the nerve centre for wildlife training, research, publication and extension in the country. This is an autonomous body registered under the Societies Registration Act, 1860. The assistance was sought from the United Nations Development Project (UNDP) and Food and Agricultural Organization (FAO) for the establishment of this institute. Before the WII came into force, the wildlife management course, however, was taught for some period at the Indian Forest College and later on it was under a separate Directorate of Wildlife Research and Education in Forest Research Institute (FRI). The WII society, the important body of this institute, is headed by the Union Minister for Environment and Forests. Other members of this body include some State Forest Ministers, nominated MPs and MLAs, officials from many Central Government Ministries and Departments, representatives from NGOs and eminent people from different walks of life. The important function of this institute is to promote the conservation of wildlife through education, research and training programmes. Nevertheless, the followings are the general objectives of WII.

- To train personnel (e.g. park managers and biologists) for protected area management and wildlife research.
- To train education and extension specialists for protected area management and wildlife research.
- To provide orientation courses for those who are involved in land use management.

- To conduct and coordinate applied wildlife research including the development of techniques appropriate to Indian conditions.

- To create a database by employing modern computerized analytical techniques for building a wildlife information system.

- To provide advisory and consultancy services to Central and State governments, universities, research institutions, other official and non-official agencies.

The institute has three faculty divisions: Wildlife Biology, Wildlife Management and Wildlife Extension. It has also two units known as Eco-development Cell and Environment Impact Assessment (EIA) Cell. Important courses offered by this institute include post graduate diploma in wildlife management, certificate course in wildlife management and M.Sc. in wildlife biology. Moreover, capsule courses like wildlife management, zoo management, and wildlife protection, law and forensic science are organized for specialists in different fields. Research is an important part of this institute. The research projects covering ecological, biological, socio-economical and managerial aspects are conducted in wilderness areas across the country. The Research Advisory Committee (RAC) screens research proposals and monitors the progress of WII research projects. Projects like conservation of Indian wolf, impact of fragmentation of the biological diversity of rain forests, small mammals and herpeto fauna of the Western Ghats mountains and development of Indian cooperative wildlife health program have been conducted with foreign collaboration.

The institute has a well developed national wildlife database centre. It also publishes manuals, books, technical reports etc regularly. Faculty members are available for consultancy services to Central and State governments and other agencies. The WII is an important regional centre in the south and south-east Asia for training and education in wildlife management and conservation.

Wildlife Preservation Society of India:

This non-government organization was established in 1958 by a group of wildlife enthusiasts. It is located at Dehradun, Uttaranchal. The society tries to promote interest and impart knowledge in the conservation of our varied and colourful wildlife. It assists the government in the formation, protection and maintenance of national parks and sanctuaries of India. The society promotes wildlife tourism in order to create an interest of our rich heritage. It also disseminates knowledge by means of journals, bulletins, monographs, films etc. The society conducts tour to local national parks and sanctuaries for students and other visitors. It brings out a bilingual journal called "Cheetal" which provides informative articles on various aspects of wildlife.

Central Zoo Authority (CZA):

The Central Zoo Authority, the highly empowered statutory body established in 1992, has been created under the provision of the Wildlife (Protection) Amendment Act 1991. It is a unique zoo organization in the world. The main purpose of creating such a highly empowered body is to give more thrust on the conservation and welfare of the zoo animals in the country. The establishment of CZA, therefore, fulfils a long felt need to the better management of the existing zoos in the country. It consists of ten members and one whole time member secretary. It is chaired by the Minister of Environment and Forests, Government of India. Two committees namely the Administrative committee and the Technical committee have been constituted for carrying out the functions of the CZA. The main functions of CZA are as follows:

- To specify the minimum standards for housing and veterinary care of the zoo animals.
- To provide technical and other assistance to zoos for their proper management.
- To identify endangered species of wild animals for the purpose of captive breeding to be assigned to a zoo.
- To restrict the power of a zoo regarding acquisition or transfer of any wild animal specified in Schedule I and II.
- To ensure maintenance of stud book (a comprehensive record of pedigree) of endangered species of wild animals bred in captivity.
- To recognize or derecognize a zoo.
- To evaluate and assess the functioning of zoos with respect to standards or norms as prescribed by CZA.
- To coordinate research in captive breeding and educational programme.

However, the CZA with the collaboration of the Department of Biotechnology (DBT) has given financial assistance to the Centre for Cellular and Molecular Biology (CCMB) for an ambitious project called "Laboratory for Conservation of Endangered Species (LaCONES)". The project will thrust on various aspects like DNA fingerprinting, establishment of gene resource bank, semen analysis, artificial insemination, in vitro fertilization, embryo transfer and cloning.

World Wide Fund for Nature-India (WWF-India):

In 1969, the World Wide Fund for Nature-India was registered as a Public Charitable Trust under the Bombay Trust Act 1950 with the aim at promoting education and research related to the conservation of flora and fauna, water, soil and other resources. Now it's headquarter is located in New Delhi. The WWF-India has 27 regional offices spreading throughout the country. Since its inception, this organization has been orchestrating a large number of conservation projects relating to wildlife. The much highlighted wildlife conservation projects assisted by this organization are the project tiger, the Gir lion ecological research project, the project hangul and the Himalayan musk deer project. The organization has also widened its capacity for monitoring of wildlife trade through TRAFFIC-India. Another important role taken by the WWF-India is to promote conservation education programme in educational institutes. The nature club of India has been set up in four different regions of the country in order to mobilize youth for conservation of wildlife. The WWF- India also publishes a quarterly newsletter containing useful information on wildlife.

Wildlife Trust of India (WTI):

The Wildlife Trust of India is a non-profit conservation organization located in New Delhi. This organization is committed to urgent action that prevents destruction of country's wildlife. The WTI works on a set of conservation related programmes in India through a team of professionals comprising biologists, veterinarians, communication specialists, lawyers and others. This organization has specific species programmes to protect endangered species like rhinoceros, elephant, tiger, sloth bear etc.

Other national organizations:

Besides the above mentioned organizations, many other societies have also been instrumental in the conservation of wildlife throughout the country. The Zoological Survey of India (ZSI) has been carrying out surveys on the faunal resources of the country. The Red Data Book on Indian animals has been published by the ZSI. Indian Institute of Forest Management in Bhopal and the Indian Council of Forest Research and Education in Dehradun are actively engaged in survey, education, research and training in forestry. The Centre for Ecological Studies of Indian Institute of Science, Bangalore is involved in studies on ecology and wildlife of the Western Ghats. Other organizations like Centre for Science and Environment Education, Botanical Survey of India, Ranthambore Foundation and National Museum of Natural History have been promoting environmental education and conservation of natural resources through various activities.

International organizations:

A large number of international organizations are involved in the conservation of wild fauna. Some notable organizations are World Wide Fund for Nature (WWF-Nature), World Heritage Sites, Green Peace International, IUCN, Whale Conservation Institute, United Nations Environment Programme, Wetlands International, the International Council for Birds Preservation, the Food and Agricultural Organization, the Smithsonian Institute, the International Wildfowl Research Bureau and World Conservation Monitoring Centre.

The World Wide Fund for Nature (WWF-Nature) was established in the year 1961. It is located at Glands, Switzerland. The main function of this organization is to provide financial support for various conservation projects. The WWF-Nature in close collaboration with the IUCN also guides by providing technical expertise on wildlife projects. It has other activities like awareness programmes against degradation of environment. The WWF-Nature has a number of centres in many countries. It makes a world wide effort to raise funds for various activities.

The International Union for the Conservation of Nature and Natural Resources (now called World Conservation Union) is widely known for its stimulating role in the conservation of natural resources. It facilitates cooperation between governments and national and international organizations for wildlife conservation. The IUCN through its Species Service Commission (SSC) also helps to protect animals threatened with extinction. Many specialist groups such as Asiatic wild cattle specialist group, Asian elephant and rhino specialist groups, lagomorph specialist group, primate specialist group and cat specialist group have been formed under the aegis of the IUCN.

The International Council for Birds Preservation (ICBP) was established by a group of ornithologists in 1992. It's headquarter is located in Cambridge, England. The main objective of this organization is to preserve the rare and threatened species and the habitat in which they live.

The Global Tiger Forum was established in 1994 with the aim at saving tigers from imminent danger of extinction. The organization was established with the following objectives:

- To work out the effective strategies for future implementation for the survival of tigers.
- To improve the habitat conditions and promotion of conservation programmes.
- To coordinate multi-pronged research efforts to generate a proper database accessible to all tiger range states.
- To campaign against the use of tiger bones and other body parts for medicinal derivatives throughout the world.
- To develop an effective intelligence system to get information about poaching, poachers and their network.

Chapter 9

Zoos and their Management

Introduction:

The zoo can be defined as a place where wild animals are exhibited in captive condition. According to the Wildlife (Protection) Act 1972, a zoo is "An establishment whether stationary or mobile where captive animals are kept for exhibition to the public but does not include a circus or an establishment of a licensed dealer in captive animals." Zoos are also known as zoological gardens or zoological parks. Usually, a small sized zoo is called zoological garden whereas a medium to large sized zoo is known as zoological park. However, a zoo has been classified by the Central Zoo Authority (CZA) as large, medium and small or mini zoo depending on the area, number and variety of animals kept and number of visitors visit the zoo (Table 9.1). Zoos can be considered as living museums. The modern zoo plays an important role in conservation and protection of many endangered wild animals. Many threatened animals like lions, tigers, primates, deer and various birds and reptiles are kept in many zoos of the country. At the time of independence, there were 22 zoos in India. There has been a phenomenal growth of zoos in recent time. The country at present boasts of having more than 300 zoos but only 164 zoos have been recognized by the CZA. However, the Government of India has made a laudable effort by establishment of the Central Zoo Authority for management of zoos more efficiently.

Table 9.1 : Classification of zoos as laid down by CZA:

Criteria	Large zoo	Medium zoo	Small zoo	Mini zoo
Area of the zoo	More than 75 hectares	50-75 hectares	20-49 hectares	Less than 20 hectares
Number of animals exhibited	More than 750	500-750	200-499	Less than 200
Animals variety exhibited(in numbers)	More than 75	50-75	20-49	Less than 20
Number of endangered species exhibited	More than 15	10-15	5-9	Less than 5
Annual attendance of visitors per year(in lakhs)	More than 7.5	5-7.5	2-5	Less than 2

History of zoo:

Since very early periods, wild animals have been kept in captivity. Domestication of animals was necessary for some obvious reasons such as food, transportation and protection. The earliest records of wild animals in captive condition can be brought into notice through some evidences such as pictures of antelopes (4000 years ago in Egypt), records of a house of deer (in China) and records of semi-domesticated elephants (in India). The first known zoo was apparently founded by Egyptian Pharaoh Thutmuse III around 1500 BC. However, the earliest zoo called Intelligence Park was established in China during the Chow Dynasty (1100 BC). It was established perhaps for the purpose of education as well as for respecting religious sentiments. The ancient history also reveals that King Solomon had a collection of 100 elephants, 4000 horses, over 100 antelopes and a large number of apes and deer. In ancient Rome and Greece, a great variety of wild animals were kept in menagerie (a collection of wild animals in enclosure especially kept for exhibition). Kings, princes, feudal lords, rulers and landholders usually used to keep animals for family entertainment. Sometimes, these animals were given to other dignitaries as a symbol of love and gift. It is said that the Mughal Emperor Akbar and Jahangir kept cheetahs as pet animals. Many other emperors also used to keep leopards, tigers, lions, wild asses and hippopotamuses for various purposes. Those privileged people in fact had kept wild animals for amusement of the society as well as to satisfy their inherent curiosities. The motive behind keeping these animals was also to establish their status in the society. Later on, however, people were interested in keeping animals for many other objectives. They were more interested in catching and keeping wild animal for the following purposes:

- to hunt or keep animals for food;
- for transportation;
- for decoration purpose;
- for companionship and curiosity; and,
- for worship (animal is believed to be god or symbol of power).

Nevertheless, the modern zoo keeping is said to have started in 1752 with the foundation of Imperial Menagerie at Schonbrunn palace in Vienna. This was opened to the public in 1765. Another zoo was established in a Royal Park in Madrid in 1775. The term "zoo" was first used in the late 19th Century as a popular abbreviation of the zoological garden in Regent's Park, London. Alfred Vance also first used it about the London zoo in 1876 in a popular song. Since the end of the World War II, there has been a rapid development of zoos throughout the world.

The concept of zoo in India is said to have started with the first domestication of elephants some 4500 years ago. Today's zoological parks and zoological gardens came into existence in the 19th Century. Many zoos in the early periods were set up next to the existing parks or gardens. Bombay zoo, for example, was established as a part of the Victoria Garden. A large number of zoos were created under the patronage of local provincial kings in Princely States. Some zoos such as Gwalior zoo, Jaipur zoo, Junagarh zoo and Baroda zoo are examples of this kind. In India, the first aviary/zoo, however, was started by Raja Rajendra Mullick in Calcutta as early as 1854.

Recently, the National Zoo Policy has been formulated by the Government of India with the aim at improving the conditions of the zoo animals. Under this policy, no new zoos can be established unless it is urgently required. Moreover, no new animals are to be acquired unless it is essential for maintaining the genetic viability of the zoo population. It has also been proposed in the above policy that no zoo shall be allowed to establish any recreation or entertainment infrastructure within its campus. The new zoo policy has given more emphasis for maintaining animals on scientifically drawn management principles. As per the guidelines of the policy, zoos would be allowed to organize their collections of endangered species into genetically and behaviourally viable groups through exchange, gifts, loans etc. Today zoos are the home of many migratory birds. The Delhi zoo, for instance, is the shelter for many winter visitors who come all the way from Europe, Siberia etc. It is said that about 700 migratory ducks including common teals, pintails, mallards and shovellers are the winter guests of Delhi zoo.

There are 370 species of wild animals found in different zoo of the country. Sizeable populations of wild animals are found in various Indian zoos. The large numbers of captive animals are now causing genetic depression and health hazards. This also becomes a financial burden to the zoo authority. Another problem faced by the zoo administration is that captive animals can not be released into the wild habitats because of the fact that the animals have lost their natural instincts and behaviour. Therefore, the CZA has recommended for reducing the number of population drastically in overpopulated zoos.

Objectives of zoo:

Zoos are established for various reasons. The purposes for which they are founded worldwide are described below:

Conservation and breeding: Many zoos in the world serve as a breeding centre for animals which are in danger of becoming extinct in the wild state. Therefore, the roles played by zoos for conservation of wild animals are immense. The endangered wild animals can be bred and reared up to maturity stage in the zoo and subsequently they may be released into their natural states. In this way, there would be a viable population of endangered animals in the captivity as well as in the wild habitats. Cheetah, the highly endangered animal listed in the Red Data Book of IUCN, is being successfully bred in a few American and European zoos in order to save this animal from near extinct. It is reported that total number of Hawaiian goose was estimated at only 50 in 1947. Realizing its immediate danger of extinction two geese were brought to the Wildfowl Trust at Slimbridge, England for the purpose of captive breeding. They bred successfully and thus saved from extinction. The captive breeding of the brow-antlered deer is taking place in India so that the present population of this most endangered mammal in Asia can be increased. There are many more examples where captive breeding of many threatened animals has been conducting. In this way, a large number of endangered species can be saved from extinction in near future. Zoos, however, also provide a refuge for animals whose natural habitats are threatened. Zoos can also replace ageing and nonproductive animals through scientific breeding programme. Today, there is a growing emphasis towards fostering the breeding of rare animals in the leading zoos of the world. To achieve this, all management facilities should be available in the zoos so that the endangered animals have a chance for survival. The emphasis should be given to three major objectives of captive breeding programmes:

(a) to propagate and manage captive populations of highly endangered animals in order to prevent their immediate extinction;

(b) to develop stable captive population of rare and endangered wild animals for education programmes; and

(c) to employ programme as a part of conservation strategies to ensure ultimate survival in the wild.

These objectives can be achieved by three important breeding levels. Firstly, the individual animal should have sufficient longevity and physical well being. Secondly, sufficient reproduction should occur at the level of breeding pairs and colonies in order to have secured continuity over the generations and finally, conservation at the population level should be firmly assured.

Study and investigation: Continuous studies and investigations are very important to get new information about captive animals. Close contact with animals by zoo researchers and observers help in collection of data on behaviour, feeding, reproduction, diseases and many other fields of activities. These observations are again properly studied and the results obtained from these studies make it easier to manage animals. The results obtained from studies on captive animals are also important to safe guard the continued existence of the animals in the wild state. In this regard, many zoos publish scientific journals and periodicals for dissemination of knowledge and information. In recent years, there has been qualitative improvement of our knowledge on various fields like breeding, feeding and veterinary care of captive animals. Moreover, for proper studies, it is important to have the basic data (e.g. individual identification, birth rate, death rate, sex, place of capture, if wild caught etc.) of each animal for the purpose of population analysis and management.

Education: Zoos provide an educational programme to the people especially children and students. It also communicates the people about the importance of animals to our life support system. Zoo educates visitors about various aspects of wild animals. Many people feel that zoo is very helpful for direct educational purpose.

Recreation: Zoo provides recreation for the leisure time of visitors. It is a place for mental and physical relaxation. Showmanship is an important aspect in zoo management. This can be best achieved by displaying the animals in an environment which is most suitable to them.

In the end, therefore, it can be said that modern zoo is an educational research institute that serves the purpose of conservation and entertainment. They play a commendable role in saving certain species of mammals, birds and reptiles which are virtually on the brink of extinction in their natural states. European bison, Pere Davis's deer and Hawaiian goose are some examples in this regard.

Zoo layout:

There may be three types of zoo layouts. A main stream or major zoo requires sufficient space and is properly exhibited by mammals, birds, reptiles, amphibians and fish. Many invertebrates are occasionally housed in a major zoo. Nevertheless, major zoos have restricted their collections to a few orders now a day. This is done due to the fact that animals from selected orders and families can be better managed with greater details. The zoo animals can be exhibited on the basis of their common principles such as taxonomy, zoogeography, ecology and ethology. The taxonomic arrangement of the exhibits is usually followed in the older zoos. In this type of arrangement, different animal groups are classified according to their taxonomic positions and are housed in separate enclosures. The layout of animals based on zoogeography is grouped according to their place of origins. Related groups may be kept in adjoining areas and they may be separated by concealed moats. Ostrich and zebra, for example, are housed together on this principle. In ecological arrangement, animals are housed on the basis of their habitats. In ethological type of layout, the collection is exhibited according to the behavioural attributes of animals. Thus, aquatic mammals such as seals, walruses, sea lions and dolphins may be arranged as per ethological principle.

The specialized zoo is concerned with animals belonging to a particular order or family. Sometimes, it displays a particular species. Animals in this zoo are classified as per taxonomic and zoogeographical themes.

Many zoos have the small mixed collections. It has a small area covering 5 acres or even less. The main objective of small mixed collection is to attract public. Chimpanzee, macaw, penguins, sea lions etc. are kept in this type of zoo layout. However, different types of zoo exhibits should have the following considerations:

- comfort of the animals;
- proper visibility to the zoo visitors;
- properly and conveniently serviced by the keepers; and
- hygienic and aesthetically pleasing to the public.

Zoo design and housing:

The designing of a zoo is the art of combination of engineering, biological and environmental sciences. The proper design and housing are vital for keeping animals healthy and productive. Zoo building should be designed and constructed in such a way so that it looks like natural settings as far as possible. The building should be comfortable to the animals. There should be adequate space for various activities such as loafing, lying down and rearing the young. It must have easy accessibility to serve and capture of captive animals. The designer must consider the behavioural aspects of the animals to be displayed in the zoo. It should also consider the proper visibility to the zoo visitors. Some important factors that need to be given due importance while designing a zoo house include height of the enclosures, provision for shelves, platforms, service doors, wall and floor surface, type of food containers, door for the cages, proper arrangements of lights, proper ventilation, control of temperature , drainage system and depth of moats. A proper plantation may be done inside and outside of the enclosures. The topography of the outer enclosure should be taken into account while designing a zoo house. The topography may be flat, sloping or undulating depending upon the types of animals to be kept in the enclosures. The other accessory structures should also be given an equal importance in order to keep the zoo most attractive to the public. They include (a) the outer boundary fence, (b) road and parking area for cars, (c) enclosures for the animals, (d) staff office, work rooms and lecture rooms, (e) store rooms for tools, machines, food etc., (f) veterinary hospital, quarantine sheds etc, (g) disposal of water and drainage system and (h) entrance, catering and toilets.

Enclosures may be closed or large open fields. A zoo house should be constructed in such a way so that it meets the basic requirements such as sufficient space, comfort, safety from predators, sufficient distance from other animals or human visitors, well visibility of the animals, easy veterinary care and easy management without putting the animals under stress. Enclosure barriers may be of fences, bars, moats or glasses. Moats are very frequently used in modern zoos. 'U' shaped moats are normally used for tigers, bears, lions, elephants, rhinoceros etc. The water moats are important for swimming purpose. Moreover, the moated enclosure makes the viewing natural. Floors may be of permanent and non-permanent types. Tiles or cement may be used for preparation of a floor. The replaceable or non- permanent floors are changed according to the need and maintenance of hygienic condition of the animals. Sand and gravel are kept for reptiles. The walls can be made of bricks, woods or glasses. It can be coloured with nontoxic paints. Utmost precaution should be taken in wall painting. According to the need of the animals many accessories such as trees, ladders, rocks and posts are kept inside the enclosure. Artificial light should be provided for nocturnal animals exhibited in zoo. Adequate sleeping quarters in the form of dens (e.g. foxes, wolves and bears) or burrows (e.g. rodents) are also provided. Some animals such as giraffes must have access to warmth and shelter during inclement weather.

Zoo nutrition:

The health of a captive animal mostly depends on the feeding status of the animal. The types and requirements of food provided to the zoo animals are based on their feeding behaviour in the wild as well as their physiological status. It is a fact that a large number of animals exhibited in the captivity are actually obtained from their natural environment. Therefore, the food given to the captive animals should be as close as possible to their natural foods. This should be maintained at least during the early periods in captivity. Here, some food materials liked by animals in their natural environment are given below:

Plants: Animals like elephant, black rhinoceros, impala, eland, giraffe, deer, camel and leaf eating monkey feed on woody (e.g. shrub, bark and twigs with leaves) and herbaceous plant materials (e.g. leaves, succulent stems, shoots, roots, flowers and fruits). Therefore, the browsing animals should be provided for plant materials that are relished by them. The grazers such as zebras, white rhinoceros, llama, rabbit and Thomson's gazelle, on the other hand, consume leafy plants with tough fibres. Animals such as bear, some apes and monkeys, and many birds are highly adapted to eat fruits and seeds for their survival.

Large mammal and fish: The natural diets of carnivores are mainly based on mammals and fish. Animals like lion, tiger, cheetah, polar bear, wolf, weasel and fish eating birds feed on larger mammals and fish.

Invertebrates: Many mammals are best thrived on the diets consisting of invertebrates. Anteaters, lizards, echidnas and many birds, for example, feed on some invertebrates such as snails, termites, insects and spiders.

Small vertebrates: Some animals subsist largely on the vertebrates like fish, frogs, lizards, small mammals and small birds and their eggs. Snakes, owls, small hunting mammals, eagles etc are the examples.

Liquid food: Some animals consume liquid food like blood (e.g. vampire bats) and nectar (e.g. humming bird). Milk is the important dietary ingredient for mammals.

So the feeding aspect should be given utmost care for the overall well being of the animals. For details, see the feed and feeding portion of different mammals, birds and reptiles in the text of the husbandry and health care section.

Management of zoo:

There are many divisions in a zoo for proper management. However, the management of zoo mainly depends on its size and kinds of exhibits present in the zoo. In small zoos, many jobs are performed by a small number of people. Large zoos, however, have more people with diversified operations. Nevertheless, the management of a zoo can be broadly operated on five specialized areas of activities.

(a) **General administration:** This division of management is concerned with the staff and financial matters of a zoo.

(b) **Husbandry of animals:** This is a specialized area of a zoo management. It is concerned with animals and related keeper's work. This also deals with veterinary health care, recording of data, research and education. Individual animal's records such as birth, death, transfer, illness, injury, disease, treatment, feeding, behaviour and mating are given due importance in this section. So, the husbandry of animals is a very important part of a zoo.

(c) **Gardening and works:** The gardening and works department is essential for a zoo. It is concerned with the maintenance of buildings, periodical plantation and proper landscaping of the zoo. The beautification of a zoo is largely depended on this section.

(d) **Public relation:** It fulfils the visitor's wishes about a zoo. Visitor's perceptions about zoo environment are very important for improvement of a zoo.

(e) **Education, information and interpretation of the animals:** This is concerned with the establishment of a good library, screening films and related works on wildlife.

Staff position of a zoo:

It depends on the size of a zoo. Each staff has specific assignment in the zoo. A good zoo should have the following staff: director or curator and administrative staff, zoo veterinarians, accountant and clerical staff, staff for designing and building new enclosures, staff for preparing animal's diet, groundsmen and gardeners, keepers, education staff, guide and graphic artists, advertising and publicity staff and book stall and café staff.

Chapter 10

Wildlife Trade and Legislation

Introduction:

Many precious flora and fauna are already extinct from this planet and a large number of existing wild flora and fauna, as reported, are threatened with extinction in near future. The reasons for extinction or dwindling of wildlife population are manifold. It is a well known fact that poaching and illegal trading in wildlife flora and fauna are the most important factors for depletion of wildlife population. It is reported that 30,000 primates, 90,000 elephants ivory, 10 million orchids, 4 million live birds, 10 million reptile skins and 15 million furs are under global trade. It is also learnt that about 250 species of birds in India are trapped for food, pet, medicine and other purposes. The annual trade is reported to be worth 20 billion US $ which is next in value to narcotics and illegal arms trade. USA is the largest consumer of wildlife.

Realizing its gravity, many legislations were framed in order to protect wild flora and fauna. During 3rd Century B.C. King Asoka made a law for preservation of wildlife and environment. In 1873, the British Government had passed the Madras Act to prevent indiscriminate destruction of wild elephants. The Constitution of India (under Article 48-A of the Directive Principles of State Policy) has clearly mentioned that "The State shall endeavour to protect and improve the environment and to safeguard the forest and wildlife of the country". The Indian Constitution has also stressed the importance of the fundamental duties of the citizens to protect wildlife and environment. It has been mentioned in part IV-A that, "It shall be duty of every citizen to protect and improve the natural environment including forests, lakes, rivers and wildlife and to have compassion for living creatures". The Constitution in its 73rd Amendment Act of 1992 on Panchayats has highlighted the importance of environmental protection and conservation.

Nevertheless, the year 1972 has been a landmark in the history of protection and conservation of wildlife in India. The Wildlife (Protection) Act 1972 was passed by the Parliament under Clause (1) of Article 252 of the Constitution. This is the most comprehensive legislation for control and management of wild animals including their habitats. This Act has been instrumental in protection and conservation of many mammals, birds, reptiles, amphibians etc. Prior to this Act, the protection of wildlife was, however, the subject to State rule. Most of these rules were framed as shooting rules under the Indian Forest Act. However, different Provinces had some special legislations to protect wild animals. The Bombay Wild Animals and Wild Birds Protection Act 1951, the Rhinoceros Preservation Act, the Elephant Preservation Act 1879 and the Game Act were some important legislations. The Wild Birds and Animals Protection Act 1912 was very significant for protection of wildlife resources. The other important acts such as Indian Forest Act (1927), Forest Conservation Act (1980), Environment Protection Act (1986) and Biodiversity Act (2002) are directly or indirectly help in protection of wild animals. The objectives for enactment of various wildlife laws are as follows: (a) to increase wildlife population particularly endangered, migratory and breeding animals; (b) to protect their habitats; (c) to assure desire and calculated wildlife harvesting; and (d) to protect interests of people (e.g. crops and orchards) from wildlife.

However, the Wildlife (Protection) Act, 1972 had some inadequacies and shortcomings and thereby it was necessary to amend this Act so that it fulfils the entire gamut of wildlife. Accordingly, it was amended by the Amendment Act, No. 44 of 1991 with effect from 2nd October, 1991. It is important to be noted that another new Chapter III-A providing the protection of specified plants has also been inserted in the Act. The other important features included in this Amendment Act include ban on wild animals hunting, restriction on transport of wildlife and plants, immunization of livestock, and enhancement of penalties. However, it was amended again and the Wildlife (Protection) Amendment Act, 2002 has come into operation with effect from 1 April 2003. It is a comprehensive revision of the Wildlife (Protection) Act, 1972. The two new categories of protected areas, viz., "Conservation Reserve" and "Community Reserve" have been proposed in the Act.

In spite of having this important legislation, poaching and illegal trading on wild fauna and flora are still persisting throughout the country. Therefore, the most important task at this juncture is the effective implementation of the existing laws. If this part is lacking, then no matter how stringent the laws, it can not achieve the desired goal for protection and conservation of our wildlife and nature. However, the National Board for Wildlife has approved a proposal to set up a Wildlife Crime Bureau on the lines of the narcotics bureau to strengthen intelligence gathering, interdiction, investigation and prosecution of wildlife crime.

Wildlife (Protection) Act, 1972 (as amended up to 2003):

This Act is more significant by saying that "An act to provide for the protection of wild animals, birds and plants, and for matters connected therewith or ancillary or incidental thereto with a view to ensuring the ecological and environmental security of the country." The addition of the words, "ensuring the ecological and environmental security of the country" is also significant. This Act is applicable throughout the country except in the State of Jammu and Kashmir. It has 6

Schedules:

Schedule I: Part I-Mammals.

                     Part II-Amphibians and reptiles.

                     Part II (A) - Fishes.

                     Part III-Birds.

                     Part IV- Crustaceans and insects.

                     Part IV-A: Coelenterates.

                     Part IV-B: Mollusca.

                     Part IV-C: Echinoderma.

Schedule II: Part I- Few mammals and lizards.

                     Part II – Beetles, snakes and few mammals.

Schedule III: Contain a few mammals.

Schedule IV: Most birds, rodents, snakes, tortoise, butterflies and moths.

Schedule V: Vermins.

Schedule VI: Specified plants.

CITES (Convention on International Trade in Endangered Species of Wild Fauna and Flora):

Many wild flora and fauna have been dwindling in their natural states. Lose of habitats, commercial utilization, pollution etc. are believed to be the important factors causing decline of many wild flora and fauna. However, over exploitation of living wild resources for trading purposes has been considered as one of the principal causes for depletion of wildlife resources all over the world. Wild animals are hunted for their flesh, hides, bones, furs, fats, oils, ivories, antlers and other by-products. It is said that about 47,000 elephants were killed annually between 1850 and 1890 to supply ivories in the London ivory market alone. The export of snake skins from India is estimated at 2.5 million pieces per year. Musk is sold at three times the price of gold. There are many more examples where illegal trades in wildlife resources are being practised in India as well as other parts of the world. Illegal trade in wildlife is said to be the world's number two largest illegal business. Report indicates that the global wildlife trade in animal and plant products is around $20 billion and one third of this is illegal. Realizing its grave situation, an international cooperation was essential to combat these illegal trades and thereby to protect certain species of wild flora and fauna against over exploitation through internal trade. Accordingly, an international treaty or convention was adopted in order to establish worldwide control over trade in endangered species and their products. This is popularly called the Convention on International Trade in Endangered Species of Wild Fauna and Flora (CITES).

On 3rd March 1973, 21 countries signed the historic Washington Convention. The convention came into force on 1st July, 1975 and India becomes party to CITES from 18th October, 1976. Its secretariat is located at Lausanne in Switzerland. The number of parties has risen from 36 in 1976 to 161 in 2003. Two-third members of this convention are the representatives from the developing world. India is one of the few countries that signed on CITES very early. Members of this convention meet once every two to three years to review the latest situations on trades in wildlife resources. The European Economic Community (EEC) is also the member of this convention. The various NGOs may also participate.

The salient objectives of this convention are as follows:
- To put a ban on trade in endangered species which are being threatened with imminent extinction.
- To monitor and regulate trade in endangered species.
- To regulate commerce in some 30,000 plants and over 2500 animal species through its system of appendices.
- To monitor illegal trade in other less endangered but potentially threatened species.

Nevertheless, the important aspect of the convention is to control the legal and illegal trade in species of wildlife in general and at the same time to ensure international cooperation in prevention or controlling trade in species threatened with extinction. The export, import and re-export of any specimen of the species listed in the Appendices (I, II & III) require the prior grant and permission to CITES. Two bodies namely the Management Authority and the Scientific Authority are very important for wildlife trade. In India, the Management Authority consists of Director General of Forests, Additional Director General of Forests (Wildlife) and the Directorate of Wildlife Preservation. The management authority in consultation with the scientific authority grants the permits and certificates for imports and exports on the biological and scientific matters of the proposed trade. The scientific authority, on the other hand, is represented by the Botanical Survey of India, the Zoological Survey of India, the Central Marine Fisheries Research Institute and Wildlife Institute of India. The function of the scientific authority is to help in examining the status of the species involved in the trade and also to assist the management authority in regulatory procedures. The issuing of licenses for export, import and re-export of wildlife, however, is under the control of the Chief Controller of Imports and Exports. Four cities namely Bombay (now Mumbai), Calcutta (now Kolkata), Delhi and Madras (now Chennai) are designated for import and export of animals. Two cities namely Tuticorin and Amritsar are designated for import and export of plants in India. The Director of Wildlife Preservation controls the trade at the centre while the Chief wildlife Warden issues the legal procurement certificate for trade at the state level.

All the member countries are required to submit their reports and trade data to the CITES secretariat. The Wildlife Trade Monitoring Unit (WTMU), a part of the IUCN Conservation Monitoring Centre at Cambridge (UK), analyses data on wildlife trade. In this regard, it is important to note that this convention does not interfere with the law of the land and it rather serves as an intermediate coordinating agency to bridge the gap in the control of wildlife trade. Moreover, CITES is not for banning trade but its purpose is to regulate international commercial trade in species threatened by trade. Therefore, the formation of CITES is a major step towards the protection and conservation of wildlife and more importantly, without this illegal trade can not be controlled.

There are three appendices in which the CITES has listed many species of animals and plants that are threatened or likely to be threatened very soon. Appendix I includes those species which are threatened with extinction and are affected by trade or may be affected by trade. The wild specimens are not permitted for international commercial trade. Appendix II includes those species which happen to be threatened unless trade is subjected to strict regulation. The species included here are endangered but a limited trade is permitted. International commercial trade in wild specimens is permitted and for which a proper documentation from exporting country is required. The Appendix III includes those species for which a range country (country of origin of species) may unilaterally seek international cooperation in the form of CITES certification for international commercial trade.

The recent proposal to transfer Southern African (e.g. Botswana, Namibia etc.) elephant population from Appendix I to Appendix II was passed by CITES. Therefore, these countries would be allowed to practice controlled trade in elephant products. Some people are of the opinion that this would benefit the rural communities and also would help in conservation and management of wildlife. The Asian elephants are, however, still in Appendix I of CITES.

TRAFFIC (Trade Records Analysis of Flora and Fauna in Commerce) – International:

TRAFFIC-International is a wildlife trade watchdog group of IUCN and WWF for- Nature. It was established in 1975 with the aim at monitoring wildlife trade worldwide. It collects and analyses data on the wildlife trade. It studies the status of species in the wild state. It also closely watches for violation of national wildlife protection laws. The head quarter of this organization is located at Cambridge, UK and its branches are located in various countries.

TRAFFIC- India was established on the first of January, 1992 as a division of WWF-India and a part of the global's TRAFFIC network. It closely monitors the wildlife trade. TRAFFIC- India also monitors the utilization of animals and plants and their derivatives. This organization also maintains database related with wildlife statistics. TRAFFIC-India supports government agencies through its legal, technical and other services.

Section 2
WILDLIFE BIOLOGY

Chapter 11

Zoological Classification of Mammals, Amphibians, Reptiles and Birds

Class : Mammalia —

The term "mammal" is derived from the Latin word *mamma* which means breast. Normally four legged animals are described as mammals. However, most reptiles are having four legs but they are not mammals. Dolphins and whales, on the other hand, are classified as mammals but they do not have legs. So mammals can be defined as a group of vertebrate animals whose young are nursed with milk which is secreted from the special secretary glands (*mammae*) of the mother. Mammary glands are only found in mammals. Moreover, there are many salient features by which we can distinguish mammals from other group of animals. The body of mammal is insulated by hair (except in some whales where they lost in adult stage). Hair become water proofed by sebum, the secretion of sebaceous glands found in hair roots. Another important feature of mammal is that its lower jaw is comprised of a single bone and is directly hinged to the skull. All mammals have a four chambered heart and only the aortic arch is present. The respiration takes place through the lungs. Another important muscular structure called diaphragm, which separates the heart and the lungs from stomach and intestines, is present in the mammals. The neck has seven cervical vertebrae (except in sloth and manatee). The mammalian brain is well developed. Males posses rudimentary nipples. Mammals have two sets of teeth: the milk or deciduous teeth and permanent teeth. Mammalian feet are adapted for walking, flying, climbing, running, swimming etc. The erythrocytes are non nucleated (the camel and the Australian marsupial koala, however, are reported to have nucleated erythrocytes in blood). Mammals are homoiothermic (warm-blooded) animals i.e. the body temperatures do not vary with their surroundings.

All mammals (except monotremes) are viviparous i.e. live bearing animals. Fertilization takes place internally. Mammalian estrous cycles are of three types: ovulation is induced by cervical stimulation during copulation and followed by a spontaneous luteal phase (in rabbits, cats, ferrets and camels); ovulation is spontaneous but the stimulus of copulation is essential to induce the luteal phase

117

(in rats and mice); and ovulation and the luteal phase are spontaneous (in primates, marsupial, cattle, sheep, goat etc.). In placental mammals, the young are developed within the mother's womb. Young are well developed (precocial). In pouched mammals the young, however, are less developed (altrical) at birth. Males have copulatory organs and majority of mammalian species have descended testes.

Classification of mammals:

The zoological classification of animals is based on their mutual relationship and affinities. All the scientifically described living mammals have been classified into a distinct class of the vertebrates, the Mammalia. Again mammals are sub-classified into three groups: Monotremes or egg laying mammals (Sub-class: Prototheria, Infraclass: Ornithodelphia); Marsupials or pouched mammals (Subclass: Theria; Infra class: Metatheria) and Placental mammals (Sub-class: Theria, Infraclass: Eutheria). The classifications are again divided into orders, families, genera and species. All the mammals have been classified into 20 orders. They are as follows:

Orders	Animals
Monotremata	platypus and echidna
Marsupialia	kangaroo, koala, wallaby, wombat, numbat, bandicoot etc.
Insectivora	mole, hedgehog, shrew , tenrecs, moonrat etc.
Tupaioidea	tree shrew
Dermoptera	flying lemur
Chiroptera	bat
Primates	ape, monkey, lemur, man etc.
Edentata	anteater, armadillo and sloth
Pholidota	pangolin
Lagomorpha	rabbit, hare etc.
Rodentia	rat, mouse, porcupine, squirrel etc.
Cetacea	dolphin, whale etc.
Carnivora	tiger, lion, cat, bear, panda, dog, hyena, weasel, mongoose, leopard, wolf, cheetah, civet, otter etc.
Pinnipedia	seal, walrus and sea lion
Tubulidenta	aardvark
Proboscidea	elephant
Hyracoidea	hyrax
Sirenia	manatee and dugong
Perissodactyla	horse, zebra, rhinoceros etc.
Artiodactyla	cattle, goat, antelope, deer, giraffe etc.

Class: Amphibia —

The word "amphibian" means double life. This is due to the fact that members in this group spend part of their lives in the water as well as on the land. Amphibians are thought to be the ancestors of all reptiles, birds and mammals. They are found throughout the world. Most species of amphibians are characterized in having long bodies, four small limbs and a large tail. They differ from the reptiles by having soft moist skins that usually lack scales. Moreover, their eggs have no hard shells. They do not have hairs or feathers. Most species of these cold blooded vertebrates are brightly coloured. The respiration takes place through gills, lungs and skin. They mostly rely on vision for food. Some are voiceless. Majority male amphibians call out to attract females during mating. Most species breed for only a limited period in a year.

Amphibians have been classified into three orders: Apoda or Gymnophiona (caecilians); Caudata or Urodela (salamanders and sirens); and Anura (frogs and toad).

Class: Reptilia —

Reptiles are widely distributed in both temperate and warm regions of the world. However, they are abundantly found in the tropics. The members belonging to the class Reptilia include crocodiles, snakes, lizards, tuataras, tortoises and turtles. Reptiles can be distinguished in having a dry and scaly skin, absence of hairs, feathers, mammary glands and presence of a few or no skin glands. The jaws are set with simple teeth. They have four limbs (except snakes) that are projected to the sides. The heart is imperfectly chambered in most reptiles and the two aortic arches arise from the ventricular region of the heart. The kidneys tend to excrete insoluble uric acid so as to conserve body water. With the exception of burrowing reptiles most species have well developed eyes. Tongue can be protruded in some reptiles.

Reptiles are cold blooded (poikilotherm) animals i.e. their body temperatures fluctuate with the atmospheric temperature. In temperate regions, they hibernate during winter months. Basking is an important behavioural function of many reptiles. They may be terrestrial, aquatic or arboreal. All male reptiles except tuatara possess copulatory organs. Females are mostly oviparous. Eggs are laid in a nest or under cover shortly after fertilization. The development of embryos occurs largely after the laying of eggs. Some species of reptiles, however, are viviparous. The number of eggs laid by reptiles varies from one to two hundred depending on the species. Many reptiles are long-lived animals.

Exploitation of reptilian resources in many parts of the globe has caused a great concern towards the conservation of reptile population. They are hunted for flesh and eggs as food; hide, skin and shell for fancy leather articles and handicrafts (e.g. belts, gloves, luggage, handbags etc.); teeth and bones as charms; fats for

medicinal properties; and venom for preparation of medicinal products. As a consequence, many species of reptiles have become extinct or are on the verge of extinction.

Classification of reptiles:

The zoological classification of reptiles is debatable and no single system of classification is accepted by all the herpetologists. The following classification, however, is acceptable to many taxonomists.

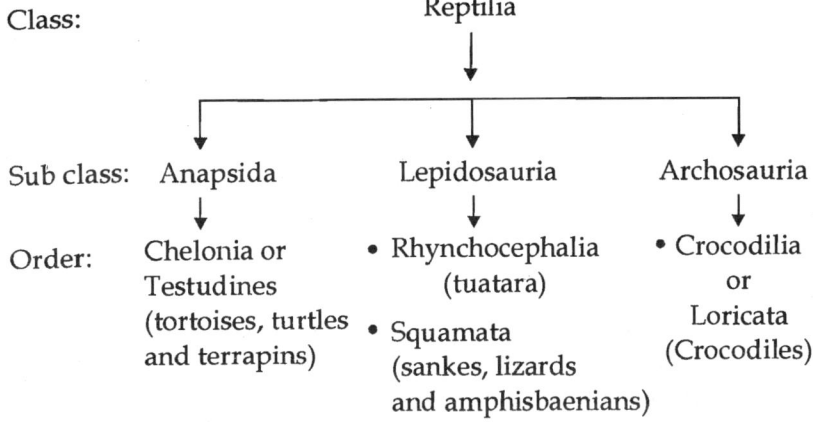

Class: Reptilia

Sub class: Anapsida   Lepidosauria   Archosauria

Order: Chelonia or   • Rhynchocephalia   • Crocodilia
          Testudines   (tuatara)   or
          (tortoises, turtles   Loricata
          and terrapins)   • Squamata   (Crocodiles)
                             (sankes, lizards
                             and amphisbaenians)

Class: Aves —

The birds have been descended from their immediate ancestors, the reptiles. Scientists believe that the earliest known bird that lived over 140 million years age is *Archaeopteryx*. Like mammals, birds are warm blooded animals i.e. they can maintain their body temperatures more or less constant and independent of surrounding atmosphere. The main characteristic feature of birds is their feathers. They grow from the epidermis and are made of keratins. They do not have sweat glands. The oil gland (preen gland) is located on the rump at the base of the tail and it keeps plumage waterproof. Normally, feet are covered with scales. The toes are tipped with claws. Typically the first toe directed backwards and the second, third and fourth toes are directed in front. The fifth toe is absent. The upper mandible which is attached to the skull is movable to some extent. The heart is four chambered. The aortic artery is lost. Living birds have no teeth. Birds have highly sight and hearing. Some members are solitary (e.g. many birds of prey) while others are gregarious (e.g. flamingo). The nests are built by many birds for laying purpose. Eggs are incubated either by female alone or by both sexes (except in megapodes where eggs are hatched by heat generated by decaying vegetation). The incubation period varies from 11 to 80 days depending on the species. Most species have nearly naked and helpless young (altrical). The life span varies according to a particular group of birds. An ostrich, for example, can live for more than 40 years, while perching birds usually have a life span of 5 to 8 years.

Birds play vital roles in maintaining the ecological balance as well as welfare of the human being. Wild birds and eggs are the food sources of human being. Feathers have been used for decoration since early times. Pigeons have been trained for carrying a message. Birds help in cross pollination in flowers. They also play an important role in controlling of pests and dispersal of seeds as well. The droppings of fish eating water birds are valued for fertilizer.

Classification of birds:

The classification of birds, as suggested by ornithologists, should be followed from the most 'primitive' order to the most 'advanced' order. The following zoological classification with salient features of each order is presented below:

Order	Salient features	Name of bird
Struthioniformes	Restricted to Africa; found in open and arid country; naked legs, neck and head; two toes; About 6-8 ft tall and weigh more than 160 kg; Egg weighs up to 2.3 kg; life span more than 40 years.	Ostrich
Rheiformes	Found in South America; habitat - preferably open grassland; long and powerful legs; foot with three toes; 4-5 ft tall and body weight about 38 kg.	Rhea
Casuariiformes	Australia, New Guinea and adjoining islands; habitat-thick scrubs forests and open plains; drooping hairy plumages with absence of tail quills.	Cassowary and emu
Apterygiformes	New Zealand; habitat- thick and swampy forests; stands 18-22 inches; short and stout legs; rudimentary wings; nostril at the tip of the bill.	Kiwi
Tinamiformes	Central and South America; habitat-include thick lowland and jungles, open pampas, sparse upland etc; strong keeled breastbone; very stout tail.	Tinamou
Sphenisciformes	Coastal regions; no flight feathers; short and thick legs; sharp pointed beak; highly social birds.	Penguin
Gaviiformes	Aquatic birds; fully webbed toes; short legs; short, narrow and tapering wings.	Diver or loon
Podicipediformes	Cosmopolitan in distribution; aquatic bird; webbed toes, short and pointed bill; weak flier.	Grebe

Pocellariiformes	Mostly found in southern hemisphere; webbed feet; tube noses; highly pelagic; musky body odour; a single white egg normally in a underground burrow.	Albatross, petrel etc.
Pelecaniformes	Temperate and tropical region; four webbed toes; hooked or straight bill; large wings; throat or gular pouches; 1-6 chalky white eggs; both sexes incubate for 21-42 days; fledging at about 35-56 days.	Pelican, cormorant, darter etc.
Ciconiiformes	Worldwide except in extreme north; wading birds of marshes or shallow water; long legs; long necks and long wings; temperate members migratory in nature.	Stork, heron, spoonbill etc.
Anseriformes	Swimming or semi aquatic birds; short legs; broad bills in true water fowl; rounded nostrils.	Duck, goose etc.
Falconiformes	Worldwide distribution; strong wings; strong fliers; hooked beaks; small clutches; long incubation period.	Falcon, vulture etc.
Galliformes	Nearly cosmopolitan in distribution; arboreal or ground living; powerful and scratching feet; good swimmers and fly for short distance.	Megapod, peacock, quail etc.
Gruiformes	World wide distribution except in polar region; habitat-dry plains, marshes, desert etc.; relatively long necks; strong or weak fliers or flightless.	Bustard, rail, crane, etc.
Charadriiformes	Worldwide distribution; most members found in marshes, beaches and coastal waters; mostly breed once a year.	Gull, tern etc.
Columbiformes	Worldwide except in extreme north; weak bills; loosely attached feathers in the skin; large crop; 1-3 eggs; young fed by regurgitation.	Pigeon, dove etc.
Psittaciformes	Found throughout tropics with some in temperate regions; stout hooked bills; brightly coloured; strong fliers.	Parrot, cockatoo etc.

Cuculiformes	Worldwide except in extreme north; long tails; strong to weak fliers; arboreal or a few ground living; some are brood parasites.	Cuckoo, coucal etc.
Strigiformes	Worldwide except in Antarctica and few oceanic islands; nocturnal ; hooked beaks and soft plumage; forward facing eyes; pronounced facial disc; strong grooming feet.	Owl
Caprimulgiformes	Worldwide except in north; nocturnal or crepuscular; very long mouth; small bills; some feed on insects caught in flight(except oil bird).	Oil bird nightjar etc.
Apodiformes	Throughout the world except in extreme north; weak feet; pure white eggs; naked and blind young.	Swift bird, Humming bird etc.
Coliiformes	Africa South of Sahara; four forward directed toes; long and pointed tails; soft plumage.	Mouse bird
Trogoniformes	Pan tropical forests except Australia; weak feet.	Trogon
Coraciiformes	Tropical and temperate regions of the world; cavity nesters.	Kingfisher, hornbill etc.
Piciformes	Forests and brush lands of tropic and temperate regions except Australia, New Zealand and Oceanic islands; arboreal; strong or weak fliers.	Honey guide, woodpecker etc.
Passeriformes	Comprised more than half of the known species of birds; distributed all over the world except polar regions and some oceanic islands; land birds; a few weaver, etc brood parasites; known as perching birds.	Swallow, robin, bulbul, finch,

The following chapters highlight the salient features of some selected animals covering three classes – Mammalia, Reptilia and Aves.

Chapter 12

Salient Features of Selected Wild Mammals

ORDER: MONOTREMATA –

The word "monotreme" means "one holed". Monotremes are confined in Australia, Tasmania and New Guinea. They have both mammalian and reptilian characters. The members may be kept as pet animals. The animals in this order includes echidna (or spiny anteater) and platypus (or duck bill).

Echidna:

Zoological name: There are two species- the common echidna (*Tachyglossus aculeatus*) and the long- nosed echidna (*Zaglossus bruijni*).

Distribution: They are confined in Australia, Tasmania and New Guinea.

Habitat: The habitats in which they live include streams, rivers, lakes, semi-arid areas and mountains.

Physical features: The physical features include short and stubby tails, long snouts without hair, long and mobile tongues, no teeth and absence of necks. The foot is turned outwards and forefeet have spatulated claws (Fig.12.1). They can be readily recognized by their spines or quills.

Reproduction: The testes are abdominal. Urine, faeces and eggs pass out to the exterior through a common opening (cloaca).The breeding season is recorded from July to October. However, it varies from place to place. One creamy yellow egg is laid at a time by the female. The egg is attached to the hair in the pouch. The young hatch after a period of 7-10 days incubation. In the wild, the female keeps the young in a burrow until their spines are developed.

Fig.12.1:

Behaviour: Ecnidnas are solitary animals. They are active at night during hot summer and diurnal during colder periods.

Life span: Echidnas have been known to live for 30-50 years in captivity with individual variation.

Platypus *(Ornithorhynchus anatinus):*

Distribution: It is found in Australia and Tasmania.

Habitat: Their habitats include freshwater lakes, lagoons and streams.

Physical features: The body weight varies from 0.5 to 2.5 kg. The flattened body is covered with dense, short and fine hair with coarse guard hair. The other salient features include dark brown hairs, longer hindlimb, both limbs armed with five claws, and flat and broad tails (Fig. 12.2).

Reproduction: The testes are abdominal like echidna. Females have no teats. Platypus is sexually mature at about five years of age. The mating is supposed to occur in water during the months of July to October. They usually lays two soft eggs and the young are hatched after a period of 7-12 days of incubation periods. Young are nursed for about 3-4 months.

Fig.12.2: Hindlimb of platypus.

Behaviour: When the animal lives under water, its eyes and ears are closed by a fold of skin. The two small internal cheek pouches are used for storing of food. It digs a burrow in the bank of lake, stream etc and builds the nest at the terminal point of the burrow. Platypus is reported to build two types of burrows - one is used for taking shelter and other one is for rearing the young.

Life span: The life span is 10-15 years or more. However, their longevity is more in captivity than in the wild.

ORDER: MARSUPIALIA

The order Marsupialia forms a diverse group of animals. The members belonging to this order are restricted to the New World and the Australian region. They inhabit a wide range of habitats ranging from desert to rain forest. Their size varies from a small marsupial mouse (e.g. planigale) to a large kangaroo. The single order Marsupialia has been classified into two sub-orders: Polyprotodonta and Diprotodonta. The animals of this order are kangaroo, wallaby, rat kangaroo, koala, wombat, glider, cuscus, bandicoot, marsupial mole, Tasmanian tiger, numbat, etc. Many marsupials are very popular pet animals. Some members are hunted locally for food and other products. Cats, foxes, wild dogs, owls, birds of prey and snakes are their natural predators. Their populations have been declined due to several factors including change of climate, forest fire and encroachment of cultivated land.

Kangaroo:

Zoological name: Kangaroos are placed into two families: the Macropodidae (true kangaroos) and the Potoroidae (rat kangaroos). The larger members in the family Macropodidae are customarily called kangaroos and the smaller species are known as wallabies. The examples are the red kangaroo *(Macropus rufus)* and the swamp wallaby *(Wallabia bicolor)*.

Distribution: They occur in Australia, New Guinea and nearby islands.

Habitat: Their habitats include open forests, open plains, woodlands, rocky slopes and cliffs.

Physical features: The length of a large red kangaroo is about 9 ft and it weighs 90 kg. Hind limbs are larger and stronger than forelimbs. The second and third digits are reduced in size and are united together in a common sheath except for the claws. Limbs are adapted for jumping. The other characteristics include heavily built tail and compartmental stomach like ruminants.

Reproduction: The sexual maturity is reached at 15-18 months. They may breed throughout the year (e.g. grey kangaroo) or become opportunistic breeders depending on the favourable conditions for breeding. Most young, however, are born during summer months. A single young is usually born after a gestation of 28-38 days. The young enter into the pouch (a flap of skin covers the nipples) after born and remain in the pouch for about 7-12 months.

Behaviour: The forelimbs are sometimes used in bringing food to the mouth and the claws of the hindlimbs are used in grooming. Rumination takes place in kangaroo but the cuds are re-swallowed without being chewed.

Life span: The life span has been reported to be 12-18 years in the wild.

Koala:

Zoological name: The zoological name of koala is *Phascoloarctos cinereus.*

Distribution: They are found in Australia.

Habitat: Koalas are almost arboreal.

Physical features: The colouration is silver to brownish grey. It weighs 4-15 kg. They are characterized in having a rounded head, brown eyes and soft fur. It has large, smooth and leathery noses, cheek pouches, a stumpy tail or virtually tailless, large paws, strongly clawed digits and a large caecum. The foot pads are granulated.

Reproduction: Both sexes attain sexual maturity at two years of age. The breeding season starts in summer months (October-February). A singly young is born usually in mid-summer after a gestation period of 35 days. The pouch life is 6-7 months.

Behaviour: Koalas spend day time in sleeping. The searching for food begins at dusk. Their apparently abstinence from water is a matter of curiosity.

Life span: the life span of koala has been reported up to 13 years in the wild (in captivity, 18 years).

ORDER: INSECTIVORA —

Insectivores are small, active and usually secretive mammals. They are distributed in most parts of the world except in Australian regions, Greenland, Antarctica and most parts of South America. Members belonging to this order include hedgehogs, moonrats, moles, shrews, solenodons, tenrecs, otter shrews and elephant shrews.

Hedgehog:

Zoological name: The well known species are desert hedgehog *(Paraechinus spp)*, the European hedgehog *(Erinaceus europaeus)*, long-eared hedgehog *(Hemiechinus auritus collaris)* and pale hedgehog *(Paraechinus micropus)*. The last two species are found in India.

Distribution: They are distributed in Asia, Europe, Africa and New Zealand.

Habitat: Their habitats include tropical rain forest, steppe, desert and wooded or cultivated land.

Physical features: Hedgehogs are brownish to almost black in colour. The other features include elongated snouts, dense spines, short tails, relatively large eyes and ears, and strong limbs with powerful claws.

Reproduction: Tropical species breed throughout the year. The usual litter size 4-7. The gestation period is 30-48 days. The young are blind and naked. Eyes and ears open in about a week.

Behaviour: Hedgehogs are largely solitary and nocturnal. Temperate species hibernate during winter months. They have the ability to curl up into a ball.

Life span: They are reported to live 3-10 years.

ORDER: CHIROPTERA —

In Latin, the word "chiroptera" means "hand wing". Chiropterans or bats have a worldwide distribution except in polar regions. They are abundantly found in tropical parts of the world. The order Chiroptera contains 925-1000 species of bats. Bats are the only unique and fascinating mammals that can fly. Their body is covered with fur and the young suckle like other mammals. These mammals are beneficial to mankind in many ways. Firstly, they consume a large number of insects and thereby play an important role in controlling the insect population. Secondly, many fruit bats are important in dropping seeds of plants in many areas and subsequently these seeds sprout into plants. Finally, many nectar eating bats help in pollination during their search for nectar. However, their droppings are also used as fertilizer. Many bats such as the flying foxes and the larger fruit eating bats are very popular exhibits in zoos. Kitti's hog- nosed bat weighing 1.5 gm is said to be the smallest mammal in the world. Hawks, owls, snakes and a variety of small mammalian carnivores are the natural predators of the bats.

The order Chiroptera is classified into two suborders: Megachiroptera and Microchiroptera. The Megachiroptera has a single family Pteropodiae that contains 165-175 species of flying foxes or fruit bats. Bats in this group are larger in size and are found in tropical and subtropical parts of the Old World. Most have a dog like face. They have large eyes and small ears. Most species have a claw on the hind digit. Their eyesight is excellent. The diet of megachiropterans principally consists of fruits and flowers.

The suborder Microchiroptera contains about 80 per cent of the entire bat population. They are found throughout the world and are usually smaller in size. They have small eyes and large ears. They feed chiefly on insects and small animals. Some well known bats in this group include mouse-tailed bats, leaf-nosed bats and vampire bats.

Indian fruit bat or flying fox:

Zoological name: The zoological name is *Pteropus giganteus*

Distribution: The Indian fruit bat, the largest Indian bat is distributed throughout India.

Habitat: The habitats include forests, cultivated plantations etc.

Physical features: They are distinguished in having dog like faces, large and bright eyes, long snouts and pointed ears. The head is reddish brown and the snout and wings are black. The second digit is armed with claw in most species. The body weight is recorded up to 1.5 kg.

Reproduction: The male has a thin-skinned scrotum. The penis has a well developed baculum. The female possesses a pair of pectoral mammae. They breed once a year and the female drops a single offspring in early February after a gestation period of 5 months.

Behaviour: The flying foxes are highly gregarious. They are nocturnal in habits. The nocturnal habit may be for two reasons. First, they want to avoid competition for foods and other factors from diurnal birds. Second, their physiological mechanism of radioactive cooling may not be advantageous in sunlight. They roost on trees and often roost in same sites throughout the year.

Life span: Fruit bats can be well maintained in the captivity and are known to live for more than ten years.

Vampire bat:

Zoological name: The true vampire bats are classified in the family Desmodontidae which contains 3 species. They include the common vampire *(Desmodus rotundus)*, the white-winged vampire *(Diaemus youngi)* and the hairy-legged vampire *(Diphylla ecaudata)* bats.

Distribution: These bats are restricted to Central and South Americas.

Habitat: They are found in forests, caves, culverts etc.

Physical feature: All vampire bats have a swollen and glandular muzzle giving a fleshy nose-leaf appearance. They weigh 15-45 g. The fur is grizzled with shades of brown. Other characteristics include well developed thumbs and hindlegs, pointed or rounded ears, absence of tail, chisel like incisor and razor-like canine teeth, muscular tongue with two lateral grooves and an elongated blind pouch stomach.

Reproduction: Sexual maturity is attained at about 10-12 months. The breeding cycle continues throughout the year. The gestation period is about 7 months. The young suckle up to 9 months.

Behaviour: They roost in the hollow trees, caves and culverts. Vampire bats are considered to be the serious pests of livestock. They are also known to transmit diseases (e.g. trypanosomiasis in cattle). This bat is also the carrier of rabies.

Life span: They are reported to live more than 5 years.

ORDER: PRIMATES —

The order Primates contains a diverse group of animals such as apes, monkeys, marmosets, tamarins, bush babies, pottos, lemurs and men. Primates are the natives of Africa, South America and some parts of Asia. They occupy in a wide range of habitats. The tropical primates are distributed in tropical forests, woodlands, grasslands and vegetation complexes. They may live in the canopy or on the forest floor. The primates live in groups. The nocturnal species, however, are largely solitary in nature except the South American night monkey which occurs in small groups consisting of two to five animals. The five major types of social organizations are found among primates: multi-male group, family group, harem group, all male group and unstable group.

The order Primates containing over 200 species is classified into two suborders: Prosimii (lower primates) and Anthropoidea (higher primates). Prosimians or lower primates are less man-like and they retain many primitive characteristics. Lower primates include true lemurs, dwarf and mouse lemurs, indris, sifakas, aye-aye, bush babies, pottos, lorises and tarsiers. They live chiefly in forests. The members belonging to the suborder Anthropoidea include capuchin-like monkeys, marmosets, tamarins, Old World monkeys, apes and men.

Ape:

The term "ape" is loosely defined as any species of higher primates except man especially one without a tail. Sometimes, this term is also used in describing certain monkeys such as Celebes black ape. However, in true sense the members belonging to the families Pongidae and Hylobatidae are called apes. The great apes include chimpanzee, gorilla and orangutan. The gibbons are the small apes.

Chimpanzee:

Zoological name: There are two species- the common chimpanzee *(Pan troglodytes)* and the pygmy chimpanzee (P. *paniscus*).

Distribution: Chimpanzees are found in Central and West Africa.

Habitat: They inhabit in grasslands, woodlands and rain forests.

Physical feature: The body weight is recorded up to 90 kg. The coat has long and sparse black hairs. Both males and females possess short white beard. The top of the head is rounded or flattened. Males bear larger canine teeth.

Reproduction: Chimpanzees attain maturity at 7-10 years of age. The wild females, however, mature a few years later. There is no particular breeding season. The oestrus cycle is recorded 25-37 days. The skin of the perineum and the posterior portions of buttocks swell to a large size during estrus like baboons and rhesus monkeys. The female may be served by half a dozen males. She gives birth to a young (rarely twins) after a gestation period of 230-285 days. The young remain with the mother for at least 5-7 years.

Behaviour: They spend the night in the shelters for which they make nests in trees. It is said that a chimpanzee builds a new nest each night. The mature male sleeps alone, while the mother sleeps with her offspring in a separate nest. They live in communities comprising as many as 60 or more individuals. The male hardly leaves the community where it is born. Most females, however, migrate to a new community during heat period. Chimpanzees are very intelligent animals and are known for their inquisitiveness.

Life span: The life span is reported to be 60 years.

Hoolock gibbon:

Zoological name: The zoological name of hoolock gibbon is *Bunopithecus hoolock.*

131

Distribution: This ape is found in Bangladesh, Myanmar and north eastern parts of India.

Habitat: Hoolock gibbon, the only ape found in India, is confined to the rain forests of north eastern parts.

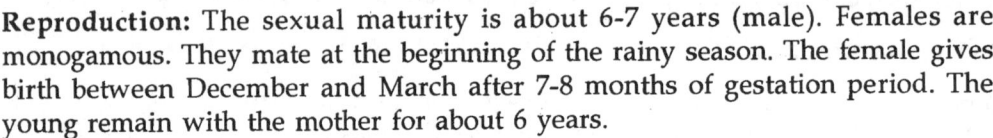

Physical feature: The body weight of this medium sized animal varies from 6-8 kg. The male is black with white brows, while the female is buff or golden colour with occasionally black patches. The young male and female are black with a silvery white band above the eyebrow. The other characteristics include rounded head and long arms twice the length of hind legs.

Reproduction: The sexual maturity is about 6-7 years (male). Females are monogamous. They mate at the beginning of the rainy season. The female gives birth between December and March after 7-8 months of gestation period. The young remain with the mother for about 6 years.

Behaviour: They are active at morning and evening and are reputed for making peculiar calls in chorus. Hoolock gibbons are timid and inoffensive. These diurnal animals live in small troops. They are known to have a special liking for spiders.

Life span: In the wild hoolock gibbons have been reported to live for 25-30 years.

Monkey

Any member except ape and man in the suborder Anthropoidea is called monkey. Members representing the two families the Cebidae and the Callitrichidae are called New World monkeys, while members in the family Cercopithecidae are known as Old World monkeys. The New World monkeys include marmosets, tamarins and capuchin-like monkeys and they are found in North and South Americas. Baboons, macaques, langurs etc. representing the Old World monkeys are found in Asia and Africa. The habitats in which monkeys live are the mountains, grasslands and forests.

Macaque:

Zoological name: There are many species of macaques such as the rhesus macaque (*Macaca mulatta*), crab eating macaque (*M. fascicularis*), lion-tailed macaque (*M. silenus*), bonnet macaque (*M. radiata*), Assamese macaque (*M. assamensis*) and pig-tailed macaque (*M. nemestrina*).

Distribution: Macaques or monkeys occur in the whole of Asian region except the barbary macaque which is found in Africa. However, some are restricted to a limited geographical area. The lion-tailed macaque, for example, confined to the Western Ghats of India. The crab eating macaque is found in the Andaman and Nicobar islands. The pig-tailed macaque inhabits in forest of northeast India.

Physical feature: The body weight ranges from 3.5 to 18 kg. The physical features in general include powerful jaws, long upper canine teeth, presence of cheek pouches, well muscled long hindlegs in arboreal species, and long and dense silky fur.

Habitat: They live in a wide variety of habitats such as swamps, semi-desert forests, mangrove areas and evergreen forests.

Reproduction: Macaques breed throughout the year. The oestrus cycle is about 28 days. The single young is born after about 5-6 months of gestation. By about four years, the young become mature.

Behaviour: They are curious and imitative in nature. Macaques are gregarious and diurnal animals. Some members such as rhesus macaques spend most of the time in fighting and playing.

Life span: Macaques are known to live more than 25 years.

Langur:

Zoological name: Some well known species include the common langur or hanuman (*Semnopithecus entellus*), the Malabar langur (*S. hypoleucos*), the capped langur (*S. pileatus*), the golden langur (*S. gee*) and the Nilgiri langur (*S. johnii*).

Distribution: The above mentioned species are found in India and neighbouring countries.

Habitat: The main habitat is the forest. However, they also make their homes in cultivated areas, dry scrubs and human dwellings.

Physical feature: The common langur has a slender body. The crown of the head is devoid of a crest. The length of head and body is about 2 feet long and the tail is about 3 feet or more. Ears, face, soles of hands and feet are black in colour. The other features are stiff and projected hairs on the eyebrows, large salivary glands, absence of cheek pouches and complex sacculated stomach.

Reproduction: Adult size is attained at about 5 years of age. The duration of oestrus cycle is about 30 days. During breeding period the oestrus females come closer to the males. The female gives birth to a single young after 6 months of gestation. The birth weight of a young may be 400 g.

Behaviour: They may remain in troops ranging from 6 to as many as 70 individuals. The mixed troops have a stable unit of related females with usually just one breeding male. Females remain in the same location but young males leave the parental troop. The territorial and grooming behaviours are seen in langurs. Agonistic behaviour for food is common among the members of langurs. In Central India, a mutual relationship between langur and cheetal has been observed in the wild.

133

Life span: Langurs are known to live 25-30 years.

ORDER: PHOLIDOTA

The pangolins or scaly anteaters belong to the family Manidae of the order Pholidota.

Pangolin:

Zoological name: There are seven recognized species of pangolins. They are the Indian pangolin *(Manis crassicaudata)*, the Malayan pangolin *(M. javanica)*, the Chinese pangolin *(M. pentadactyla)*, the giant pangolin *(M. gigantea)*, the white-bellied tree pangolin *(M. tricuspis)*, the long-tailed pangolin *(M. tetradactyla)* and the cape pangolin *(M. temmincki)*. The former three species are called the Asian pangolins and the latter four species are known as the African pangolins. In India, both Indian and Chinese pangolins are found.

Distribution: They are found in many parts of Asia and Africa.

Habitat: Their habitats range from the dense tropical rain forest to arid sparsely wooded savannah.

Physical feature: The body weight varies from 4 to 27 kg. The colouration ranges from light yellowish-brown through olive to dark brown. The most striking anatomical feature of this animal is its horny body scales. These prominent fish like scales are composed of cornified epidermis that overlaps in an imbricate manner as we see tiles on a roof. The other body features are short and conical head, short and powerful limbs with five clawed, longer forefeet, absence of teeth and elongated tongue. Hairs are present at the base of the body scales in Asian pangolins while they are absent in their African counter parts.

Reproduction: There are two pectoral mammary glands in female. The puberty is attained at two years of age. The female gives birth one to three young depending on the species. The litter size is usually one in African species. Births generally occur between November and March after a gestation period of about 140 days. The new born weighs 200 g. The weaning period is normally stated to be four months.

Behaviour: Pangolins are specialized in feeding of ants and termites with the help of their tongues. They are nocturnal in habits and sleep during the day in the burrows except the small white bellied pangolin. The location of prey is achieved by smelling. Another important behaviour of the pangolins is that they can roll more or less into a ball. Pangolins also dig a hole in the ground with their powerful claws. Some are terrestrial (e.g. Indian pangolin) while others are arboreal (e.g. Chinese pangolin).

Life span: They are known to live as long as 14 years.

ORDER: LAGOMORPHA —

Lagomorphs have a worldwide distribution. These animals are found as native species in many parts of the globe. Lagomorphs have been introduced by many countries such as New Zealand, Australia and Madagascar for various purposes. They live in a wide variety of habitats such as desert, grassland, swamp, tundra and tropical forest. In the wild, lagomorphs are hunted for food, fur, sport etc. These small to medium sized animals are commercially reared for meat, skin and biomedical research. Lagomorphs were classified earlier in the order Rodentia. Later on, they have been given the status of a separate order Lagomorpha. However, lagomorphs can be distinguished from rodents by having two pairs of incisor teeth in the upper jaw (in rodents-one pair) in which the small and peg-like second pair teeth are located behind the long and constantly growing chisel-like anterior teeth. At birth, there are 3 pairs of incisors but the outer pairs are soon lost. There is a gap between the incisors and premolars. Lagomorphs can also be distinguished in having larger ears, fully furred feet (some rodents may have furred feet), eyes set high on heads and less developed jaws muscles. The testes are permanently external unlike the rodents.

The order Lagomorpha has two families:
(a) Ochotonidae- pika
(b) Leporidae- rabbit, hare and cottontail

Foxes, bobcats, wolves, weasels, owls and hawks are the natural predators of lagomorphs. A conservation and management programmes of lagomorphs including habitat management, stocking programmes, harvest, control and protection have been taken up all over the world.

Rabbit:

Zoological name: The present day domestic breeds of rabbit have been derived from the European wild rabbit (*Oryctolagus cuniculus*). The examples of wild rabbits are the Assam rabbit or hispid hare (*Caprolagus hispidus*), the Sumatran short-eared rabbit (*Nesolagus netscheri*) etc.

Distribution: They are found almost every part of the world. In the wild state, the Assam rabbit occurs along the Himalayan foot hills ranging from Uttaranchal to Assam.

Habitat: Rabbits live in a wide variety of habitats such as grasslands, woodlands, prairies, marshes, swamps and deserts. However, they prefer to live in open woodlands and scrublands.

Physical features: The mature body weight ranges from 2 to 5 kg. Their forelimbs are shorter than hindlimbs. Rabbits have grey, white, brown or black hairs. The Assam rabbit, however, has dark brown upper part while brownish white under part. The salient body features include five toes on the forefeet and four on the hind feet, toes equipped with powerful claws, 28 teeth, gap between the incisors and premolars, and short tail. The most striking feature of the rabbit (also other lagomorphs) is its well developed caecum.

Reproduction: The testes descend into scrotum at about two months of age. The testes may, however, be withdrawn into the abdominal cavity through an open inguinal canal when rabbits are frightened or exposed to cold. Males are sexually matured at six or seven months of age in the wild while in captive condition, they may be used for breeding purpose at about 20 weeks of age. Females attain puberty at an earlier age than males. In the wild state, the sexual maturity is achieved at 3-6 months. The wild rabbits are seasonal breeders and they breed between February and September in northern Europe. The domestic rabbits, on the other hand, breed almost throughout the year. Rabbits are not spontaneous ovulators. Ovulation takes place only after coitus. The naked and blind young are born after a gestation period of 28-32 days. The young weigh about 40-50g or more.

Behaviour: Wild rabbits are nocturnal. Their territories are made by the odoriferous substances produced from the anal and inguinal glands. Rabbits dig burrows. However, the Assam rabbit shelters in burrow left by other animal. The maternal and nest making behaviours are noticed a few days before kindling (the act of parturition). Normal nest building activity starts 3-5 days before the expected date of kindling. She makes a nest with fur and other materials.

Life span: In the wild, the life span has been recorded about 10 years. In captivity, they may live as long as 13 years.

Hare:

Zoological name: The well known species of hare are the European hare *(Lepus europaeus)*, the Indian hare *(L. nigricollis)*, the snow shoe hare *(L. americanus)* and the cape hare *(L. capensis)*. The Indian species has 3 varieties: the desert hare *(L.n .dayanus)*, the rufous-tailed hare or northern form of Indian hare *(L.n.ruficaudatus)* and the black-naped hare or southern form of Indian hare *(L.n.nigricollis)*.

Distribution: Hare inhabit in Europe, Africa, Asia and North America. The desert hare found in Rajasthan and adjoining areas. The rufous-tailed hare has a wide distribution ranging from the south of Himalaya to the river Godavari. The black-naped hare occurs in peninsular region.

Habitat: Most species prefer to live in open grassy areas. However, some species have been reported to live in different habitats such as open areas (e.g. cape hare) and boreal forest (e.g. snow shoe hare). The Indian hare inhabits in open bush country.

Physical feature: The average adult body attains more than 4 kg (1.3-7 kg). The general colour pattern is grayish brown to yellow upper coat and lighter or white under coat throughout the year. The fur colour of northern altitude species is usually greyish brown in summer and white in winter. The desert hare is sandy yellow in colour. The rufous-tailed hare has black patches on the back and face. The black-napped hare has a dorsal blackish brown patch on the back running from neck to shoulder, and the upper part of this is black. The morphological features of these animals include slender bodies, long hind legs, long ears and short tails. Most species have black ear tips. The total number of teeth is 28 (I: 2/1; C: 0/0; PM: 3/2; M: 3/3).

Reproduction: Sexual maturity attains at 7 to 8 months or more. The litter may consist of 1-9 young (leverets) depending on the species. The young are born after a gestation period of 40 to 42 days. The fully furred young have their eyes open at birth. The black-napped hare breeds mainly between October and February.

Behaviour: Most hares are solitary and nocturnal. They live on the surface.

Life span: The life span is recorded up to 12 years.

Although both rabbit and hare have some common features, they differ from each other by certain characteristics. They are as follows:

Rabbit	Hare
(a) Rabbits inhabits in open grass land, cultivated crops, swamp, rocky desert and sub-alpine valley.	Most species of hares prefer to inhabit in more open areas.
(b) Rabbits are well domesticated mammals.	The domestication is difficult.
(c) They are usually gregarious animals and most rabbits dig a burrow.	Hares are usually solitary animals (except European hare) and they do not dig burrows (except snow shoe and arctic hares).
(d) Rabbits are generally smaller The hind legs are weaker.	Generally larger in size. The well developed hind legs are stronger.
(e) Rabbits have shorter ears and ear tips are of same colour. They have vertical grooves on the upper surface of the anterior incisors that are located in the upper jaw.	Hare have longer ears and in some species the tips of ears are black in colour. Hare have no such vertical grooves.
(f) Sexual maturity is attained at an early age (3-5 months).	Most species attain puberty at 7-12 months of age.
(g) Young are born in burrow. Newly born young are devoid of hairs. The eyes and ears are closed at birth and young start eating solid food after 18-21 days of age.	The female drops her young in the open Young are well developed and have hairs. Eyes are ears are open at birth. They start feeding solid food after 3-10 days of age.
(h) The gestation period varies from 28-32 days and the litter size is 5-8 young.	The gestation period is higher (40-42 days) and the usual litter size is 2-5 young.
(i) Superfoetation (two simultaneous pregnancies at two stages of development) does not occur in rabbits.	Superfoetation occurs in hares.

ORDER: RODENTIA—

The members belonging to the order Rodentia account for half of the mammalian species living today. Rodents are widely distributed throughout the globe except Antarctica and some islands. They inhabit in almost every habitat. The order Rodentia is divided into three suborders: Sciuromorpha, Myomorpha and Caviomorpha. The classification is based primarily on the arrangement of the jaw muscles especially the deep masseter muscles. The members belonging to the suborder Sciuromorpha (squirrel-like rodents) include squirrels, beavers, spring hares, pocket mice, kangaroo rats, scaly-tailed squirrels, pocket gophers and mountain beavers. Rodents belonging to the suborder Myomorpha (mouse-like rodents) are rats, mice, dormice, jumping dormice, jerboas, birch and desert dormice. Some notable members in the suborder Caviomorpha (cavy-like rodents) are cavies, porcupines, capybaras, coypus, pacas, chinchillas, chinchilla rats and agoutis.

Most members of this order have short limbs. The forelimbs are slightly smaller than the hindlimbs. All species possess long and sensitive whiskers. Some are nocturnal (e.g. porcupines), while others are active during day light hours (e.g. squirrels). Rodents may be semi aquatic (e.g. capybara), aquatic or semi aquatic (e.g. beavers) and arboreal (e.g. New World porcupines). Some are burrowing animals (e.g. voles), while others are grazing rodents (e.g. capybara). Most rodents are small in size. However, the largest rodent, capybara may weigh up to 65 kg. In the wild they are known to live less than two years, whereas, in captivity the life span may be up to five years.

Rodents are useful as well as harmful animals. It is useful in the sense that some rodents are highly valued for their fur (e.g. chinchilla and beaver) and biomedical research (e.g. guinea pigs, rats, mice and hamsters). Their harmful effects have also attracted a worldwide attention. Some rodents are responsible for damaging agricultural crops and seeds of grains. They consume lot of grains causing a great loss to the farmers. The drainage systems in the cities and agricultural fields are also damaged by these animals. Rodents are widely known to be the reservoirs of many dreadful diseases (e.g. bubonic plague and tularemia). Some species are also known to kill domestic chickens and occasionally eat their eggs.

Rodents have common characteristics with lagomorphs. The following points, however, are important for distinguishing each other:

Rodent	Lagomorph
(a) Rodents have one pair of incisor teeth in the upper jaw.	(a) Lagomorphs posses two pairs of incisors in the upper jaw. The second set of incisors are located directly behind the upper front incisors.
(b) The scrotum is posteriorly located to the penis.	(b) The scrotum in lagomorphs is anteriorly located to the penis.
(c) Rodents have an os penis	(c) Os penis is absent.

Rats and mice:

The rats and mice are classified in the family Muridae of the suborder Myomorpha. Many a time it is very difficult to differentiate between rat and mouse, as there is no sharp dividing line of these two rodents. Nevertheless, the word "mouse" is loosely applied to almost any of many small rodents usually having rounded ears and long skinny tails. On the other hand, rats are generally larger than mice and most rats are characteristically having scaly tails. In this text, a general description of rat and mouse is given below:

Among all the rats, the best known species is the brown or Norway rat *(Rattus norvegicus)*. It is widely distributed in all over Europe and some parts of United States. Rats are devoid of a gall bladder. They are primarily nocturnal. Rats are able to breed at three to four months of age and can produce upto 7 litters a year. Each litter comprises of six to twenty two young. The life span is about four years.

The house mouse *(Mus musculus)* is extensively used in research purpose. They are found in farms, warehouses and are also frequently entered into houses. Wild mice are nocturnal in habits, but they are active both day and night in the laboratories. They make a nest from plants and animal fibres. Mice have a short oestrus cycle, short gestation period and long breeding season. The size of a litter is as many as 12 young. The gestation period is three weeks. They are known to live for about three years.

Indian rats and mice can be classified into three groups according to their habitats in which they live. The first group comprises those wild species which are found in scrubs and forests and they are mostly arboreal in nature (e.g. Indian bush rat-*Golunda elliot*, and white-tailed wood rat-*Rattus blanfordi*). The second group is the field rats that are common pests to crops (e.g. *Mus booduga*). The final group is the habitual household pests (e.g. common house rat or black rat-*Rattus rattus)*.

Chinchilla:

Zoological name: The chinchillas may be considered as a single species *Chinchilla laniger* or two species-the long-tailed chinchilla *(C. laniger)* and the short-tailed chinchilla *(C. brevicaudata)*.

Distribution: They are the natives of South America. However, these rodents are commercially reared for valuable fur in many parts of the globe.

Habitat: In the wild, chinchillas live in rocky and barren mountains or in burrows.

Physical feature: The body weight varies form 0.5 kg to 1 kg. The dense fur is bluish-grey above and yellowish below. The other features include large ears, long and strong hindlegs and tufted tail. There are 20 teeth (I: 1/1; C: 0/0; PM: 1/1; M: 3/3).

Reproduction: They become sexually mature at 5-9 months. The female produces as much as 6 young in a single litter in the wild. They have usually two litters in a year. The pregnancy period is recorded for 105-118 days. The young are fully furred at birth and suckle for about 45-60 days. Young are weaned at 6-8 weeks old or more.

Behaviour: Chinchillas are active at dawn and dusk. They are found in colonies. The female does not build nest.

Life span: The life span is 8-10 years or more.

Hamster:

Zoological name: The golden hamster (*Mesocricetus auratus*) and the common or black-bellied hamster are the two examples of many reported species. The golden hamsters are widely used as laboratory animals.

Distribution: Hamsters are distributed in Europe, China and Middle East of Asia.

Habitat: They live in various habitats such as rocky mountain slopes, steppes, cultivated fields and arid and semi-arid regions.

Physical feature: The body weight may reach up to 900 g depending on species. The coat colour of the golden hamster is light reddish brown above and white below. The general morphological features of most hamsters include rounded bodies, short limbs, small tails, large dark eyes, prominent ears, sharp claws and long whiskers. They have thick and soft fur. The distinctive feature of hamster is its cheek pouch. The cheek pouches are used for carrying food materials to their shelters.

Reproduction: Hamsters breed throughout the year in captivity, while they are reported to breed once or twice in wild. Maturity attains at 6-10 weeks of age. Estrus lasts 4-5 days. The females (e.g. common hamster) become receptive to males at 6 weeks of age and drop young at 8 weeks. The number of young produced by a female varies from 6 to 12. Young are blind and naked. The normal lactation lasts for 3-4 weeks.

Behaviour: Hamsters are solitary and nocturnal in the wild. They are very docile and gentle pet animals, though their wild counterparts are reported to be exceptionally aggressive in nature. They usually live in burrows. Hibernation takes place during winter months. Territories are formed by the secretions produced from the glands. Another important behaviour is that these animals use their front paws for bringing food materials into the cheek pouches.

Immediately after mating, the female builds a nest in the burrow. Cannibalism of pups is common especially in first litters among hamster.

Life span: They are known to live for 2-4 years.

Porcupine:

Zoological name: Some well known species of porcupines are the Indian porcupine *(Hystrix indica)*, the Himalayan porcupine *(H. hodgsoni)*, the Asiatic brush-tailed porcupine *(Attherurus macrourus)*, the prehensile-tailed porcupine *(Coendou prehensilis)* and the North American porcupine *(Erethizon dorsatum)*. Porcupines are also known as "quill pigs".

Distribution: They are found Asia, Africa, South, Central and North Americas.

Habitat: Porcupines live in a wide variety of habitats such as dense forests, semi-deserts, open grasslands and deserts.

Physical feature: Some members have elongated bodies and their short legs are covered with short, flat and chocolate coloured sharp bristles. The most salient feature is the presence of stout and sharp quills. The crested porcupine, for example, has black and white spines on two third upper parts and flanks of the body. Most species have a crest comprising long and coarse hairs. The body weight ranges from 13 to 27 kg. The other features include pig like eyes and short and sturdy legs.

Reproduction: The mammary glands are located on the sides of the body. Sexual maturity is attained at about two years of age. The female gives birth one to two young after a gestation of 90-120 days. Only one litter is produced annually. The young weighing 300-330 g is fully furred with eyes open. Young are nursed for 13-19 weeks.

Behaviour: The Old World porcupines are nocturnal and terrestrial.

Life span: The life span varies from 8 to 20 years in the wild.

ORDER: CARNIVORA –

The order Carnivora comprises a very distinctive group of animals numbering 274 living species. The members belonging to this order are called carnivores and have a worldwide distribution. The term "carnivore" broadly applies to any flesh eating animal. Although carnivores are principally meat eaters, some members in this group (e.g. pandas) feed mostly on vegetable matter. Most carnivores are ground living mammals, though aquatic (e.g. sea otter) and arboreal (e.g. palm civet) members are also found. Their size ranges from the smallest weasel to the largest polar bear. The most important anatomical character by which carnivores can be distinguished from other mammals is their shearing cheek teeth or carnassials. The fourth upper premolar and first lower molar are collectively known as carnassials teeth. They are fit together perfectly to provide a shearing surface to cut meat. Incisor and canine teeth are used for seizing, holding and biting of prey. The forefeet and hindfeet usually bear four and five digits respectively. Digits are armed with sharp claws. The claws in extended and retracted position are given in Fig 12.3 and 12.4. Some carnivores possess retractile claws, whereas, others have non-retractile or semi-retractile claws. Most members have odorous skin glands. Some carnivores walk on soles of their feet called plantigrades (e.g. bears), while other species walk on their toes called digitigrades (e.g. dogs). Sweat glands are restricted to their pads of the feet. The male copulatory organ of many carnivores have an elongated bony structure called the baculum (os penis). Most members in this order lead a solitary life except in breeding season. In the wild, these animals are spread over a large area.

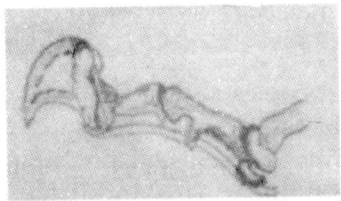

Fig.12.3: Extended claw   Fig.12.4: Retracted claw in carnivores.
in carnivores.

The order Carnivora contains two super-families and each super-family has four families. The zoological classification is as follows:

Super-families:

Feloidea	Canoidea
The paroccipital process is not so prominent; cowper's gland is absent; most species have retractile or semi-retractile claws.	The paroccipital process is prominent; most cases, the cowper's gland is very large; claws are non-retractile.
Families:	Families:
(a) Felidae- tiger, lion, leopard etc.	(a) Ursidae- bear
(b) Hyenidae- hyena	(b) Procyonidae- panda
(c) Protelidae- aardwolf	(c) Canidae – dog, wolf, fox etc.
(d) Viverridae-civet, mongoose etc.	(d) Mustelidae- weasel, marten etc.

Felids:

The members belonging to the family Felidae are called felids. Felids have a worldwide distribution except in few places. They live in a diverse range of habitats. Felids are broadly classified into two groups: big cats (cheetah, tiger, lion, leopard, clouded leopard, snow leopard and jaguar) and small cats (bobcats, wild cats, lynx etc. in the genus *Felis*). The basic difference between the big and the small cats is that some big cats can roar, while small cats can purr but can not roar. The big cats have a round pupil in their eyes, whereas, it is generally vertical in many small cats. The size of felids ranges from a small black footed cat weighing 1-2 kg to Siberian tiger that weighs more than 350 kg. Short and rounded heads with flat face, large and well developed brains, and short legs are the general features of felids. With the exception of Pallas's cat and lynx, all the felids have 30 teeth (I: 3/3; C: 1/1; PM: 3/2; M: 1/1). They are generally nocturnal though some animals may prowl by day. The majority species are solitary, extremely secretive and are not easily accessible. Felids when surrounded with their young become dangerous. Some members are good swimmers.

Felids have always been fascinated by human beings for their elegant looks. Many members are very popular zoo exhibits.

The salient features of some felids are given below.

Cheetah/Hunting leopard:

Zoological name: The zoological name of cheetah is *Acinonyx jubatus*. There are mainly two subspecies- the African cheetah (*A.j.jubatus*) and the Asiatic cheetah (*A.j.venaticus*).

Distribution: The fastest land animal, the cheetah is found in Africa, South Asia and Middle East. In India, the last cheetah in the wild was found in 1947.

Habitat: They inhabit in most habitats such as open woodlands, grasslands, scrub and open plains. However, they prefer to live in open semi-desert country.

Physical feature: A cheetah measures about 7 ft in length including tail. The body weight varies from 50 to 80 kg. The colour is tawny to buff or grayish white, and the coat is spotted with small and solid round black spots almost throughout the body. A loose and slim built, small and round head, high-set eyes with round pupils, small and round ears, long and thin legs and a deep narrow chest are the distinctive features of a cheetah. Unlike other felids, the feet are furnished with hard pads. The well developed claws are blunt and is not covered by a sheath when retracted, and remain exposed and extended (i.e. non-retractile claws). However, sometimes it becomes semi-retractile. The most salient feature is its "tear stripe" or a black line which runs down from the corner of each eye to the upper lip. Its nasal passage is larger than that of other big cats thereby allowing an increased air intake which enables the cheetah to maintain a relentless suffocating bite.

Reproduction: Sexual maturity is attained at 20-23 months, though some individuals may attain this at 14-16 months. They are seasonally polyestrous.

Some authors have reported that there is no regular breeding season and cubs may born in all the months. Estrus lasts for 1 to 3 days or more. They remain together for a day or two so as to mate several times during this period. The dominant male usually mates with the female. She gives birth one to eight (average three) cubs after a gestation period of 84-95 days. The new born cubs weighing 250-350 g have a coat colour of blue grey on the back and brown with dark spots on the rest. Young are cared by mother for about one year of age.

Behaviour: Cheetah hunts in open plain predominately by day. They kill most prey by strangling. In the wild, it drags the prey to a safe place before eating. Females are more solitary but they are also found in a family unit comprising a single female and her young. Cheetahs defecate on tops of logs, rocks, or in shady areas. Most salient behaviours observed among mature cheetahs include social grooming, licking the fur of other family members, and capturing and eating flies and ticks. Young are reputed for their playfulness.

Life span: Cheetahs are known to live 10-12 years in the wild, though they have been reported to live up to 17 years in captivity.

Lion:

Zoological name: The zoological name of lion is *Panthera leo*. There are two sub species- the Asiatic lion *(P.l. persica)* and the African lion *(P.l.leo)*.

Distribution: Lions are found in Africa and Asia. The Asiatic lions are now only confined to Gir forest of Gujarat state in India.

Habitat: Lions prefer to live in open grasslands.

Physical feature: An adult lion may reach up to an average length of 9 ft including tail and weighs up to 240 kg. It stands 4 ft at the shoulder. The coat colour is sandy buff with whitish on belly and inner side of legs. The back of the ears is black. The male has well developed manes. Eyes are yellow in colour. The female has four nipples. The upper surface of the tongue is coated with backward curved horney papillae. However, the Asian lion can be distinguished from the African lion by the following points:

- The Asiatic lion is the cousin of the African lion and is slightly smaller than African lion.
- The Asiatic lions have a thicker coat and have a longer tassel of hair at the end of the tail. Moreover, they bear smaller and scantier manes.
- The Asiatic lion has a more pronounced belly fringe and a more prominent tuff of hair on its elbow.

Reproduction: Sexual maturity occurs at three to four years of age. Wild animals exhibit an early maturity than the captive specimens. Males start breeding between four and six years. In captivity, lions may breed every year, but in the wild they usually breed once in two years. The cubs are born round the year. A pair may copulate several times a day. Usually two to three cubs (range 1-6) are born by the female after a gestation period of 92-119 days. The new born cub weighs about 450 g. The fully furred young have a dark spotted coat marking and this usually disappears with the advancement of age. The teeth start to appear at about three weeks of age. The young become independent when they are 1.5 to 2 years of age.

Behaviour: Among all the felids, lions are the most social animals. They live in family prides of 2-4 dominant males, several adult females, a number of sub-adults and cubs. They maintain their territories by urine markings, patrolling or by roaring. They are known to hunt in packs comprising mostly of females, and sometimes the pack may go up to thirty individuals in one encounter. Lions are mostly hunt during the day. An interesting behaviour of the lion is that an established adult male in a pride usually behaves friendly towards the females and the cubs. However, a newly introduced male, on the other hand, is reported to kill at least some of the cubs in the pride where he takes it over. A lion normally eats the hind quarters first followed by the forequarters and finally the head of the prey. When many lions are present to one kill, the kill is torn to pieces.

Life span: In the wild lions normally live for 15 years, while in captivity the life span may reach up to 25 years.

Leopard/Panther:

Zoological name: The zoological name of leopard is *Panthera pardus.* There are 14 recognized subspecies of leopards all over the world. The Indian subcontinent is blessed with five subspecies such as the Indian leopard (*P.p.fusca*), the Persian leopard (P.p.saxicolor), and the Sikkim and Nepal leopard *(P.p.pernigra)*.

Distribution: These carnivores are the natives of Africa and Asia. Leopards are the most widespread members in the Felidae family and this is largely because of their highly adaptable hunting and feeding behaviours. Moreover, they can adapt to a wide variety of habitats.

Habitat: As mentioned earlier that leopards are adapted to a wide variety of habitats ranging from the tropical rain forests to almost urban dwellings. Leopard is equally comfortable in marshy land, semi-desert terrain and low lying scrub. The black panther is found in the high rainfall area of eastern India and Western Ghat.

Physical feature: The size ranges from about 5 ft 4 inch to 9 ft in length and weighs 30-90 kg. Like most members in the cat family, males are larger and heavier than females. The ground colour is usually tawny yellow and is marked with rounded black or dark brown spots. These spots are hollowed unlike cheetah. The black colour is due to melanin pigment.

Reproduction: The attainment of sexual maturity in leopards is about two and a half year. Over most of their ranges, they breed throughout the year. Females are sexually receptive at three to seven weeks of interval. Usually three blind and fully furred cubs (range 1-6) are born after a gestation period of 90-105 days. The new born cubs weigh 430-570 g and their eyes are open after about ten days of birth. Cubs remain with the mother up to 18-24 months. These animals can be crossed with the lions in captivity and the resulting crossbreds are called "leopons".

Behaviour: Leopards are least social animals and have no adult social groups. Territories are marked by urine spraying onto logs, branches and tree trunks. They hunt singly generally at night. Another interesting fact is that leopard often drags its prey up tree where it can eat and store so as to avoid any disturbance from scavengers. Leopards enjoy climbing trees and have a habit of basking on the branches in the sun. When there is shortage of its natural prey base in the wild, the leopard comes in to the periphery of human habitat and preys on the domestic livestock. The black panther is bolder and pugnacious.

Life span: The life span is 12-15 years in the wild, though it has been reported a long life span (20 years) in captivity.

Tiger:

Zoological name: *Panthera tigris* is the zoological name of tiger. Among eight recognized sub-species, three have become extinct (the Caspian tiger, the Bali tiger and the Javan tiger). The living subspecies are the Indian tiger *(P.t.tigris)*, the Indo-Chinese tiger *(P.t. corbetti)*, the south Chinese tiger *(P.t.amoyensis)*, the Siberian tiger *(P.t.altaica)* and the Sumatran  tiger *(P.t.sumatra)*. The south Chinese tiger is said to be extinct very soon, as reported that a few tigers are left in the wild.

Distribution: Tigers are distributed solely in Asia. They are thought to have originated in Siberia and the New Siberian Islands and subsequently they extended their territories over the greater parts of the Asian continent where conditions were favourable for establishing themselves.

Habitat: Their habitats include tropical rain forest, mangrove swamps, snow-covered coniferous and deciduous forests, scrub and grasslands.

Physical feature: The average length of a tiger (head to tail tip) ranges from 9 ft 6 inches to 10 ft 3 inches and it stands 3 ft to 3 ft 6 inches high at shoulder. The body weight may reach up to 384 kg. The ground colour is reddish fawn and is broken at intervals by dark vertical stripes. The male normally appears to have fewer strips than the female. The other salient features include longer hindlimbs, sharp and long retractile claws in forepaws, and rudimentary clavicle. The back of the ears is black with a prominent white spot in the centre. The black markings in the patch of white hair above each eye are found to be so distinctive that individuals can readily be distinguished by them. The white tiger, a recessive mutant of the Indian race, is found in India. It is characterized by chalky white colour marked with ash or light black stripes, pink nose, palm pads and icy blue eyes. They are generally bigger in size.

Reproduction: Sexual maturity is reached by three to four years of age. The onset of breeding season varies from place to place. The tropical tigers breed throughout the year, while in the north the breeding is restricted to the winter months. The majority of cubs in India, however, are born in April and November and in the cold regions, the female drops young in the spring. The estrus lasts for 3-10 days. They mate as many as hundred times over a period of two to three days. After each coitus the male is thrown off by the female and he tries to get out of her reach as quickly as possible. However, after a short period of rest, the male again is attracted towards the female for copulation. The usual litter size is two to three young, though as many as six young may be born. The young, each weighing about 1-1.5 kg, are born after a gestation period of 103-115 days. The new born cubs are blind and helpless. The young are cared by the mother alone. They come out from the den at about eight weeks of age and follow the mother. The lactation period continues for 4-6 months. Nevertheless, cubs remain for 2-2.5 years under mother's care and guidance. In the wild, the interval between two successive pregnancies is reported to be three years. Tigers and lions are closely related and hybrids are produced in captivity. The hybrid cubs are the "tigon" (male tiger and female lion) and the "liger" (male lion and female tiger). Both crossbreds are generally larger and darker than either of their parents.

Behaviour: Tigers are ordinarily solitary and independent in the wild except in mating time. However, the basic social unit is mother and young. In the wild state, both male and female demarcate their territories. The home range can be maintained by a variety of methods. One such method is that urine which is mixed with anal gland secretions is sprayed onto trees, bushes and rocks along the trails. They also mark scrapes on the ground and tree trunks. The black tarry colour faeces is left in conspicuous places. They are good swimmers. Unlike leopards, the tigers seldom try to climb trees. Male tigers are reported to kill smaller cubs when encounter with them. The mother secretes her newborn cubs in a den normally a cave or rock overhang, among dense reeds and bushes. In the wild tigers and leopards are said to avoid each other.

Tigers hunt alone and their hunting technique is unique. They usually attack from the sides or from the rears. A prey is killed by grabbing the head between forepaws and twisting its neck with violent jerk and finally a biting to the throat or neck. The death is most likely caused by suffocation as well as snapping of the spinal cord. It often drags its dead prey into cover in the vicinity of water in the wild. A tiger drinks frequently during meal. During summer months, they spend much of the day time resting near water bodies and occasionally lie or stand in water for cooling.

Sometimes, tigers become man eaters. The Bengal tigers in Sundarban areas of West Bengal are known to be human flesh eaters. The reasons for man eaters are manifold.

- Some tigers are very old or they get disabled or wounded and thereby unable to stalk and kill their natural preys. When their canine teeth are broken, they are also tempted to attack human beings.
- When tigers killed a man inadvertently, tasted the meat and apparently found it to their liking. There might be the other reason that tigers first scavenged on unburied human corpses and later transferring their attention to living prey.
- The availability of natural preys in the habitats.
- The destruction of habitats may also be an important factor for tiger to be turned into a man eater. The scarcity of freshwater resource in Sundarban areas is believed to be one of the reasons for development of this habit.

Many a time, it is very difficult to ascertain whether a prey is killed by a tiger or a leopard in the wild. Therefore, the following points may be helpful in identifying the killer:

Tiger	Leopard
(a) The size of the canine punctures and the gap between them are larger and wider.	(a) In case of leopard it is smaller and less wider.
(b) Tiger drags the kill further into the bush and usually stays near the kill for a few days. The kill is covered with twigs, debris and earth so that carcass can be concealed from other competitors such as jackals. It may feed on its kill for 4-5 days depending on the size of the kill.	(b) Leopard normally brings its kill into tree in order to protect from other predators and scavengers.
(c) Tiger are clean feeders and start feeding normally from hind quarters.	(c) Leopards are considered to be messy feeders and usually begin to eat from the chest or stomach.

Life span: Tigers have been known to live for about 15 years in the wild, while in captivity the life span has been reported more than 20 years.

Snow leopard:

Zoological name: The zoological name of snow leopard is *Panthera uncia.*

Distribution: They are found in Central Asia and Indian subcontinent. In India, it occurs in the higher altitudes of the Himalayan range (from Kashmir to Arunachal Pradesh).

Habitat: Its homeland is the mountain steppe and coniferous forest scrub.

Physical feature: The snow leopard is about 4 ft 6 inches length (head to body) and tail is about 3 ft. It is smaller than a leopard and weighs 23-41 kg. The ground colour is pale whitish grey often with a yellowish tinge and white below. They differ from other big cats in having short muzzles, elevated foreheads and vertical chins. The unbroken spots in snow leopard are distinct on the heads, napes and lower parts of the limbs. The body is covered with medium brown blotches ringed with black or dark brown. A black streak runs from the middle of the back to the root of the tail. The surface of its paws is furnished by a cushion of hair that protects the soles from cold. The ears are white with black edge.

Reproduction: Snow leopard is able to breed at two to three years of age. The mating season begins between January and May. The heat period lasts for about 4-8 days. The litter size is usually one to four. The gestation period is 93-105 days. The blind young are born in spring or early summer in a well-hidden den lined with the mother's fur. The young become active by two months. Female usually breeds every alternate year.

Behaviour: The snow leopard is nocturnal. It leads a solitary life except mating time.

Life span: The life span is reported to be 15 years in captivity.

Small cat:

Zoological name: Small cats belong to the genus *Felis.* There are 28 species of small cats. Some well known species of small cats are leopard cat *(F. bengalensis)*, jungle cat *(F.chocus)*, fishing cat *(F.viverrina)*, Asiatic golden cat *(F. termincki)*, Pallas's cat *(F.manual)*, puma *(F.concolor)*, lynx *(F.lynx)* and wild cat *(F.silvestris).*

Distribution: Small cats are distributed in America, Europe, Asia and Africa. Many small cats such as Pallas's cat, marbled cat, leopard cat, rusty-spotted cat, jungle cat and fishing cat are found in India.

Habitat: Their habitats range from arid regions to cool temperate forests.

Physical feature: The body weight varies from a small black cat weighing 1-2 kg to a large puma that weighs up to 103 kg. Many species of small cats have spots or stripes. The forefeet have five digits and four on the hind feet. Small cats walk on the toes. The black tear stripes are usually present. Some species have rounded ears with tufted tips. The iris of eyes may be orange, yellow or green. However, the small cats can be distinguished from the big cats by the following characteristics:

- As mentioned earlier that small cats do not roar but they can purr. This is because the arms of hyoid apparatus which is formed of an uninterrupted chain of small bones hold the larynx close up to the skull and thereby restrict its movement. So, the fully ossified hyoid bone in the vocal apparatus prevents the small cat from roaring.

- During resting, the small cat draws forepaws under its body by bending them at the keen or carpal joint and the tail is wrapped round the body. The big cats on the other hand, place their paws in front of their bodies and the tails are extended straight behind them. Small cats feed in a crouched position while big cats feed in lied down position.

Reproduction: They mature at about 10 months of age. The size of a litter may go up to eight kittens. The gestation period ranges from 56-68 days or more depending on the species.

Behaviour: They mostly sleep during day and become active towards evening.

Life span: Small cats may live up to 15 years.

Hyena:

Zoological name: There are three species in the family Hyaenidae: the spotted or laughing hyena (*Crocuta crocuta*), the striped hyena (*Hyaena hyaena*) and the brown hyena (*H. brunnea*).

Distribution: Hyenas are the natives of Africa and Asia. The striped hyena is found in India.

Habitat: They inhabit in open plains, brush lands and open forested areas. The striped hyena inhabits preferably in open habitat or light thorn bush country.

Physical features: The body weight is about 30- 40 kg (striped hyena) or 50-80 kg (spotted hyena). The forequarters are heavier than the hindquarters. The other features include longer forelimbs, non-retractile claws, large and rounded ears, bushy tail pointed at tip, massive head with powerful jaws, large eyes, sharp and elongated canines, dorsal mane and well developed carnassials. The anal pouch lies between the rectum and the base of the tail and this can be turned inside out. Both sexes are very difficult to differentiate. The female sex chromatin is "drumstick" appearance.

Reproduction: The female has a large clitoris. There may be four teats (female spotted hyena) or 6 teats (female striped hyena). The striped hyena breeds throughout the year. Litter size is 1-6 young. The gestation period varies form 90 to 110 days. The young weigh about 1.6 kg. Young are nursed for about one year.

Behaviour: Hyena has earned bad reputation for its habit of eating carrion. They are master scavengers and eat any item that remains untouched by other animals. The sense of smell is well developed. Hyenas are nocturnal, though they may be active by day. They are solitary animals. However, hunting takes place in packs. Hyena is famous for its "laughing call". In India, the striped hyena is reputed for its habit of predating of livestock and occasionally lifting of child.

Life span: The life span has been reported up to 20 years or more in the wild depending on the species.

Viverrid:

Viverrids are the animals of the Old World. They inhabit a wide variety of habitats such as desert, semi-desert, forest, woodland and savannah. Civets, genets, mongooses and linsangs are the well known representatives of viverrids. Long and slender bodies, elongated heads, short legs, long and bushy tails, small and rounded ears and tapered muzzles are the general features of viverrids. Most species are terrestrial. Some are arboreal. Most viverrids produce a yellowish substance from the perineal glands called civet. The active substance of the civet is known as civetone. It has a pleasant musky odour. Civetone is used for preparing medicine and perfumes.

Mongoose:

Zoological name: There are many species of mongoose. The banded mongoose (*Mungos mungo*), crab eating mongoose (*Herpestes urva*), Indian brown mongoose (H. fuscus), Egyptian mongoose (*H.ichneumon*), small Indian mongoose (*H.javanicus*) and broad-striped mongoose (*Galidictis fasciata*) are some well known species of mongoose.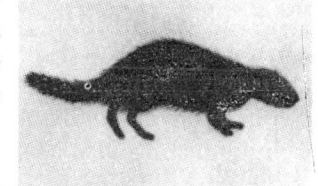

Distribution: They are found in Africa, Asia and southern Europe.

Habitat: They inhabit in forest, cultivated areas, desert etc.

Physical feature: The body weight is attained up to 5 kg depending on the species. The hair is coarse and rather long. Unlike civets and genets, mongooses have no spots. There may be one, two, or three pairs of teats depending on the species.

Reproduction: The puberty is reached when they are about two years of age. The small Indian mongoose may, however, breed when it attains nine months old. They may breed at almost any time of the year or in a particular season. The litter size usually ranges from two to six offspring. The gestation period is normally two months.

Behaviour: Most species are solitary, though some live in pairs or in small groups. Mongoose may be nocturnal or diurnal. They are reputed for stealing eggs and killing snakes.

Life span: The life span may be four to ten years in the wild.

Bear:

Bears are well distributed in Asia, Europe and North and South Americas. Their habitats range from arctic coasts to tropical forests. They are distinguished in having a compact body, massive head, short and powerful limbs, and short tails. Each foot bears five toes. Claws are long, curved and non retractile. The carnassial teeth are poorly developed and have lost shearing function. They have a long and shaggy coat with predominantly black or some shades of brown or white.

The family Ursidae contains seven species: the American black bear (*Ursus americanus*), the grizzly or brown bear (*U. arctos*), the polar bear (*U. maritimus*), the spectacled or Andean bear (*Tremarctos ornatus*), the sloth or honey bear (*Melursus ursinus*), the Malayan sun bear (*Helarctus malayanus*) and the Asiatic or Himalayan black bear (*Selenarctos thibetanus*). The sun bear is the smallest among all the recognized species of bears and weighs 25-65 kg. The following text highlights some salient features of the Asiatic black bear and the sloth bear.

Asiatic black bear:

Distribution: The Asiatic black bear occurs in southern and eastern Asia.

Habitat: It is found in forest and brush cover.

Physical feature: The salient features include large tufted ears, short tanned muzzle, black claws and a large characteristic white crescent marking on the chest. The coat colour is typical jet black, though brown and reddish brown individuals are also seen. The body weight ranges from 42 to 120 kg.

Reproduction: Females are sexually receptive at about two years of age and the male attains sexual maturity a couple of years later. They breed in the late autumn. The gestation period is about 210- 240 days. The litter size is usually one or two young. The birth weight varies from 350-400g. The mother with its cubs comes out from the den in the month of May. The young remain with the mother for about two years.

Behaviour: They are good climbers and the males are notorious for killing females. Asiatic black bear are solitary and nocturnal in nature. They go for a period of hibernation though many individuals remain active during winter months.

Life span: The Asiatic bears are known to have a life span of 20-24 years in the wild.

Sloth bear:

Distribution: They are found in India and Sri Lanka.

Habitat: The sloth bears inhabit in forests.

Physical feature: The ground colour is normally black occasionally mixed with brown, grey or rusty red. Sloth bears are characterized by having long white curved claws, naked and mobile lips, short tails, long snouts, extensive tongues, small molar teeth and long and shaggy hairs. The flesh is dark in colour. It stands about 4-5 ft at the shoulder and weighs 90-150 kg. It has a characteristic whitish 'V' shaped mark on its chest.

Reproduction: Southern populations breed throughout the year, while their counterparts in the northern area breed in June. The female usually gives birth two young (range 1-3) in a ground shelter after a gestation period of 210 days. Cubs leave the den after two to three months of birth. Young remain with the mother for about 2-3 years. Females usually breed only once in three years in the wild.

Behaviour: Sloth bears usually sleep during day time and hunt at night. They are very fond of termites and honey. It digs up 3- 4 ft deep at the base of termite mound leaving an indication of its presence in a particular area. The sloth bear travels a long distance in search of food. They do not have winter dormancy periods, though sloth bears shelter in dens for seclusion and protection.

Life span: It may live as long as 40 years in captivity.

Procyonid:

Procyonids are reputed for their active and curious nature. Many members are very popular pets and zoo exhibits. The members belonging to the family Procyonidae include pandas, raccoons, kinkajous, coatis, olingos, cacomistles and ringtails. They occur in tropical and temperate parts of North, Central and South Americas and eastern Asia. The body weight ranges from 0.8 kg (ringtail) to 160 kg (giant panda). The members are solitary in nature except ringtails. Most species are nocturnal. They are arboreal and move around in trees. They prefer to make their dens in hollow trees or rocky areas or they shelter in dens left by other animals. A brief description of pandas is given below:

Giant panda:

Zoological name: The zoological name of giant panda is *Ailuropoda melanoleuca*.

The giant panda is sometimes called bamboo bear.

Distribution: It is found in Szechuan, Shensi and Kansu provinces of Central and Eastern China.

Habitat: Giant pandas are found in cool and dense bamboo forests at elevations of 5000 to 13000 ft.

Physical feature: The length is about 160-180 cm including the tail and the body weight is 75-160 kg. The giant panda has a dense white coat marked with deep black on the limbs, ears, muzzles, shoulders and around the eyes. Total number to teeth is 42 (I: 3/3; C: 1/1; PM: 4/4; M: 2/3). The carnassial teeth are adapted to both slicing and crushing.

Reproduction: The male panda is unique in having separate scrotal sac for each testicle. The female reaches sexual maturity at four to six and a half years of age. The male probably takes a few years more to reach at reproductive state. Mating starts in the months from mid-March to May. The peak of estrus lasts for about 1-3 days. The female bears a litter of one or two young. Yellowish and pink young are blind, almost naked and helpless. They weigh about 90-150 g. Young are born in a cave or in a hollow tree after completion of 97-163 days gestation. They are able to move at three months of age. At about six months, young start to eat bamboo leaves. The independent life begins at about one year of age when young attain about 35 kg.

Behaviour: In the wild, the giant panda is solitary except mating season and pre-weaning period. They communicate by sight, sound and smell. It is very difficult to locate the giant pandas in their natural habitats.

Life span: They can live about 14 years in the wild. However, in captivity, the life span has been recorded more than 30 years.

Red panda/Lesser panda:

Zoological name: The zoological name of red panda is *Ailrus fulgens*. It is also commonly called bear cat, common panda or Himalayan raccoon. It has two recognized subspecies: *A.f.fulgens* and *A.f.styani*. The former race is found in the temperate forest of Nepal, Sikkim and Himalayan region of eastern India, while the later is found in South China and Myanmar.

Distribution: It is more widely distributed than giant panda and its homeland ranges from the eastern Himalayas to Western China.

Habitat: They inhabit in temperate forest zone of the Himalayan ecosystem where bamboo forests are available.

Physical feature: The length from head to body is 51-63.5 cm and the tail is 28-

48.5 cm. They weigh about 3-5 kg. The red panda has a thick fur. The colour is reddish brown above and black underneath. Face and ears are white to buff. The face has a red brown stripe which runs from each eye to the corner of the mouth. The other salient physical characteristics of red panda include a rounded head, short and broad face, pointed snout, pointed ears, semi retractile claws and long and bushy ringed tail. The soles are covered with hair.

Reproduction: Reproductive maturity occurs at one and a half to two years of old. The litter size ranges from one to four young (usually two). The gestation period is 90-150 days. The long gestation period may be due to delayed implantation. The young weighing about 105-131 g are blind, helpless and fully furred. It is readily adapted to the captivity, though breeding is not easily taken place.

Behaviour: The red panda is very curious and gentle. It is fairly nocturnal and may live alone, in pairs or in family groups. In the wild, they are reported to be most active at dawn, dusk and at night. The ideal social grouping is mated pair and the dependent offspring. They probably make territories. Faeces is deposited in well defined latrine areas.

Life span: The red panda can live up to 14 years in the wild.

Canid:

The family Canidae forms a diverse group of members and they include wolves, foxes, jackals, coyotes and dogs. Canids have a worldwide distribution except for New Zealand, West Indies etc. The domestic dogs, however, have been introduced in many places where wild canids are naturally not found. The members are adapted to a wide variety of habitats ranging from desert to arctic tundra region. Their size varies from the smallest fennec fox weighing about 1 kg to the largest grey wolf weighing 80 kg. Canids are easily recognized by their long muzzles, erected and pointed ears, strong and blunt non-retractile claws, more or less bushy tails and well developed jaws. Canids walk on the toes of the feet. The male has a well developed os penis. All the canids are reputed for their opportunistic and adaptable behaviours. They tend to hunt in pack. Some canids such as arctic fox and timber wolf are important for their fur production.

Wild dog:

Wild dogs are found in Asia, Africa, South America and Australia. They include Asian wild dog (dhole), African wild dog (cape hunting dog), bush dog, raccoon dog and dingo. The domestic dog (*Canis familiaris*) is thought to be descended from wolf and golden jackal. The following text highlights about the Asian wild dog or dhole.

Dhole:

Zoological name: The zoological name of dhole is *Cuon alpinus*. In India, three subspecies namely the tawnier peninsular form *(C.a.dukhunensis)*, the deeper red Himalayan form *(C.a.primaevus)* and the paler trans-Himalayan form *(C.a laniger)* are existed.

Distribution: They are found in eastern and central Asia.

Habitat: Dholes are exclusively inhabitants of forests.

Physical feature: Dhole is characterized in having a wolf-like body, short leg, a thick set muzzle, and a bushy tail. The average body weight is 18 kg. They differ from wolves, jackals and domestic dogs by having 12 cheek teeth in the lower jaw (others have 14). The total number of teeth is 40 (I: 3/3; C: 1/1; PM: 4/4; M: 2/2). The characteristic red coat varies in shades of yellow or white according to the localities and seasonal changes. The tail is black. The female bears six or seven pairs of teats.

Reproduction: Dholes are sexually mature when they are about one year old. In India, mating occurs between September and January. The female usually gives birth four or five pups in a den during November-April. Dens are made by the females on the banks of stream bed or among rocks. The gestation period in dholes is stated to be 60-63 days. During early periods of live, the young are fed on regurgitated food. At about five months of age, young actively follow the pack.

Behaviour: Dholes are pack-hunters and hunt mainly in the day, though occasionally it happens at night. The smell plays an important role in locating a prey. They occasionally jump high in the air or stand briefly on their hind legs to spot prey if it is hidden by grass. Dholes are efficient killers. Two to three animals are able to kill a deer of 50 kg within a few minutes. In true sense, dholes do not bark but yap or utter. They communicate each other by whistling cries.

Life span: In the wild, they can live for 10-18 years.

Mustelid:

Mustelids have a worldwide distribution except in Australia, Antarctica, Madagascar and some oceanic islands. The members occupy almost every habitat including fresh and salt water. These small to medium-sized animals have a long body with short legs. Members belonging to the family Mustelidae include weasel, polecat, mink, marten, wolverine, stoat, ferret, sable, skunk, otter, badger and honey badger. A thick oily, yellow and strong smelling substance called musk is secreted from the glands. Mustelids are mostly solitary animals, though they may be found in small groups. Most species of mustelids are active both day and night. The majority of animals are terrestrial. However, they are also occurred in semi arboreal (e.g. Marten), burrowing (e.g. badger) and amphibious (e.g. otter) states. The otter is described in greater detail here.

Otter:

Zoological name: The Eurasian river otter *(Lutra lutra)*, Oriental or Asian short clawed otter *(Aonyx cinerea)*, Indian smooth-coated otter *(Lutrogale perspicillata)* and sea otter *(Enhydra lutris)* are some well known species of otter.

Distribution: Otters occupy throughout the world except in Australia, Madagascar and polar region.

Habitat: Otters are truly amphibious. They are found in marshes, coastal mangrove, swamps etc.

Physical feature: The body weight may vary from 4 to 12 kg. The salient morphological features include streamline bodies, small and powerful legs, webbed toes (in most species), flat rounded heads, small and rounded ears, longer hind feet, fully haired tails, stiff vibrissae around nose and snout, and non-retractile claws (in most species). Their body is covered with water proof underfur and long guard hairs. The coat colour varies from black brown to a pale grey.

Reproduction: Otter attains puberty at about two years of age. The breeding seasons starts during October or December in case of the Indian smooth coated otter. The litter size is one to five young. The gestation period is usually 60-70 days.

Behaviour: They are docile and gregarious for most part of the year. Although otters are nocturnal, they may come out during day time. Some species are well known for their manual dexterities. Otters are good swimmers and can travel under water for sometime without surfacing for air.

Life span: In the wild, they are known to live up to twelve years.

ORDER: PROBOSCIDEA —

Elephant

Elephants have been domesticated for at least 4000 years. However, today's elephant evolved from small pig-like ancestor called *Moeritherium* about 50 million years ago. The elephants are placed in the family Elephantidae of the order Proboscidea.

Zoological name: There are two species of elephants: the Asian elephant *(Elephas maximus)* and the African elephant *(Loxodonta africana)*. The Asian elephant again has four subspecies: the Indian elephant *(E.m.indicus)*, the Ceylon elephant *(E.m.maximus)*, the Sumatran elephant *(E.m sumatranus)* and the Malaysian elephant *(M.m. hirsutus)*. The African elephant, on the other hand, has two subspecies: the bush or savannah elephant *(L.a.africana)* and the smaller forest elephant *(L.a. cyclotis)*.

Distribution: Elephants, the largest living terrestrial mammals, are distributed in African and certain Asian countries.

Habitat: They are found in savannah grasslands and forests. In India, they are found in the areas of tropical and some limited subtropical forests except in the dry scrub lands and saltwater mangrove forest. However, elephants prefer to live in undulating hilly terrains with thick forests. Some factors such as availability of water, food, environmental temperature, humidity, light, and predators play an important role in shifting their habitats.

Physical feature: The body weight varies from 4000-7000 kg depending upon the species involved. The adult height is recorded 10.8 ft or more. The height is reported to be approximately twice the circumference of forefoot. The body length ranges from 20 to 24.5 ft. The body surface is covered with sparse and coarse hair. The soles of the feet are covered with a soft and fatty cushion of white elastic fibres. The proboscis or pendant trunk is the elongation of the upper lip and nose. It enables the animal to feed and drink from the ground. The other functions of the trunk include breaking off branches, picking of leaves and shoot, feeding from trees and shrubs, drinking water and throwing dust. Another salient anatomical feature of an elephant is its tusks. They are actually modified and elongated upper incisor teeth. It is composed of ivory which is a mixture of dentine, cartilaginous material and salts. Tusks first appear at about 2-3 years of age and grow throughout life. However, they attain full growth in about 25 years of age. The average weight of a male tusk may be 60 kg. or more. The average length of a tusk may range from five to eight feet. The tusks are used for digging of roots, removing the bark of trees etc. Tusks also serve as a defensive weapon. The total number of teeth is 26 (I: 1/0; C: 0/0; PM: 3/3; M: 3/3). The premolar and molar teeth replace each other in succession throughout life. It is said that the first milk molar teeth appear at the age of two weeks and shed about two years of age. In this way, the second pairs shed at the age of six; third pairs at nine; the first adult teeth or the fourth pairs between twenty and twenty five year; the fifth pairs at sixty and the last pairs remain for the rest of the life. So, this is an important tool for estimation of age of an individual animal with great accuracy.

Elephants have 20-21 pairs of ribs. Ears serve like a radiator of a car. Each elephant has a characteristic pattern of blood vessels in the ear. The ear cartilage of old animal becomes hard and the fringe of the ears becomes torn and ragged. The kidney is having five to seven lobes. Another salient anatomical feature is that the lungs are adhered to the chest wall by fibrous connective tissue resulting in the absence of pleural cavity. However, the pleural cavity is found in the young elephant. Elephants are non ruminants with a simple stomach. There is no gall bladder.

The Asian elephants can be distinguished from the African elephants by the following points:

Asian elephant	African elephant
1) They are smaller in size (up to 5000 kg or more) and average height up to 3.2 m.	1) Larger in size (up to 6000 kg or more) with 4 m height.
2) Trunk has less rings, more rigid and one finger life projection at the trunk tip.	2) Trunk has more rings, less rigid and two finger like projections at the trunk tip.
3) Males only have tusks. There are also tuskless males.	3) Both sexes have tusks.
4) Each forefoot bears five nails.	4) Forefoot has four nails.
5) Ears are small and do not cover shoulders. The shape of the back is convex. The highest point of the body is the top of the head.	5) Ears are large and cover the shoulders, the shape of the back is concave. The highest point of the body is top of the shoulder.
6) Forehead has two humps.	6) Forehead is curved smoothly.

Reproduction: The testes are located within the abdominal cavity. The large penis may weigh about 50 kg. The captive male attains sexual maturity at 8-10 years of age, while the female may reach at as early as 6 years. The length of estrus cycle ranges from 120-130 days. There is a seasonal variation in breeding. Usually the female drops a single calf after a long gestation period of 20-22 months. The newborn calf is about 3 ft height and weighs about 90 kg. The newborn is able to walk after an hour of birth. The young is nursed by its mother for about four years. The male leaves the group at about 10-15 years of age. A pair of teats is located in between the anterior legs and the young suckles by its mouth within two hours of birth. The elephants are capable of reproducing calves up to 60 years or more with an average calving interval of four years. So, a female may deliver 10-12 calves in her life time.

Behaviour: Elephants are very intelligent animals. They are found in herds. The herd comprises of the females and their young. Asian herd comprises 10-20 or more elephants.The dominant female leads the herd. However, males sometimes join the groups. They use their sense of smell to find out each other. The older bulls usually live singly. The large family unit may split into small groups during scarcity of food. In the wild, elephant sometimes digs a hole with its forefeet to get water. They also take dust bath in absence of water. Elephants migrate seasonally for food, water etc. They roam sixteen to eighteen hours for feeding in wild. Elephants have a strong liking for salts and minerals. They often have a mud bath (wallowing). The mud plaster provides a protection against insect bites and overheating. Elephants sleep for two to four hours normally before dawn.

Another important physiological and behavioural state of male elephant called musth is observed in elephants. It is a phenomenon where male elephants, both domesticated and wild, on attaining maturity, are sometimes subject to excitement and misbehaviour and this seems to have some correlation with sexual function. This aspect of biology has been observed in Asian elephants since ancient times but this has been recognized in African elephants more recently. The onset of musth can be observed from swelling of the temporal glands and the secretion of fluid streams down the cheeks of the elephants. The temporal glands which are located between the eyes and the ears, secret a thick and dark brownish in colour with a pungent smell oily substance called musth fluid. The duration of musth may last from a few days to months depending on the age and condition of the bull. Males come into musth first between the ages of 15 to 20 years. However, the average age of onset of musth in African elephants has been reported as 29 years. Although it is not properly understood about the physiological links between body conditions and musth, there is enough literary information regarding the role of good nutrition and body conditions for successful expression of musth in bull elephants. Besides oily secretions from enlarged temporal glands, the other observations seen in musth include increased aggressive behaviour, dribbling of urine, decreased response to commands of handlers (mohouts), and reluctant to take feed and water. Musth seems to have some correlation with the sexual behaviour. The high level of testosterone (20-40 ng/ml) has been observed in musth bull as compared to 0.2 to 1.4 ng/ml in nonmusth bull.

Life span: They may live up to 70 years in the wild. The life span in captivity may be more.

ORDER: PERISSODACTYLA —

The members belonging to the order Perissodactyla are called odd-toed ungulates (ungulate is a general term denoting a group of mammals having hooves). Perissodactylids are medium to large sized mammals. These simple stomach animals have a large caecum. Nasals are broad at the posterior end. The true bony growths in terms of horns and antlers are not present. Two mammary glands are located in the inguinal region. The number of toes present in the perissodactylids varies according to the species. The tapirs have four toes on the forefoot and three on the hind. The rhinoceros, on the other hand, bear three toes on each foot. The horse, however, possesses one functional toe on each foot. With the exception of tapirs, all members of this order, therefore, either have one or three toes (hence called odd-toed ungulate) on each foot.

All the members belonging to the order Perissodactyla are placed into the following three families:

(a) Tapiridae - tapirs.

(b) Rhinocerotidae - rhinoceros.

(c) Equidae - horses, asses and zebras.

Rhinoceros:

The term "rhinoceros" is derived from two Greek words *rhis* meaning nose and *keros* meaning horn. Thus, the literally meaning of rhinoceros is "nose horn". It is said that rhinoceros have been roaming on the earth for more than 60 million years. Rhinoceros are the second heaviest ground living mammals. They have been hunted for horns and skins which are believed to have medicinal values. In India, rhinos are threatened by poaching, habitat destruction and natural disaster.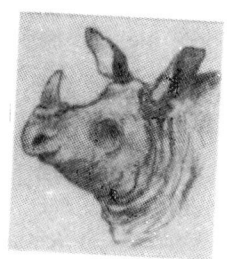

Zoological name: There are five living species of rhinoceros. They include the great Indian one-horned rhinoceros *(Rhinoceros unicornis)*, the Asiatic two-horned or Sumatran rhinoceros *(Dicerorhinus sumatrensis)*, the lesser one-horned or Javan rhinoceros *(Rhinoceros sondaicus)*, the square-lipped or the African white rhinoceros *(Ceratotherium simum)* and the hook-lipped or the African black rhinoceros *(Diceros bicornis)*.

Distribution: The present day living species are found in Africa and southern parts of Asia. In India, they are restricted to Assam (mainly) and West Bengal.

Habitat: Rhinoceros prefer to live in open grassland or forest interspersed with swamp, lake and marsh. The Indian rhinoceros prefers to live in forest mixed with swampy grassland.

Physical feature: Rhinoceros are characterized in having short and stout limbs and thick-skinned bodies. The body weight varies from 800 kg. (e.g. Sumatran rhinoceros) to 2500 kg (e.g. white rhinoceros) depending on species. The skull is broad at the base. Each foot is furnished with three toes with a hoof-like nail. The sole is tough and horney. The important anatomical feature is its horn or horns. The African species have two horns, while the Asian species with the exception of Sumatran rhinoceros have a single horn. The horns lack a bony core and are composed of collagen fibres. Both horns are located one behind the other in the middle line of the snout. The anterior horn is longer than the posterior one. The horn grows throughout the life and it begins to grow immediately once it is broken. The horn is loosely attached at its base. Females have smaller horns than males. The number of teeth varies from 24 to 30 depending on the species. Incisors and lower canine teeth are absent in the African species, while one or two upper incisors in each half of the jaw are present in the Asiatic species. The Javan and the Sumatran species have one canine tooth on each side of the lower jaw. Except the African white rhinoceros, all have a pointed upper lip. Like horse, they have a large caecum and colon. The gall bladder is absent like elephant. Asiatic rhinoceros (except Sumatran rhinoceros) have more folds of skin as compared to African rhinoceros.

Reproduction: The male rhinoceros attains sexual maturity at 4 to 8 years of age, while the females attain this at 3 to 4 years. Testes do not descend into a scrotum like elephants. The penis is pointed backwards when retracted. As a result, the urine is squirted behind between the legs. Two teats are located in between the hind legs. The oestrus cycle in rhinoceros has been reported to be 35 to 60 days. The oestrus lasts for about 2-3 days and is characterized by frequent urination, restlessness, squeaking and squealing sounds. Coitus usually lasts from about half an hour to more than an hour. The peak conception is reported during rainy season. The female drops a single calf (exceptionally two) after a gestation period of 15-16 months. The pink colour calf weighs 40-65 kg. depending on the species. The calf usually remains with the mother for 3-4 years. The calving interval varies from two to four years and female is capable of producing calves up to 40 years of age.

Behaviour: Rhinoceros are basically solitary animals. They may be grazer (e.g. Indian rhinoceros) or browser (e.g. black rhinoceros). Rhinoceros are active both in day and night. These herbivorous animals often wallow in the mud to regulate their body temperature as well as for protection of skin against biting flies. The male makes a territory which is marked with faeces and urine. In the wild, the rhinoceros often rubs its body against a sloping tree trunk leaving an indication of its presence in a particular area. The rubbed area is often coated with mud. Like a few other animals (e.g. blue bull), the rhinoceros have the characteristics of "dung piles". Rhinoceros are also known for their erratic behaviour like chasing vehicles and even butterflies.

Life span: The life span is reported to be 30 to 50 years or more.

Wild ass:

Zoological name: There are two living species of wild ass: the African wild ass *(Equus asinus)* and the Asiatic wild ass *(E. hemionus)*. The African wild ass has 3 subspecies: the north African wild ass *(E.a.atlantis)*, the Nubian wild ass *(E.a.asinus)* and the Somali wild ass *(E.a.somalicus)*. The Asiatic wild ass, on the other hand, has four living subspecies: the Kulan of Mongolia *(E.h.hemionus)*, the Indian wild ass *(E.h.khur)*, the Persian onager or ghorkhar *(E.h.onager)* and the Kiang of Ladakh, Tibet, Nepal and Sikkim *(E.h.kiang)*.

Distribution: Wild asses are found in dry areas of Asia and Africa. The Indian wild ass inhabits in Little Rann of Kutch in Gujarat.

Habitat: They live in the dry areas.

Physical feature: The African wild ass is characterized in having grayish coat colour with white belly, narrow dark stripes along the back, larger ears and normally a shoulder bar. The body weight is recorded up to 275 kg.

The Asiatic wild ass, on the other hand , is distinguished by reddish brown coat (in summer) to lighter brown coat (in winter), white belly, broad round hoofs, usually no shoulder bar, smaller ears, and a brush like tassel. The mature weight of an Asian wild ass is about 290 kg.

The Indian wild ass is reddish grey to fawn in colour. The lower parts of the body are almost white. Ears are short. The body weight varies from 200 to 250 kg.

Reproduction: Mating occurs in August-October. The female gives birth to one young after a period of 11-12 months of gestation.

Behaviour: They graze in the shrubs during day time and are found in groups of 10-30 individuals. However, ass may be found as solitary animals. These animals have a great ability to conserve water in the body and may remain without water for many days in harsh and dry land areas.

Life span: The longevity ranges from 20 to 25 years.

Zebra:

Zoological name: There are three living species: the plain or common or Burchell's zebra *(Equus burchelli)*, Grevy's or imperial zebra *(E. grevyi)* and the mountain zebra *(E. zebra)*.

Distribution: Zebras are found in Africa.

Habitat: They inhabit in dry areas.

Physical feature: Zebras are characterized in having a long face, sensitive lips, and grass cutting teeth. Like other equids, each foot bears only one functional toe. The most striking feature of these animals is their coat stripes. Like human finger prints, each species has distinct stripe patterns.

Reproduction: The female matures at the age of two and a half year. They tend to mate seasonally and the foaling usually takes place in the night after a pregnancy period of 11-12 months. The captive female may give birth to a foal at an interval of 12-15 months.

Behaviour: They are exceptionally social animals living in small to large groups. Zebras form a herd during rainy season. The herd, however, breaks off in the dry season. Zebras are normally nocturnal feeders. The males are very hostile to each other. Zebras have the habits of biting and kicking.

Life span: The life span is recorded up to 35 years.

ORDER: ARTIODACTYLA

The members of the order Artiodactyla are widely distributed almost all over the world. These even-toed ungulates are large and medium sized animals that provide meat, milk, hides and wool. Many artiodactyls are very popular exhibits in the zoos. Members of this order have either two or four toes (hence called even-toed mammals) on each foot. The first toe is absent. The second and fifth toes are either absent or reduced in size. The most important is the third and fourth toes which are large and equal in size and support the body weight. Each toe is encased in a horney hoof and a single hoof cleft is present in between the third and fourth toes (hence the name cloven-hoofed mammals). Many members have horns or antlers. They are ground living mammals except hippopotamus. With the exception of swine and peccaries, all the members of this order are almost completely herbivorous animals.

All members have been placed into three suborders: Suiformes, Tylopoda and Ruminantia. The taxonomic details are described below:

- Suborder: Suiformes (not true ruminants; two or three chambered stomach; low-crowned cheek teeth with simple cusps; tusk like canine teeth; incisor teeth on both jaws; and absence of horns or antlers).

Families:

  (i) Suidae- pig.

  (ii) Tayassuidae - peccary.

  (iii) Hippopotamidae - hippopotamus.

- Suborder: Tylopoda (three chambered stomach; two toes on each foot; padded foot; incisor and canine teeth on both jaws; and absence of antlers or horns).

Family: (i) Camelidae - camel, alpaca, llama, guanaco and vicuna.

- Suborder: Ruminantia (true ruminants; four chambered stomach except chevrotain; and horns or antlers present).

Families: Tragulidae - chevrotain.

Moschidae - musk deer.

Cervidae - deer.

Giraffidae - giraffe and okapi.

Antilocapridae- pronghorn.

Bovidae- antelope, yak, wild buffalo, gaur etc.

The salient features of some well known artiodactyls are described below:

Deer:

Cervids or deer belong to the family Cervidae. The musk deer has been placed in the separate family Moschidae. There are more than 40 species and 190 subspecies of deer distributed in Europe, Asia, South and North Americas and North western Africa. Deer, however, have been introduced to almost all the land areas inhabited by human beings. They are predominantly found in Asian countries as antelopes in Africa. These fascinating creatures are characterized by having elongated bodies, slender legs and necks, angular heads, short tails, and large and round eyes. The size ranges from the largest Alaskan moose deer measuring 90 inches at the shoulder with a body weigh of 800 kg to the smallest pudu deer which measures a shoulder height of 14.5 inches with 4 kg body weight. Generally, males are larger than females. The body colour may be brown, red or grey. The mature animals are generally lighter or more reddish in summer. The body is almost covered with hair. Another anatomical feature of deer is that they have a large fissure in the skull just in front of eye socket. Lateral hooves are well developed. All species (except musk deer and mouse deer) have no gall bladders. Preorbital or facial glands are present in all deer (except roe and musk deer). The upper jaw has no incisor teeth. Most species lack canine teeth in upper jaw. The upper canine teeth develop into tusks in males of musk deer, Chinese water deer, the tufted deer and muntjacs. The premolars and molars are used to grind fibrous materials. So, the total number of teeth is either 32 or 34 (I: 0/3; C: 0/1 or 1/1; PM: 3/3; M: 3/3). In warm climate, deer may have several ruts annually. However, in temperate climate they breed in particular season.

The most striking physical feature of these animals is their antlers that are shed and renewed each year. Males usually have antlers while females do not possess antlers. The males of musk deer and Chinese water deer, however, lack antlers. The females of reindeer (caribou), on the other hand, bear antlers. Females of roe deer occasionally may have antlers. Antlers are used for marking territories, weapons in fight etc. It is also believed that antlers have mystic healing properties and this is one of the reasons why deer populations have been dwindling in their natural habitats of many parts of the world.

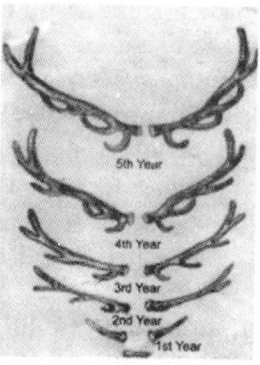

Fig.12.5: Antler growth of five year old red deer.

Antlers are solid bones without a marrow. They grow directly from the permanent bony growths of the frontal bones of the head called pedicles. The development, maturation and casting of antlers are closely related to the physiological and reproductive activities. Under the influence of hormones, the mesodermal cells of the pedicle proliferate and thereby start new antler growth. In young deer, antler looks a simple spike. However, it takes a familiar form of an antler when the animal becomes older. The newly grown antler is covered with a highly vascular and finely hairy thick soft skin called velvet. Depending on the physiological conditions and other factors, an antler may attain to its full growth with in three to five years of age (Fig.12.5). The newly developed antlers are tender and are highly subjected to trauma. When the antler is reached to the limit of growth, a ring of bony matter forms at the point where the antler unites with its pedicle. The bony ring gradually constricts and the blood supply cuts off to the solid bone and velvet. As blood vessels dry up, the velvet shrinks, dries and begins to peel off. Interestingly, deer rub their antlers against trees, buildings, or any other convenient projections for clearing the velvets during this period. The cleaning and hardening of antlers generally coincide with the onset of the breeding season (rutting season). This period is also characterized by maximum production of spermatozoa. The clean and hard antlers are shed after the completion of breeding season. Shedding of antler causes a small haemorrhage and shortly afterwards, a layer of cicatrical tissue appears in the form of a scar and the cycle of development of new antlers starts. The time for shedding of antler, however, varies with age, locality and feeding conditions of the animal. Young males shed their antlers earlier than older animals. The period of antler growth coincides with the seasons when the food is most abundant. High dietary protein and calcium are important for antler growth.

As stated earlier that the antler cycle in deer is closely associated with the seasonal rhythm of reproduction. The cleaning of velvet occurs when the testosterone levels are high. The casting of antlers, on the other hand, takes place when the level of testosterone is very low. When the antler cycle is disturbed, it may result in premature casting, failure to shed velvet or retention of antler. In short, there are four important phases of antler cycle under the control of sexual and growth hormones. They are as follows:

(i)   Phase-I: It is related to the development of pedicles under the control of testosterone.

(ii) Phase-II: Velvet starts developing during phase II when there is increased secretion of insulin like growth factor and decreased testosterone level. During this period the antler becomes live, soft, sensible and highly vascularised.

(iii) Phase-III: In this phase, there is again rise of testosterone level causing mineralization and occlusion of blood vessels. The velvet becomes dry and peels off.

(iv) Phase- IV: This is the period when the antler becomes hard, bare and dead. The casting of antlers occurs in this phase. During this time the testosterone level declines below the critical level and simultaneously the osteoclastic activity of pedicle is activated and leads to casting of antlers. The bare pedicles are soon overgrown by skins.

In this regard, both antlers and horns can be distinguished each other by the following points:

Antler	Horn
• Antlers are solid bones which are regenerated each year.	• Horns are not cast and regrown each year except pronghorn antelopes. They are the permanent structures and lost for ever once they are broken.
• Antlers have branches except pudu deer.	• Horns do not have branches (except pronghorn antelope).
• The antler is covered by a skin on which hair grow. Velvet is peeled off.	• Horns have a permanent bony sheath that covers the bony core.

The biological data relating to reproduction are well studied in deer. The female conceives as early as her first year. The length of oestrus cycle in polyestrous species lasts for about 18-21 days. The litter size usually varies from one to two. The triplets, however, are common in water deer. The gestation period ranges from 25 weeks to 40 weeks depending on the species. The general rule is that the larger the animals the longer the period of pregnancy. The young are nursed by the mother for several months.

In warm climates, deer may have several ruts annually. Deer in temperate climates, on the other hand, breed in particular season. The following is the breeding cycle of deer in the temperate regions.

Deer are found in small herds during the months beginning from December to April.

They break up in smaller groups or harems in May and July. During this period, the male deer thrashes its antlers and the female gives birth one to two young after about eight months of gestation.

The smaller groups comprising a dominant male and many females are formed during August-September. New antlers are grown in males.

Then it is the period of rut (October-November). Velvets are removed. The oestrus cycle starts. Both male and female unite for mating. During the period of rut, male becomes dangerous and aggressive and the fighting is common with other males.

Finally the smaller groups break up and the normal small herds are formed in the month of December-April as usual.

It has already been mentioned that deer are found in almost every part of the globe. They are usually known as Asian deer, North American deer, South American deer, and European deer. Some widely known species of deer are sambar (*Cervus unicolor*), cheetal (*Axis axis*), hog deer (*Axis porcinus*), sika deer (*Cervus nippon*), muntjacs (*Muntiacus spp.*), fallow deer (*Dama dama*), roe deer (*Capreolus capreolus*), rein deer (*Rangifer tarandus*), swamp deer (*Cervus duvauceli*) mule deer (*Odocoileus hemionus*), moose deer (*Alces alces*), pampus deer (*Ozotoceros bezoarticus*), marsh deer (*Blastocerus dichotomous*), water deer (*Hydropotes inermis*), red deer (*Cervus elaphus*),white-tailed deer (*Odocoileus virginianus*), pudu deer (*Pudu spp.*), thamin deer (*Cervus eldi*) and rusa deer (*Cervus timorensis*).

Deer found in India include swamp deer, sambar, thamin deer, hangul, hog deer, muntjac, cheetal, musk deer and mouse deer. The salient features of some deer species found in India are summarized below:

Swamp deer/Barasingha:

Zoological name: The zoological name of swamp deer is *Cervus duvauceli*. There are three races of swamp deer. They include the tarai race (*C.d.duvauceli*), the central race (*C.d.branderi*) and the eastern race (*C.d.ranjitsinhi*).

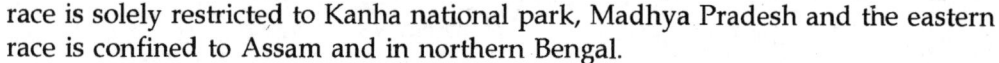

Distribution: They are found in a few isolated pockets of India. The tarai race is confined to tarai region of Uttar Pradesh (mainly in the West Kheri Forest Division near the Sarda river). The central race is solely restricted to Kanha national park, Madhya Pradesh and the eastern race is confined to Assam and in northern Bengal.

Habitat: They live in grasslands, river banks, vicinity of swamps and inside the forests. The tarai race and the eastern race are found in swampy dry grassland, while the habitat of central race is open hard-ground grass land.

Physical feature: This large and graceful deer stands 44 to 46 inches high at the shoulder. The mature male weighs 200-260 kg. and the female may weigh 135-140 kg. The coat colour is generally brown with yellowish shades below. During hot season, stags become reddish brown and the females turn yellowish brown like sambar. A dark brown band bordered by white spots runs down the length of spine. Fawns are spotted with white. Manes are present on the neck of the male. The tarai race has characteristic broad flat and splayed-out hooves that help in moving over swampy land. The central race has solid and pointed hooves. Antlers begin to grow shortly before or at the beginning of rains and usually have 10-14 points, though 12 points are common (hence called barasingha). The antlers of fully adult animals are having the characteristics of main beam sweeping upward for over half of the length before branching repeatedly in the distal third. Antlers are shed between February and March.

Reproduction: Puberty is attained at about two years of age. The onset of rut is marked by two important behaviour patterns- wallowing and bugling. They usually mate between November and March. The female drops normally one young after a period of 240-250 days of gestation.

Behaviour: They are highly gregarious animals and graze till late in the morning and again in the evening in the wild. The sense of smell is excellent. On being alarms, swamp deer will dash off with a loud chorus. They are less nocturnal than sambar. Barasingha visits a water hole soon after daylight and in the late afternoon during summer months. A prominent behaviour is ear-waving that displays during the hot season. The majority of animals roam with an area of about four square miles for seven to eight months of the year.

Life span: The life span is twenty years or more.

Sambar:

Zoological name: The zoological name is *Cervus unicolor*. Out of six recognized subspecies, the *C.u.niger* (India), *C.u.unicolor* (Sri Lanka) and *C.u.malaccensis* (Mayanmar, Thailand, Indo-China etc.) are well known.

Distribution: It is the largest deer in south-east Asia. Besides India, sambar also makes its home in China, Myanmar, Sri Lanka, Indonesia, Sumatra etc. They are widely distributed in India.

Habitat: The habitats of this deer include mountains, coastal forests, swampy lands and agricultural fields.

Physical feature: A full grown male may reach up to 320 kg in body weight. The stags stand 48-56 inches high at the shoulder. The length of body is 6-7 feet and tail length is 12-13 inches. It has a coarse and shaggy coat. Colouration is uniformly dark-brown with yellowish tinge under the chin, tail and inside the limbs. The winter coat is grey-brown to dark brown while the summer coat is brown to chestnut brown. Adult males are darker than does and young. The tip of the tail is black. The other salient features include erect ears, bushy tails and large infraorbital glands.

Antlers are branched with long brow tines at acute angle and have forwarded pointing terminal forks. The antlers are generally about 20-35 inches long. It attains full growth at about four years of age. In central and southern India, shedding of antlers takes place between March and April. The growth of new antlers begins towards the end of May and ceases about the end of September.

Reproduction: Sambar sexually mature at about 16 months of age. The rutting season begins in the months of October-December. One or two fawns are born by the female shortly before rains. Gestation period is recorded about eight months. Usually the female gives birth once a year. Fawns remain with mother for about two years.

Behaviour: They are predominantly a forest animal coming out into the open occasionally at dusk and during the night. They are mainly active at night. On being alarms, the male has a loud and whistling call. Sambar are essentially not gregarious animals and the typical herd is composed of not more than six animals. The male lead a solitary existence except rutting season. Occasionally, stags stand up on their hindlegs like a goat called "preaching". Sambar have regular stamping grounds in the forest and as well as in the open ground. In the early part of the rut the "sore spot" becomes noticeable on the ventral surface of the neck of sambar. Wallowing is a prominent behaviour of stags during rutting season.

Life span: The longevity ranges from ten to twenty five years.

Cheetal/Spotted deer:

Zoological name: *Axis axis* is the zoological name of cheetal. There are two recognized subspecies: *A.a.axis* and slightly smaller *A.a.ceylonensis*.

Distribution: They are found in India and Sri Lanka. Cheetals have also successfully been introduced in many parts of the world.

Habitat: The grasslands and forests form the ideal habitats of these strikingly beautiful and social animals. They are normally found in dry and moist deciduous forests. However, cheetal occupies the evergreen and thorn forests peripherally.

Physical feature: The male may weigh up to 85 kg. This medium sized deer stands about 35-38 inches high at the shoulder. The coat has a bright rufous fawn with white spots. Cheetal has a dark dorsal stripe that runs down the back from the nape to the tip of the tail. The underparts are white. An adult carries three-tined antlers, each composed of a beam forked at the summit and a brow tine that grows at right angles with the beam. The shedding of antlers may take palace in any time of the year. However, antlers normally drop in the month of August and September in south and in central India. The hindfeet of cheetal have well developed interdigital glands.

Reproduction: Cheetals are highly prolific animals. Sexual maturity is occurred at 16-18 months of old. Pairing is reported to take place during winter months, though they are known to breed round the year in north India. The oestrus cycle lasts for 21 days. The litter size varies from one to three. Gestation period is 210-240 days. The interbreeding between cheetal and hog deer is known to occur.

Behaviour: Cheetals are less nocturnal than other deer and feed till late in the morning and again in the afternoon. An unusual association between the cheetal and the langur is commonly seen in the wild in central India. The cheetals browse on the leaves on the floor which fall from the trees by feeding langurs. This symbiotic relationship also helps in early detection of predators by responding alarm calls to each others. Both the species are benefited from the good sense of smell of cheetal and acute vision of langur. This unique association in the wild state, however, dramatically declines during the rainy season when other food materials are abundantly available. Cheetal are considerably more benefited from this association.

They drink usually twice a day (in the early morning and late afternoon) during the hot season. Cheetal normally bark at the sight of a tiger. Cheetals are found in herds. The usual herd size is 5 to 10 animals though as many as 50 to 70 are not uncommon. However, the herd size depends on the availability of food and water. The female becomes solitary a few days before parturition and remains in the vicinity of dense brush and high grass. The young nibble on grasses when they are 15 months old.

Life span: They are reported to live for 15 to 20 years.

Muntjac / Barking deer:

Zoological name: The five species of muntjacs have been recognized. They include the Indian muntjac *(Muntiacus muntjac)*, the black muntjac *(M. crinifrons)*, Fea's muntjac *(M. feae)*, Roosevelt's muntjac *(M. rooseveltorum)* and Reeves's muntjac *(M. reevesi)*.

Distribution: The muntjacs are widely distributed in India, southern China and other parts of southern Asiatic regions.

Habitat: Muntjacs usually prefer to live in thickly wooded bamboo-clad ravines.

Physical feature: The male may attain a body weight of 20-35 kg. The female has a body weight of 13-15 kg. The colour ranges from greyish brown or reddish to dark brown. The winter coat becomes darker than summer coat. The salient anatomical feature of the animal is that the skull possesses a curious facial rib (hence the alternative name of muntjac is "rib-faced" deer). The male has a small antler and usually have a single branch. They are shed in the hot season. Males have tusk like movable upper canine teeth.

Reproduction: Muntjacs seem to breed at all seasons, though rut mainly takes place in winter months. One or two lightly spotted white fawns are born after a gestation period of 6-7 months.

Behaviour: They are solitary and fairly diurnal in habit. Muntjacs, however, may remain in pairs. It cries like a dog (hence called barking deer).

Life span: The life span is about sixteen years.

Musk deer:

Zoological name: The zoological name of musk deer is *Moschus moschiferus*. The typical form *M. m. moschiferus* is found in India.

Distribution: They are found in India, Nepal, Russia etc. In India, musk deer are found in the Himalayan ranges.

Habitat: The ideal habitat of this deer is the dense vegetation, especially hills with rocky outcrops in coniferous, mixed or deciduous forests.

Physical feature: A mature animal may reach a height of 28 inches at the shoulder and weighs 7-17 kg. Females are heavier than males (unusual among the deer). The coat is greyish-brown to golden in colour. The salient features include small heads, longer hind legs, long and pointed central hoof, absence of antlers, presence of a pair of movable and elongated canine teeth in the upper jaw, spatula-shaped incisor teeth in lower jaw, and small and naked tail with a tuft of hair at the tip. Musk deer have a gall bladder.

The most unique feature of musk deer is the musk glands for which these precious animals are recklessly hunted. The musk glands are located in between the genitals and the umbilicus of the male. It secretes a jelly-like oily substance which is reddish brown in colour with unpleasant smell. The secretion from musk gland can be collected by inserting a spoon into the aperture of the sac. When the secretion dries, the powdery mass becomes black and it emits the scent of musk. The odorous principle of musk is muscone (3 methyl cyclopentadecanone). This gland develops when the animal reaches puberty and on an average 6 to 8 g musk can be obtained from each adult animal annually. The musk is used for preparation of perfumes. It is also believed that musk has some medicinal values.

Reproduction: The copulatory organ of the male has a thread like extension. Unlike other deer, the female possesses one pair of teats. The sexual maturity is reached at about 18 months of age in female while male attains this at two to three years. The rutting begins in November and it lasts up to January. The oestrus cycle exhibits for about a month and during this period the female is chased by one or two males. The litter size is one to two. The gestation period is 150-180 days. The new born has a body weight of 600- 700 g with a characteristic striped and spotted coat. The young remain with the mother for two months.

Behaviour: Musk deer are shy and solitary animals. They are rarely seen in pairs. Their sense of hearing is very keen. These animals are active at dawn and dusk. However, they are also active at night. They spend most of the day resting in dense undergrowth or bamboo thickets or in shades. Musk deer has one or two "latrine sites" where they defecate. Musk deer are highly territorial.

Life span: They are known to live for about 13 years in captivity.

Chevrotain/Mouse deer:

Zoological name: There are four species: the Indian spotted chevrotain *(Tragulus meminna)*, the larger Malayan chevrotain *(T. napu)*, the smaller Malayan chevrotain *(T. javanicus)* and the water chevrotain *(Hyemoschus aquaticus)*.

Distribution: They inhabit in southern Asia and Africa. The water chevrotain is distributed in Africa. The rest three species are found in southern Asia.

Habitat: Mouse deer or chevrotain lives in the dense vegetation and forests. The Indian chevrotains inhabit in grass-covered rocky hillsides or in forests.

Physical feature: The body weight may reach up to 13 kg. They are closely related to pigs. The anatomical features include four-chambered stomach, lack of upper incisor teeth, presence of only three premolar teeth, well-developed toes, absence of antlers, dagger-like upper canine teeth in males and premolars with sharp crowns. The Indian chevrotain has lighter spots and stripes. The gall bladder is present.

Reproduction: The male copulatory organ is spiral shaped at the tip with lobed on the sides. The female bears four nipples. Sexual maturity is reached when animals become 8-10 months of old. Pairing takes place between June and July. Copulation lasts for longer periods like pigs. The litter consists of one to two offspring. The gestation period varies from 5-9 months depending on the species involved.

Behaviour: They are nervous, solitary and nocturnal animals. The female is known to ingest the placenta after birth.

Life span: The life span is reported to be 10-14 years.

Antelopes:

There is no precise zoological definition of antelopes. The term "antelopes", however, are implied to those members in the family Bovidae which are slenderly built, swift moving, graceful and plain dwellers. In this regard, it is to be noted that the pronghorn in the family Antilocapridae is also called an antelope. Nevertheless, the antelopes inhabit in the habitats of desert, steppe or in open plain.

It is said that Africa is the land of antelopes. Some widely known antelopes of Africa include bushbuck *(Tragelaphus scriptus)*, bongo *(Taurotragus eurycerus)*, eland *(Taurotragus oryx)*, waterbuck *(Kobus ellipsiprymnus)*, reedbuck *(Redunca arundinum)*, oryx *(Oryx gazella)*, hartebeest *(Alcelaphus buselaphus)*, black wildbeest *(Connochaetes gnou)* impala *(Aepyceros melampus)*, gerenuk *(Litocranius walleri)* and Thomson's gazelle *(Gazella thomsoni)*.

India boasts of four antelopes. They include the blue bull, the blackbuck, the Indian gazelle and the four-horned antelope. The salient features of Indian antelopes are described below:

Blue bull / Nilgai

Zoological name: The zoological name of blue bull is *Boselaphus tragocamelus*.

Distribution: Blue bull, the largest antelope of Asia, is distributed in peninsular India.

Habitat: They are found in the hilly grasslands or lightly wooded regions. However, blue bull prefers to inhabit in dry deciduous and thorn forests.

Physical feature: The horse-sized antelope, the blue bull stands about 52-56 inches high at the shoulder and weighs 200-270 kg. The body colour is dark bluish grey or bluish black in adult males. Young males and females, on the other hand, are tawny colour. The other salient features of this animal are smooth upright horns in males, long and pointed head, and a white ring below each fetlock with two white spots on each cheek, longer forelegs and short dark manes. Another distinctive character of the male is that it has a tuft of stiff black hair on the throat. Bulls have short smooth horns of 6 to 8 inches long.

Reproduction: The puberty is reached at about 30 months. One to two offspring are born after a pregnancy period of 8-9 months. They have no regular breeding season. Most of the births occur shortly before the rains.

Behaviour: They are usually found in small groups consisting of 10-15 individuals. Blue bulls are much diurnal in habits. An important behaviour is that the blue bull has the habit of regularly defaecating on one particular spot like rhinoceros and four-horned antelope. They establish territories and form breeding herds during the rut.

Life span: The life span is about 15 years.

Blackbuck:

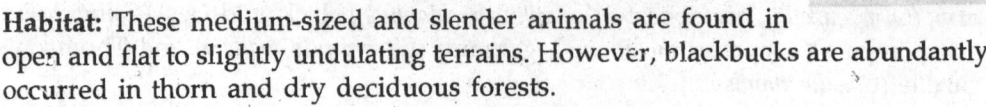

Zoological name: *Antilope cervicapra* is the zoological name of blackbuck. It is also known as *Krishna mriga*.

Distribution: Blackbucks are typically distributed in India with a few numbers in Nepal. In India, blackbucks are largely restricted to the states of Gujarat and Rajasthan.

Habitat: These medium-sized and slender animals are found in open and flat to slightly undulating terrains. However, blackbucks are abundantly occurred in thorn and dry deciduous forests.

Physical feature: It stands about 29-33 inches high at the shoulder and the total length is about 49-58 inches. It weighs 25-35 kg. The coat colour of the females and immature males is light fawn with white chins, eyes and bellies. However, the neck and upper parts in male turn into dark brown to black after about three years of age. Pure white blackbuck is also occurred. Horns are large, black and spirally twisted. Young below one year of age, however, have horns without a spiral. Females are usually hornless. There are prominent glands below the eyes in males.

Reproduction: The reproductive maturity is occurred at 19-23 months of age. However, maturity may attain at early or late depending on the physiological and nutritional status of the animal. The rut begins throughout the year with peaks between March-April and August-October. One to two fawns are born after 180 days of gestation. Young are usually concealed in bushes.

Behaviour: They are found in herds consisting of 20-30 animals and the herd may consist of mixed animals, breeding herds and buck. The animals may graze till mid day and again in the late afternoon. The irregular deposition of droppings in common heap is an important behaviour of blackbuck. It is said that blackbucks persist for days and even weeks without drinking.

Life span: Blackbuck lives for about 15 years.

Chinkara / Indian gazelle:

Zoological name: The zoological name of chinkara is *Gazella gazella*.

Distribution: Chinkara are distributed in north-western, central and some parts of southern India.

Habitat: These small to medium sized antelopes are found in the habitats of open scrub forests, ravines, low foothills and hilltops at the edge of open or cultivated ground.

Physical feature: An adult male measures 26 inches at the shoulder and weighs 15-20 kg. The body is chestnut in colour. The chin, breast and the underparts have white markings. Each side of the face of this animal has an alternative pattern of dark and light stripes. The short tail is black and haired. Horns are present in both sexes and are tightly annulated almost to the tip. The hooves are pointed.

Reproduction: Chinkara are sexually matured at one and a half year. One to two fawns are born at all seasons after a gestation period of five months.

Behaviour: They are found in small groups comprising three to seven animals in the wild. Chinkara have also a nocturnal habit.

Life span: They are known to live for ten years in the wild.

Four-horned antelope / Chausingha:

Zoological name: The zoological name is *Tetracerus quadricornis*.

Distribution: They are exclusively found in Indian subcontinent.

Habitat: Four-horned antelope occurs in undulating or hilly country.

Physical feature: The mature animal stands about 2 ft at shoulder. The body weight varies from 15 to 25 kg. The coat colour is dull light brown with shades of white on the lower parts. The most distinguishing anatomical feature of this animal is its horns. It is the only antelope that are blessed with two pairs of horns (hence the alternative name "chausingha"). The male only bears horns and the anterior pair is shorter than the posterior one and sometimes is hidden by the hairs.

Reproduction: The puberty is reached at about 24 months. Rutting takes place in June-July. The female drops one to three young. The gestation period is 225-240 days.

Behaviour: Four horned antelope live in pairs or solitary. They preferably spend most of the times near the water bodies as these animals have the regular drinking habits.

Life span: The longevity is about ten years.

Wild bovids:

The family Bovidae contains yak, mithun, gaur, banteng, bison, wild sheep, wild goats etc. besides antelopes. Many members of this family are found in domesticated ones. However, some are existed in their wild states. The following text highlights salient features of some well known wild bovids:

Yak:

Zoological name: The zoological name of yak is *Poephagus grunniens*. They are also called grunting ox.

Distribution: Yaks are confined to the central Asia and high regions of Tibet. In India, they are found in Arunachal Pradesh, Sikkim, Himachal Pradesh and Jammu & Kashmir.

Habitat: Yaks prefer to inhabit an area with an elevation of 14,000-20,000 ft.

Physical feature: It stands about 6 ft and 8 inches at the shoulder and the body weight is recorded 250 to 550 kg. This animal has shaggy fringes of coarse hair coat with dense undercoat of soft hair. The colouration of wild yak is very dark brown or black. The horns are more or less rounded with outward and upward directions. The other salient features include drooping heads, short and sturdy legs and high humped shoulders.

Reproduction: Sexual maturity is attained at about five years of age. The rutting season starts during the late autumn. The bull accompanies four to five breedable cows during this period. The female comes in heat after 21 days and estrus lasts for 24-72 hours. The single young is born after a period of 258-270 days gestation. The reproductive life of a female yak is reported to be 15-16 years.

Behaviour: Bulls are generally solitary. However, they may be found in small groups of two to three animals. Cows and calves, on the other hand, form a large herd comprising a large number of animals. The grown-up bulls make smaller groups of their own. They graze both in morning and in evening.

Life span: Yaks are known to live for twenty five years.

Mithun/Gayal:

Zoological name: The zoological name of mithun is *Bos frontalis*. They are semi wild in nature and can be maintained as tamed animals. They can be bred with cattle with some limitations. There are two reported breeds: sally and gamba. Sally breed is known to be more productive while gamba breed is larger in size but less productive.

Distribution: Mithuns are distributed in north-east India.

Habitat: The live in the hilly regions. They prefer to live in cool forests ranging from 300 to 2700 meters altitude.

Physical feature: The coat colour varies from black with a coffee brown shade to dark brown. The extremities may be white or black. The other salient features are short tail, well developed dewlap, white marking on the forehead in some animals and absence of ridge in the median line of the skull between the horns.

Reproduction: The female matures at three to four years of age and she usually gives birth to one calf yearly.

Behaviour: Mithuns are generally found in small groups. They are comparatively hardy and thrive well on coarse fodder of the forest. Mithuns are docile animals. It has the ability to graze on slopes of hill.

Life span: The life span is reported to be 15-20 years.

Indian bison/Gaur:

Zoological name: *Bos gaurus* is the zoological name of Indian bison. There are three recognized subspecies: *B. g. gaurus* (India and Nepal), *B. g. readi* (Myanmar, Thailand and Indo-China), and *B. g. hubbacki* (Malaysia). The gaur is the largest animal among all the wild cattle.

Distribution: Indian bison is restricted to India, Myanmar, Vietnam, Nepal and Malaysia.

Habitat: Indian bison is found predominantly in the hills. Their ideal habitats are characterized by (a) large, relatively undisturbed forest tracts, (b) hilly terrain below an altitude of 5000 to 6000 ft, (c) availability of water, and (d) an abundance of forage in the form of coarse grasses, shrubs and trees.

Physical feature: A mature animal may stand 64 to 72 inches high at the shoulder and the total length is 11-12 feet including tail. It weighs 700-1000 kg. Males are having dark brown to black coat colour, whereas, the cows and young tend to be reddish. The salient physical features of these animals include large and massive upward horns, huge heads, convex roman noses, deep massive bodies with sturdy limbs and short tails with a small black tassel. There is a ridge in the middle of the skull between the horns. It has a prominent dorsal ridge formed by the extension spinous processes of the third to eleventh vertebrae terminating abruptly near the middle of the back. There is a small dewlap below the chin.

Reproduction: They are sexually mature at three years of age. Rutting takes place between November and March. Usually, a single calf is born after completion of about 9 months gestation. The newly born calf is light golden-yellow in colour. Calf remains with the mother for about two years.

Behaviour: Indian bisons forage in the forests. However, they come out onto the meadows only to eat and drink during the hot season. They are generally found in small herds comprising eight to twelve animals. Bull usually leads a solitary life or associate with other males in bull herds except rut. The "lateral display" is the striking behaviour pattern of the Indian bison and appears to one of the important means by which bull establishes the rank without actual physical contact.

Life span: Gaur may live up to 25 years of age.

Wild buffalo:

Zoological name: The zoological name is *Bubalus bubalis*.

Distribution: In India, they are mainly restricted to Bastar region of Chhattisgarh and in some protected areas of Assam.

Habitat: Their ideal habitats include tall grass jungles and reed brakes in the neighbourhood of swamps. However, dry hard ground scattered with trees and grasses also makes the habitat of wild buffalo.

Physical feature: The body weight may reach up to 900 kg. The wild buffalo is sleeker, heavier and more robust looking than the domestic buffalo. The colour is slaty black.

Reproduction: Breeding season starts at the end of rains. Although calving may take place throughout the year, the female usually gives birth between March and May. The gestation period is about ten months.

Behaviour: Wild buffaloes are exhibited in small herds of 10-15 animals comprising a bull, few breeding females and their young.

Life span: They can live 20-25 years.

Wild sheep:

Wild sheep are found in many parts of the world. Mouflon *(Ovis musimon)*, bighorn *(O.canadensis)*, blue sheep *(Pseudois nayaur)* and urial *(O.orientalis)* are well known examples of wild sheep. Blue sheep or bharal is found in the higher altitudes of the Himalayan region. The salient characteristics include short legs, long ears and long tails. The male has no beard. It has a light salty blue coat colour with shades of pale fawn. The rounded with upward and outward directions horns have a semicircular sweep. The body weight may reach up to 70 kg.

The urial or shapu is found on the steep grassy hill slopes of the Laddakh region.

Wild goat:

Wild goats are distributed in many parts of the world. They are adapted to rugged and rocky terrains and are found in bands comprising 5-20 animals. The notable wild goats are the Siberian ibex *(Capra ibex sibirica)*, the Spanish ibex *(C. pyrenaica)*, the Himalayan tahr *(Hemitragus jemlahicus)*, the Nilgiri tahr *(H. hylocrius)* and the markhor *(C. falconeri)*.

The Himalayan tahr is usually reddish brown to dark brown in colour. It has long robust limbs and narrow erect ears. The neck has a full mane. Horns are curved backwards. The Himalayan tahr is ranged from Kashmir to Sikkim. They are forest loving animals and prefer to live in herds on steep rocky slopes covered with oak and other trees. Mating occurs in winter and a single young is born during summer.

The Nilgiri tahr or Nilgiri ibex is a dark yellowish brown animal and are restricted between Nilgiri hills and Ashambu hills of Tamil Nadu and Kerala. Females and young, however, are grey. It can be distinguished from the Himalayan tahr by having a distinct lighter patch on the back of the body which is mostly dark brown in colour. The female Nilgiri tahr possesses black carpal marks that are absent in the Himalayan tahr. They are true browsers and can be found in small groups of six to twelve individuals or more. Breeding takes place between April and August. The gestation period is reported to be 174-201 days. The female is reported to produce two young once at a birth.

Goat antelope:

Goat antelopes are found in south-east Asia. They hold an intermediate position between goat and antelope. These animals occur in small herds in the mountain regions. They have a more or less goat-like build, goat-like teeth and short tails. Goral, takin and serow are the goat antelopes.

Gorals, the stockily built goat like animals inhabited throughout the Himalaya at an elevation of 3000-14000 feet. Both male and female have conical, backwardly curved horns marked with rings. They are found in parties consisting of 4-8 animals. Old males, however, are solitary.

Wild pig:

Zoological name: There are nine species of wild pig. They include the wild boar *(Sus scrofa)*, the pigmy hog *(S. salvanius)*, the bearded pig *(S.barbatus)*, Celebes wild pigs *(S. celebensis)*, Javan warty pig *(S. verrucosus)*, the warthog *(Phacochoerus aethiopicus)*, the giant forest hog *(Hylochoerus meinertzhageni)*, the bush pig *(Potamochoerus porcus)* and the babirusa *(Babyrousa babyrussa)*.

Distribution: They are confined to Asia, Africa and Europe. These animals have also been introduced in other parts of the world. The wild pigs *(S .scrofa)* are distributed in many parts of India. The pigmy hog is confined to the state of Assam.

Habitat: The habitats in which the wild pigs are found include forests, woodlands and grasslands.

Physical feature: The wild pigs are distinguished by having a barrel-like body with a large head. The snout is prominent. Each foot has four digits. The body is covered by bristly hair. The upper canines are well developed and turn upward.

Reproduction: The puberty is attained at about one year of age. The females are highly prolific animals. The normal litter size varies between six and eight. The gestation period is 115-120 days.

Behaviour: Wild pigs are mainly active at night and live in groups. The adult male leads a solitary life.

Life span: The longevity is 10-20 years in the wild.

Camelid:

All the members in the family Camelidae are popularly called camelids. There are Old World camelids and New World camelids. The Old World camelids (camels) belong to the genus *Camelus*, whereas, the New World camelids (llama, alpaca, guanaco and vicuna) are placed in the genus *Lama*. The four species of the New World camelids are also called lamoids or cameloids. Among all the lamoids, the llama and the alpaca have been domesticated, while the remaining two species are still wild in nature. All the camelids are well adapted to walking on sandy deserts or on rough mountain terrains. The members of this family are characterized by having fatty fibroelastic sole pads, two toes with broad nails on each foot, three-chambered stomachs, forked lips, no horns, absence of gall bladder and elliptical red blood corpuscles. However, camels differ from lamoids by the following characteristics.

Camel	Lamoid
(a) Camels are larger in size and are less aggressive.	(a) Lamoids are smaller in size and are very aggressive by nature.
(b) The total number of teeth is 34.	(b) The total number of teeth in lamoids is 30.
(c) The pads of the toes are not so deeply widen.	(c) The pads of the toes are deeply widen as compared to camels.

Camel:

Zoological name: The genus *Camelus* contains two species: the Arabian or dromedary camel *(Camelus dormedirius)* and the Bactrian camel *(C. bactrianus)*. Both the species can be interbred and offspring produced from such crossing are fertile.

Distribution: They are found in Central Asia, South West Asia and North Africa.

Habitat: Camels are adapted to deserts.

Physical feature: The body weight ranges from 400 to 750 kg. The males are larger than females. The colour varies from brown to grey. Legs are elongated. The long curved neck, the split upper lip, small nails, long hairs on top of the heads, and horny callosities on the chests and joints are the other characteristic features. However, both the species can be distinguished by the following points.

Arabian camel	Bactrian camel
(a) Arabian camels live in deserts of north Africa and south west Asia.	(a) Bactrian camel are adapted to cold and rock deserts of Central Asia (e.g. Mongolia)
(b) The coat colour varies from white to medium brown.	(b) The coat colour is uniformly light to dark brown. The winter coat becomes thicker.
(c) The single hump is the salient feature of Arabian camel.	(c) The Bactrian camel has two humps.
(d) The Arabian camel carries its head high. The legs are comparatively long.	(d) This animal holds its head lower. Legs are shorter.
(e) The Arabian camels are probably not found in true wild condition. However, they may be seen in feral state.	(e) The bactrian camels occur in the the wild.

Reproduction: Mating takes place all the year round. However, a peak of births coincides with the season of vegetation growth. The female drops a single foal on every second year after 390-410 days of gestation. The young may be fully grown at 5 years of age. The crossbred "Rama" resulting from the crossing between a camel and a llama has been successfully produced for the first time at the Camel Production Centre at Dubai.

Behaviour: Camels are known for high adaptability to extreme cold and hot conditions. They are found in small herds comprising four to 6 animals. The male protrudes a red pouch (soft palate pouch) in rut.

Life span: In captivity, camel may live for 25-30 years.

Giraffe:

Zoological name: The zoological name of giraffe is *Giraffa cameloparalis*. There are several races of giraffe based on the different types of markings and their spots.

Distribution: They are found in Africa.

Habitat: Giraffes inhabit in woodlands and wooded grasslands. They are occasionally found on grass plains.

Physical feature: Giraffe, the tallest animal on the earth, measures a height of 16-19 ft and weighs 550-1930 kg. The coat colour is pale brownish chestnut with large blotches of a darker tint. Individual animal has a distinct coat marking that distinguishes from others. The forelimbs are longer than the hindlimbs. Both sexes possess permanent horns that grow continuously throughout life. The frontal bony protuberance is located on the forehead. The other features of giraffe include thin and mobile upper lip, long neck, extensile and prehensile black tongue, two to three lobulated canine teeth, large eyes with eyelashes, and short mane on the neck. The gall bladder is present (however, there are reports of not having any gall bladder during dissection of some animals). The female has four mammary glands.

Reproduction: The male is sexually matured at three and a half year old or more, whereas, the female conceives at four to five years of age. A single offspring (rarely twin) is born at any time of the year after a gestation period of 14-15 months. A female may produce 10-12 calves in her life time. The new born weighs about 95-120 kg.

Behaviour: Giraffes are found in small family troops comprising two to six individuals or more in the wild. It is said that females usually drop calves on particular calving grounds. They often ruminate for short period while lying, standing and walking.

Life span: The average life span is thiry years.

Hippopotamus:

Zoological name: There are two living species: the common, river or Nile hippopotamus *(Hippopotamus amphibious)* and the pigmy hippopotamus *(Choeropsis liberiensis)*. They are distantly related to pigs.

Distribution: These animals are distributed in Africa.

Habitat: The ideal habitats of common hippos are short grasslands, lakes and rivers. The pigmy hippos may remain in lowland forests, swamps etc.

Physical feature: The barrel-shaped body of a common hippo may reach up to a length of about 12 ft. and an individual may reach up to a body weight of 4000 kg (usually 1600-3200 kg). The other salient features are smooth skin, short and stumpy legs, four toes on each leg, presence of incisor and canine teeth, flat tail, devoid of sweat glands, and no gall bladder. The penis is directed backward.

Reproduction: The male attains puberty at about 8-10 years. The female, however, attains maturity at about 6 years. She gives birth usually to a single calf after a gestation period of 8 months. The parturition may take place in water or on the beach. The birth weight of new born varies from 42 to 50 kg. The young are weaned after one to two years of age.

Behaviour: The males are found solitary or in bachelor herd. A small territory is marked with faeces in the wild.

Life span: The common hippo may live up to 45 years.

Chapter 13

Salient Features of Selected Wild Reptiles

As mentioned earlier that reptiles are widely distributed in both temperate and warm regions of the world. The members in the class Reptilia include crocodiles, snakes, lizards, tuataras, tortoises and turtles. Many reptiles are useful in controlling the agricultural pests like insects and rodents. The salient features of some selected reptiles are described below:

Turtle:

Turtles along with tortoises and terrapins are placed in the order Chelonia. The aquatic species are usually called turtles and the terrestrial members are known as tortoises. Turtles are normally slow moving and non-aggressive animals. They do not have teeth. Instead, jaws are armed with horny plates that help in chewing food. The body is covered with a protective bony shell. The upper portion of the shell is called carapace and the lower portion is known as plastron. The limbs are flattened laterally and toes are webbed. Some species of turtles have paddle-like flippers with two or three protruded claws. The fore flippers are used in propulsion, whereas, the hind flippers serve for digging nest. Males can be identified from females by their usually concave plastrons.

The majority of turtle species are found in or near the tropics. Breeding takes place in water and it happens usually once annually. The female goes on the land for laying eggs. For that purpose, a hole is made by the hind legs in the ground near the bank. During lying time, the female hangs her tail in the hole and lays egg one by one. With the help of hind legs, she picks the eggs up and meticulously allows them to slide to the ground. After depositing the egg, the hole is filled up and she returns to the water. The eggs are hatched by the heat of the soil alone.

The freshwater turtles have a flattened disc-like shell covered with soft skin. The other characteristics include retractile head and neck, paddle-like limbs and three claws on each limb. Most of the Indian freshwater turtles are distributed in the larger river systems of the Indo-Gangetic plain. The examples of freshwater turtles are the Indian mud turtle (*Lissemys punctata*), chitra turtle (*Chitra indica*) and ganges softshell turtle (*Aspideretes gangeticus*).

Marine turtles remain in water except in breeding season. They have paddle-shaped limbs. The front limbs are longer than hind limbs. Unlike freshwater turtles, the marine turtles can not retract their heads and necks into the shells. Out of seven species, India is blessed with five species of marine turtles. They are the loggerhead turtle *(Caretta caretta)*, olive ridley turtle *(Lepidochelys olivacea)*, hawksbill turtle *(Eretmochelys imbricate)*, leather back turtle *(Dermochelys coriacea)* and green turtle *(Chelonia mydas)*. All marine turtles have nasal glands that serve for excretion of salt. The details of olive ridley turtle is given below:

Olive ridley turtle:

Distribution: They are found in the tropics of Indo-Pacific and the East Atlantic Ocean. The olive ridley turtles are commonly distributed along the sea coasts of India.

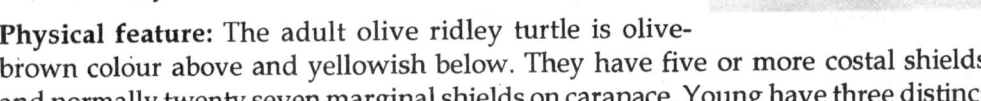

Habitat: They are found in the sea water.

Physical feature: The adult olive ridley turtle is olive-brown colour above and yellowish below. They have five or more costal shields and normally twenty seven marginal shields on carapace. Young have three distinct keels on carapace. The length may be up to 1 m and weight up to 150 kg.

Reproduction: The sexual maturity may attain between four and six years. They breed throughout the year, though peak periods may be observed during the month of January. They dig a nest of a half to one metre deep with their hind flippers. The total number of eggs laid by the females varies from 40 to 150. She may nest several times in one season before returning to her feeding area which is perhaps thousands of kilometers away. After 50 to 70 days of incubation, the hatchlings emerge *en masse* at night.

Behaviour: The olive ridley turtles are nocturnal nesters. The hatchlings locate the sea by orienting to the brighter horizon created by reflection of moon, sand dunes etc. and they then orient to wave direction, swim offshore and gradually get imprinted with the earth's magnetic field. Mortality is reported to be high in this period. It is reported that less than one in-a-thousand is believed to survive from hatchling to adulthood.

Another interesting fact is that the adults migrate back to nest on their natal beaches, where they were born. Olive ridley turtles are particularly known for their mass nesting behaviour. Thousands of turtles come ashore simultaneously to nest. The Gahirmatha coast of Orissa is one of three places in the world where this phenomenon occurs (the others are in Mexico and Costa Rica). It is reported that 1, 00,000 female turtles visit Gahirmatha coast area for the purpose of laying eggs. Loss of nesting habitats, incidental capture in fishing nets, lighting on the coast and non human predation are some of the reasons for reducing the turtle population.

Life span: The life span may be up to 100 years or more.

Tortoise:

Any terrestrial member of the family Testudinidae in the order Chelonia is known as tortoise. Tortoises inhabit temperate and tropical regions. The majority, however, are found in Africa and Madagascar. Their habitats range from dry and hot desert regions to the humid jungles. They are characterized by having usually rigidly ossified and high domed shells and hard-scaled forelegs. The head and neck 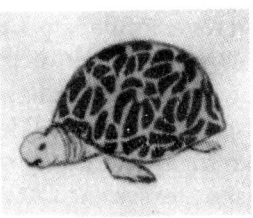 are completely retractile. The hind limbs are club and column shaped. Feet are armed with horny claws. Some tortoises have partly ossified strong scales on the outside of both fore and hind legs. Tortoises are very slow moving animals and are able to breed at about six years of age. The well known tortoises are the East Asian tortoise *(Indotestudo elongate)*, starred tortoise *(Geochelone elegans)* and eastern hill tortoise *(Manouria emys)*.

In this regard, tortoise and turtle can be distinguished by the following points:

(a) Tortoises mainly inhabit on land. Turtles, on the other hand, chiefly inhabit in water. They come to the land for laying eggs.

(b) The limbs of the tortoise are equipped with claws which enable them to move on the ground. Turtles have flattened limbs with webbed toes that facilitate swimming.

Terrapin:

The terrapins or freshwater tortoises are highly prized as delicacies. The majority of the species occur in the large river systems of the Indo-Gangetic plain. They have axillary and inguinal scent glands. The commonly known terrapins  are Indian pond terrapin *(Melanochelys trijuga)*,Indian tent terrapin *(Kachuga tentoria)*, Assam roofed terrapin *(Kachuga sylhetensis)*, spotted black terrapin *(Geoclemys hamiltonii)*, river terrapin *(Batagur baska)*, and brahminy terrapin *(Hardella thurjii)*.

Crocodile:

The order Crocodilia contains the alligators and caimans (family Alligatoridae, the true crocodiles (family Crocodylidae) and the gharials (family Gavialydae). They inhabit the warmer parts of the world. Crocodiles are adapted to fresh and marine water as well as terrestrial life. The body is covered with hard horny scales or scutes and the skin is soft between the scales. The other characteristics include short and powerful legs, five toes on each forefoot and four on the hind foot. The eyes of crocodilians are protected by a flap of skin. The broad and thick tongue can not protrude. Each jaw has a row of sharp teeth. Crocodilians have well developed smell, sight and hearing. The penis of the male is located at the cloaca. The copulation takes place in water. The female normally mates only once in a year. She lays hard shelled eggs in a nest. The nest is prepared by the female along the bank of the river, lake or in other suitable place. The chalky-shelled eggs are oval in shape and their number varies from 20 to 100 or more depending upon the species. The female remains in the vicinity of nest and helps the emerging young by breaking open the eggs. The mother also assists in carrying the hatchlings into the water.

The alligators are found in North America and China. There are two species: the American alligator *(Alligator mississippiensis)* and the Chinese alligator *(A.sinensis)*. The alligators can be distinguished from the true crocodiles by their arrangement of teeth. In the alligator, the fourth tooth of the lower jaw is contained within the mouth and does not show externally. However, the fourth tooth of the lower jaw of true crocodiles fits into a notch on the side of the upper jaw and is visible when the mouth is closed.

True crocodiles as mentioned earlier are placed in the family Crocodylidae. The freshwater and the estuarine crocodiles are found in India. The gharial or long-snouted crocodile in the family Gavialydae is also found in India. The detailed descriptions of these three species of crocodiles are described below.

Gharial / Gavial:

Zoological name: The zoological name of gharial is *Gavialis gangeticus.*

Distribution: The gharials are distributed in the rivers Ganga, Mahanadi, Indus, Brahmaputra and their tributaries of Indian subcontinent. Today they are mostly found in the Chambal, Girwa, Rapti and Narayani rivers of the Ganga river system in India. They are also distributed in the Irrawaddy and Arakan river systems in Myanmar.

Habitat: They are found in deep pools at river junctions and the deep gorges in hilly country.

Physical feature: The length is usually 12-15 feet, though individual specimen may reach up to 20 feet or more. An adult gharial may be dark olive or brownish olive in colour. The young, however, are greyish brown with five irregular transverse bands on the body and nine on the tail. Gharials are distinguished by having very long and narrow snouts and sharp-toothed jaws. It has small legs and a large tail base. The long jaws are armed with more than hundred teeth. There are 27 to 29 undifferentiated teeth on each side of the upper jaw and 25 to 26 teeth in lower jaw of each side. The first 3 teeth of the lower jaw fit into notches on the upper jaw. The male develops a big outgrowth at the tip of its snout. It is more pronounced during breeding season. In India, this outgrowth is compared to an earthenware pot called "ghara" and hence, it is called gharial.

Reproduction: Sexual maturity is attained in females at about eight to ten years of age. However, male attains reproductive maturity at 13-14 years of age. Mating takes place in December and January. After about one month of breeding, the female searches for a suitable nesting site. She digs a nest with her hind legs in a sandy river bank. Once the nest is made, the female puts her hind legs inside the nest and starts laying eggs. During laying, eggs are first held in the paws and then gently drop at the bottom. The eggs are laid at an interval of about a minute. The nest is filled with sand after laying of eggs. On one clutch the female gharial, on an average, lays about 40 chalky-shelled eggs. The total number of eggs may be up to 100.

The female guards the nest in close proximity. The optimum incubation temperature lies between 32^0C and 34^0C. The young hatch after 72-92 days of incubation period. The young make a croaking sound signaling the mother about the emerging out from the eggs. The newly hatched young are about 15 inches long and weigh about 75- 97 g. The hatchling is carried by the mother on her back or mouth in order to release it into the water.

Behaviour: Basking during winter months is an important behaviour of gharials. This is seen in midstream islands and sandbanks. Gharials are also found near river banks during the high-water monsoon months. Young animals make a groaning noise when disturbed.

Life span: Gharials are reported to have a life span of 29 years in captivity.

Marsh crocodile / Mugger / Freshwater crocodile:

Zoological name: The zoological name of marsh crocodile is *Crocodylus palustris.*

Distribution: Marsh crocodiles are distributed in India, Pakistan, Sri Lanka, Myanmar, Nepal and Iran. They are widely distributed in India.

Habitat: Marsh crocodiles inhabit in rivers, ponds, lakes, reservoirs and other large water bodies in the plains and up to 600 m in the hills.

Physical feature: An adult may reach up to 4 m and weigh more than 200 kg. The colour is olive brown with speckles of black. The below is white or yellowish. They are distinguished in having broad snouts. Another important physical feature by which it can separate from estuarine crocodile is that marsh crocodile has four prominent occipital scales just behind the head. Moreover, the back is armoured with 16 or 17 transverse and 6 (occasionally 4) longitudinal series of bony plates embedded in the skin. The toes are webbed.

Reproduction: Mating has been observed from mid January to March. It normally occurs in water. The female lays eggs in a hole dug by her in sand, earth or gravel on a stream, river, lake etc. The number of eggs laid by the female varies form three to forty in a clutch. The incubation period varies from 60 to 90 days.

Behaviour: Marsh crocodile is an excellent swimmer. Adult marsh crocodiles occasionally roar. During night in summer months, they migrate to more permanent water bodies or they aestivate in transitory waters. Marsh crocodiles have also the habits of basking on the river banks. They make a burrow on the banks of lake, river, at the foot of tree and below rocks by the side of stream for habitation.

Life span: The life span is reported to have more than 35 years.

Estuarine crocodile / Salt water crocodile:

Zoological name: *Crocodylus porosus* is the zoological name of estuarine or salt-water crocodile.

Distribution: Estuarine crocodiles are distributed in India, Sri Lanka, Thailand, Indonesia, Myanmar etc. In India, it is now mostly confined to the coastal mangrove areas.

Habitat: They inhabit in tidal estuaries of the larger rivers, marine swamps and coastal brackish water lakes.

Physical feature: It may reach up to 5 m in length and weigh more than 400 kg. The body colour is dark olive or brownish olive above interspersed with yellow. Young have black markings above. The estuarine crocodile has a longer snout than marsh crocodile. They have six to eight longitudinal series of bony plates or scutes. Unlike the marsh crocodile, a strong ridge in front of eyes is present in estuarine crocodile. Another distinguishing feature by which the estuarine crocodile can separate form the marsh crocodile is that it has no enlarged post occipital scales. However, the following are the summaries to distinguish estuarine crocodile from mugger crocodile.

Mugger	Estuarine crocodile
1) Less than 5000 are today found in India.	1) Restricted to Bhitarkanika Wildlife sanctuary, Sundarbans and Andaman & Nicobar Islands.
2) Medium sized crocodile hardly attains four metres in length.	2) It reaches to seven metres or more
3) They are found in freshwater ponds lakes, rivers and streams in Goa, it is found in brackish water lagoons and mangroves.	3) They live in mangrove swamps, lagoons, estuaries and rivers with thick vegetation.

Mugger	Estuarine crocodile
4) Sometimes, found in sewage treatment plants and temple tanks. They tend to bask out in open.	4) They are normally cryptic and shy of basking in the open.
5) They have a row of four very horny post occipital scales behind the head. It has a wider snout.	5) They do not have post occipital scales; the back of the neck just behind the head looks like the skin of a jackfruit. It has a relatively slender snout.
6) Makes a nest by digging a hole in the ground in dry season.	6) They make a nest in the wet season by scraping vegetation and soil together to form a great big heap.
7) Muggers are quite social capable of living in congregation.	7) Both males and females are very territorial, frequently battling for turf.
8) They are tolerant to human beings and attack on humans are rare.	8) Large estuarine crocodiles are known to prey on humans.

Reproduction: Sexual maturity is attained at about 8-10 years or more. Mating occurs in dry seasons. The female makes a nest which is composed of mound of vegetation and mud. The clutch size varies from 20 to 72 eggs. The weight of eggs varies from 91 to 137 g. The young hatch after an incubation period of 80-90 days. The mother prepares one to four guard wallows near the nest. The length of a hatchling is 25 to 30 cm. They are looked after by the mother for about 10 weeks.

Behaviour: Estuarine crocodiles are more aquatic as compared to marsh crocodiles. Sometimes, the young crocodiles spend considerable time on land inside the forest. It is said that the estuarine crocodile is capable of consuming a prey even as large as a whole deer.

Life span: The life span is reported to have 35-40 years or more.

Snake:

Snakes occupy throughout the world except many islands and near the polar regions. The majority of snakes have a distribution in the tropics. They are well adapted to deserts as well as fresh and salt water bodies. Snakes are more closely related to lizards and can be distinguished by many distinctive features. The jaws are usually flexible and are loosely attached to the skull. With the exception of a few primitive snakes (where rudimentary hind limbs are present), all the snakes lack limbs. The body is covered with scales. The eyes are small and have no eyelids. The outer ear and tympanum are absent. There is no diaphragm. There is only one functional lung. The left lung is absent or vestigial in most snakes. Pythons, however, have developed left lungs. The long and cylindrical tongue is forked. Successor teeth lying below immediately replace the lost teeth.

Teeth help in holding preys and assisting in swallowing. The poisonous snakes are characterized in having the highly specialized teeth in the upper jaw called fangs. The fangs are shed periodically and may be found in the faeces or on the bottom of the cage when they are kept in captivity. Reserve fangs are always present behind each functional fang. The fang has a poison gland (modified salivary gland) at its base and the venom produced by this gland flows through a groove or canal running through its centre. Another unique feature of the snake is the absence of urinary bladder. The cloaca holds the excretory products before voiding.

Most species of snakes are non poisonous. The teeth of non poisonous snakes are short and solid and their bites leave a row of smooth marking. Snakes that come out at night are usually poisonous. They have large belly scales. The scales on the head, however, are usually small (except cobra and krait). Most poisonous snakes possess only right lung. The poisonous snakes have compressed tails, long, grooved and canulised fangs and their bites leave an impression of two fangs. The venom consists of two types of toxins: neurotoxin and haemotoxin. Nevertheless, there may be some toxic bites from the harmless snakes.

The hearing or sight is poorly developed. The chemical sense in most snakes is mediated by a specialized organ called vomeronasal organ or Jacobson's organ. Although snakes have no external ears, they are sensitive to vibrations of the ground. Their locomotion involving lateral undulations of the body is commonly called "serpentine locomotion". A healthy snake may shed its skin at intervals of two months or less.

The copulatory organ in males is paired (hemipenes) and it lies at the base of the tail. The females in many species do not have the left oviduct. Courtship in snakes is initiated by the male and copulation may last from two to twenty hours. Sperm may lie dormant in the oviduct. Most snakes lay eggs. Eggs are white or yellow in colour and are usually laid under natural vegetation cover. Many snakes lay eggs in holes for which they scoop out sand or soft earth with their snouts. The incubation period usually varies from 30 to 70 days. Some species are viviparous and give birth to live young. Many snakes reach sexual maturity at two years of age. In larger species it may take four or five years.

The following text highlights the salient features of some well known snakes.

Indian python:

Zoological name: The zoological name of Indian python is *Python molurus*. There are two races: *P. m. molurus* and *P. m. bivittatus*.

Distribution: They are distributed in India, Sri Lanka, Pakistan, Nepal and Indo-China.

Habitat: Indian pythons found in dense and open forests with rocky outcrops.

Physical feature: The massively built python may reach up to 5.85 m and weigh 90.7 kg or more. The adult may be greyish, whitish or yellowish. A dark streak running from eye to nostril is present in young. The other salient features include flattened head, long snout, large nostrils, small eyes, iris flecked with gold, short and prehensile tail and rudimentary hind limbs as curved claw-like processes on either side of anus. A distinct dark and oblique band runs from eye to gape.

Reproduction: The Indian pythons able to breed at about five years of age. Mating occurs in the months of December, January and February. However, in north they are reported to lay eggs in March-June. The size of clutch varies from 8 to 100 soft and white eggs. The female broods the eggs by coiling around them. The incubation period is reported to be two months. The mother takes no interest in her brood after hatching.

Behaviour: These non poisonous snakes are lethargic and slow moving. It has a peculiar method of movement i.e. the body moves in a straight line like a millipede. They usually remain near water bodies. Indian python swims well and also climbs in search of prey. They hibernate during colder months. Indian pythons are both diurnal and nocturnal.

Life span: The longevity of Indian python has been reported more than 22 years.

Russell's viper:

Zoological name: The zoological name of Russell's viper is *Vipera russelli (Daboia Russelli).*

Distribution: Russell's vipers are widely found in India, Sri Lanka, Thailand, Indo-China and Indo-Australian archipelago.

Habitat: They normally live in plains.

Physical feature: The usual length is 1.2 m. The ground colour is brown with varying shades with three series of large oval spots. The spots are brown in centre and margined successively by black and white or buff. There is a dark stripe running from eye to lip. The head is covered with small scales. The neck is constricted.

Reproduction: Sexual maturity may reach in three years of age. The breeding takes place throughout the year. They are viviparous i.e. live-bearing snakes. The young may be born in a caul. Sometimes, the unfertilized eggs are also voided along with the young. About thirty to forty young are born between May and November. The gestation period is reported more than six months. However, the peak period of birth is reported in June and July.

Behaviour: It is a highly poisonous snake. They strike when get irritated. Russell's viper is mainly nocturnal. The young are reported to be cannibalistic.

Life span: The life span is reported to be 10-15 years in captivity.

Indian cobra:

Zoological name: On the basis of the hood pattern three species have been recognized: the spectacled or binocellate cobra *(Naja naja)*, the monocellate cobra *(N. kaouthia)* and the black cobra *(N. oxiana)*.

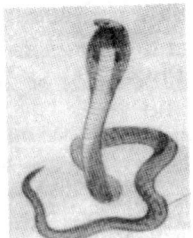

Distribution: The cobras are distributed in India, China, Nepal, Philippines, Sri Lanka etc. The spectacled cobra occurs in the peninsular India. The monocellate cobra is commonly distributed in eastern part of India and the black cobra is restricted to the extreme northwest of India.

Habitat: They are found in all habitats (e.g. deep forests, open cultivated fields, near water bodies and human dwellings) except arid deserts and in the hills above 1,800 m.

Physical feature: The usual size varies from 1.3 m to 1.6 m. The monocellate cobra is generally olive, brown or black in colour. The spectacled cobra is yellowish, brownish or black above. An adult black cobra may be black or brown. The most distinct feature is the presence of a small 'cuneate' scale between the 4th and 5th infralabials. However, in rare cases it may be absent or there may be two small 'cuneate' scales. Another important character is its hood which is formed by the elongated ribs. The monocellate cobra has a single yellow or orange 0-shaped mark on the hood. The other features include short and rounded snout, depressed head, large nostrils and glossy shields head.

Reproduction: The breeding occurs in the month of January and the majority of eggs are laid in April and May. The female may lay eggs numbering 12 to 22 or more at a time. Eggs are guarded by one or both the parents. Incubation is done by both sexes. The eggs hatch in about 48-69 days. Like adult, the hatchlings are also highly poisonous.

Behaviour: Cobras are usually active at night. They are normally not aggressive reptiles. However, when disturbed, they become very aggressive. It spreads hood when alarmed. During hood, the cobra sways backwards and forwards. Occasionally they eject poison as a spray by the forceful thrust. Cobras are good swimmers.

Life span: In captivity they are known to live 10-12 years or more.

King cobra:

Zoological name: The zoological name of king cobra is *Ophiophagus hannah.*

Distribution: The king cobra occurs in India, Pakistan, Philippines etc. In India it usually occurs in Western Ghats, Andaman Islands and eastern part of India such as West Bengal, Orissa and Assam.

Habitat: They inhabit in dense forests, hills, plains, grasslands and estuaries.

Physical feature: An adult king cobra may reach up to 5.5 m and weigh 12 kg or more. Adults are blackish brown or light brown. Hatchlings are deep black and have pure white bands round the body. They have a pair of occipital shields and costals. The adult has 32-43 lighter bands round the body and 11-13 bands round the tail. The other salient features are glossy scales, flat heads and rounded snouts. The hood is relatively less dilatable than the Indian cobras.

Reproduction: The mating occurs in March and she lays eggs in the month of April. However, eggs have been found up to July. The number of eggs laid by the female may be up to 51. The average weight of an egg is 40.85 g. It is one of the few nest-building snakes. The nest which is made of leaves or vegetable rubbish and humus is built by the female. Both male and female stay in the vicinity of the nest and usually the female is coiled up on the eggs. The incubation period varies from ten to eleven weeks.

Behaviour: The highly poisonous snakes, king cobras are reputed for their aggressiveness and courage. They are largely diurnal. King cobra erects its forebody for about one-third its length and spreads the hood like Indian cobras. As mentioned earlier that it is one of the few nest-building snakes and a nest may be built with in three days. They are ophiophagus.

Life span: The life span of king cobra has been reported for over 12 years in captivity.

Common Indian krait:

Zoological name: The zoological name of common Indian krait is *Bungarus caeruleus*. It is one of the deadliest snakes among the poisonous snakes of the world. It is reported that krait venom is 15 times more virulent than the cobras.

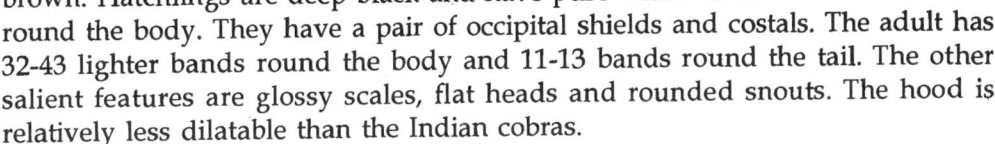

Distribution: They are widely found in peninsular India, Pakistan and Sri Lanka.

Habitat: The Indian common krait is found in human dwellings, fields and low scrub forests. They are frequently occured near or in water bodies.

Physical feature: The usual length is 1.2 m. It is lustrous black or bluish black above. They have narrow white crossbars on the back more or less prominently in pairs. The other features include long and cylindrical body, glossy scales, black iris and short tail.

Reproduction: Sexual maturity is reached at about three years of age. Pairing takes place in February and March. The number of eggs laid by the female varies from six to fifteen. The female stays with the eggs. Eggs hatch between May and July.

Behaviour: The Indian common krait is basically nocturnal. There are many instances of people sleeping being bitten by this snake.

Chapter 14

Salient Features of Selected Wild Birds

Birds are found throughout the world. A detailed zoological classification has been described earlier. Here the detailed descriptions of selected birds are given below:

Ostrich:

Zoological name: The zoological name of ostrich is *Struthio camelus*. There are five subspecies. The Masai ostrich *(S. c. massaicus)* of southern Kenya and Tanzania, the Somali ostrich *(S. c. molybdophanes)* are well known. All can be distinguished partly by the colour of bare skins on heads, necks and thighs.

Distribution: Wild ostriches are now restricted to Africa.

Habitat: The largest living birds, ostriches are found in open and arid country.

Physical feature: An adult ostrich stands 6-8 ft tall and weighs more than 160 kg. The male is somewhat larger than the female. The male is black with white shades on wings and tails. The females and young, however, have pale brown bodies. Legs, feet and bills are horn coloured. Ostriches have large brown eyes with thick black eye lashes. Legs, neck and head are naked. It has only two toes.

Reproduction: Ostrich matures at 3-4 years of age. The breeding season ranges from August to March, though it varies according to the region where they live. She lays about 10 to 20 fertile eggs shortly after mating. Eggs are laid in a scrape made by the female. A single nest may be used by many females for laying purpose. Each female lays egg usually every alternate day. The white to yellowish egg may be 6-8 inches long and weighs up to 2.3 kg. In the wild state, eggs are often partially covered with sand in day time. The incubation is carried out by both sexes and the sturdy and dappled chicks are hatched after 5-6 weeks of incubation period.

Behaviour: Ostriches are diurnal birds and forage during day light hours. They usually travel in bands comprising five to fifty birds. The male fights for a harem of three to five hens. The male sounds a hollow booming call to attract the female.

Another interesting behaviour of ostriches is that they congregate with zebras, antelopes and other large grazing mammals for mutual benefits.

Life span: They are said to live 50 years.

Pelican:

Zoological name: The well known species are the brown pelican *(Pelecanus occidentalis)*, the American white pelican *(P. erythrorhynccus)*, the dalmatian pelican *(P. crispus)*, grey pelican *(P. philippensis)* and rosy pelican *(P. onocrotatus)*.

Distribution: Pelicans are distributed in temperate and tropical regions of the world. In India, three species namely rosy pelican, Dalmatian pelican and grey pelican are found.

Habitat: They are found on or near water bodies such as rivers lakes, mangrove swamps, lagoons and estuaries.

Physical feature: The body weight may reach up to 13 kg. Pelicans are characterized in having large wings, long beaks, small nostrils and pneumatic skeletons. All four toes are webbed. The most characteristic feature of the pelican is its large, bare and elastic pouch (gular). The throat pouch can hold food particles more than twice the stomach capacity. The pouch acts as a net for scooping up fish. It also serves as a cooling device.

Reproduction: Maturity takes place at three to four years. Pelicans breed in large colonies numbering 50 to 10,000 or more. The nest is built on the ground lined with sticks or vegetation or it may be a loose platform of sticks and reeds in low trees, bushes or mangroves. The nesting season of grey pelican begins in November-April. The number of eggs laid by the female varies from one to four. Eggs are chalky white in colour. The incubation is done by both sexes and the young hatch after an incubation period of 28 to 35 days. The young are naked, blind, ugly and helpless. They are fed by both parents through regurgitation. Young grow rapidly and are able to fly at about 60 days.

Behaviour: Pelicans usually travel in small flocks.

Life span: Pelicans have been known to live 30 years or more.

Flamingo:

Zoological name: There are four living species of flamingos: the greater flamingo *(Phoenicopterus ruber)*, the lesser flamingo *(Phoeniconaias minor)*, the Andean flamingo *(Phoenicoparrus andinus)* and James's flamingo *(Phoenicoparrus jamessi)*.

Distribution: Flamingos are distributed in the Old World (except Australia), North and South Americas and some oceanic islands.

Habitat: The habitat of flamingos is restricted to the saline and alkaline lakes and lagoons of dry regions.

Physical feature: Flamingos are characterized by slender bodies, large wings, long legs, short tails, and long and thick necks. The legs and necks are proportionately longer as compared to their bodies. There are three front toes with fully webbed. The bill is sharply bent at the middle. The bill is uniquely adapted for filter feeding. The tongue serves as a pump to suck and eject the semi liquid while straining food. Flamingos have mainly pink plumages tinged with light vermilion.

Reproduction: Flamingos breed in colonies on remote areas. They may breed two or three times in a year. The nest is built up by soft mud. The clutch size varies from one to two large, elongated and pale blue or chalky white eggs. Both the male and the female incubate the eggs for 27 to 32 days. Both parents partake in rearing the young. Young forage with in three weeks. They fly at 65 to 70 days.

Behaviour: Flamingos are seen in large flocks. In flight, flamingo's neck stretches out in front and the legs trail behind. Flamingos do well in captivity.

Life span: The life span is recorded more than 20 years.

Stork:

Zoological name: There are twenty species of stork. Some well known species are the Oriental white stork (*Ciconia boyciana*), black stork (*C. nigra*), painted stork (*Mycteria leucocephala*), adjutant stork (*Leptoptilos crumeniferus*), black-necked stork (*Xenorhynchus asiaticus*) and Asian open billed stork (*Anastomus oscitans*).

Distribution: Storks inhabit in temperate and tropical zones of the Old and the New Worlds.

Habitat: They are adapted to a life in shallow waters, marshes and dry uplands. The painted storks, for example, are found in coastal mud flats, lakes, marshes, inundated paddy fields and occasionally in river banks.

Physical feature: Storks are long-legged, long-necked, long-billed birds with broad, rounded wings, and short tails like other wading birds. Toes are rather short and are partially webbed. The rear toe is elevated above the other three toes. The physical characteristics of painted stork include red-brown legs, heavy, orange coloured bills with slightly decurved at the tips and eyes surrounded by bare yellow and red skin. The plumage colour pattern of storks is usually black and white with a metallic sheen, though there is a species variation. For example, the painted storks are white with black in colour and the innermost secondary feathers are pink. Legs and bills are brightly coloured.

Reproduction: They build nests or large stick platforms in trees or on rocky cliffs and buildings. The nesting materials are comprised of twigs, grasses, sticks, rags, paper and lumps of earth. Males normally bring all these nesting materials, while females arrange these to form a rough platform. The female lays three to six white eggs. Both sexes incubate the eggs. Young hatch after a period of 29 to 35 days of incubation. The naked chicks become downy in later periods. Young are fed by both parents. They remain in the nest for about two months before flight.

Behaviour: Storks are diurnal birds. They fly with alternatively of flapping and soaring. Like many other birds (e.g. cranes), storks hold their necks straight out in flights and their long legs trail behind. It is said that stork comes back to the same rooftop to nest year after year and young tend to return to the same place for nesting. The northern species are migratory in nature and usually travel in large flocks. The painted storks are normally found in pairs or in small parties. During breeding season they congregate at favourite heronries.

Life span: Storks have been reported to live about twenty years in the wild.

Crane:

Zoological name: There are fourteen species of cranes. Common crane (*Grus grus*), sandhill crane (*G. canadensis*), Siberian white crane (*G. leucogeranus*), Japanese crane (*G. japonensis*), whooping crane (*G. americana*), sarus crane (*G. antigone*), white necked crane (*G. vipio*), black necked crane (*G. nigricollis*) and demoiselle crane (*Anthropoides virgo*) are some examples of cranes.

Distribution: Cranes are found throughout the globe except South America, New Zealand and some other parts of the world. Most of the cranes in the genus *Grus*, however, are found in Asia. The Siberian and whooping cranes are the most endangered crane species in the world.

Habitat: They occupy in the habitats of marsh lands, wet plains and prairies. Cranes are sometimes seen in sandy flats and seashores. The common crane, for example, inhabits in very wet area with occasionally in or near wooded country.

Physical features: Cranes may attain a size of 5 feet tall. They are characterized in having long legs, long necks, short and wide tails, long and wide wings, and long bills. The front toes are connected by a membrane at the base and the small hind toe is considerably elevated. They have slaty grey, brown or white plumages. The adults may have a partly bare head. The most salient anatomical feature of the crane is its wind pipe or trachea. The strongly convoluted wind pipe looks like the coils of a trumpet and it pierces through the breast bone and the flying muscles of some cranes. The wind pipe is responsible for utter a loud and penetrating call especially in male.

The common crane is a widely known species. The body colour is lead grey with blackish head and neck in most cases. Eyes are red. There is a red patch on the crown.

Reproduction: Crane builds its bulky nest on ground in marshy area. The same nest may be used year after year for laying purpose. Cranes breed only once a year. They are monogamous and mate for life. The female lays one to four (usually two) olive grey spotted with brown eggs. The egg colour of a crowned crane, however, is pale blue. Chicks hatch after a period of 28-36 days of incubation. Both sexes share for incubation. The mottled red brown to greyish downy young can run shortly after hatching. Both male and female share the responsibility of caring the young. They are able to fly at 70 day.

Behaviour: Cranes are migratory in nature. They are seen in large flocks during winter months. In flight, the neck is extended straight out in front. Cranes are also reputed for their remarkable courtship and group displays. Unlike storks and herons, cranes hardly perch on trees. Their loud and penetrating calls are more pronounced during migration. As mentioned earlier that cranes are monogamous and mate for life. Sometimes, cranes are found in agricultural fields causing considerable damages to the crops.

Life span: The longevity of cranes has been reported more than 50 years. There is a report that a Siberian white crane died at the Washington National Zoo at the age of 62 years.

Jungle fowl:

Zoological name: Four species of jungle fowl have been recognized. They are red jungle fowl *(Gallus gallus)*, Ceylon or La Fayettii's jungle fowl *(G. lafayettii)*, grey or Sonnerat's jungle fowl *(G. sonneratii)* and green jungle fowl *(G. varius)*.

Distribution: The jungle fowl are confined in South and Southeast Asia. The red jungle fowl is the widely known species. They occur in India, Bangladesh, Thailand, Malaysia, China and many other parts of Southeast Asian countries. The Indian red jungle fowl *(G. g. murghi)* are distributed mainly in the Himalayan foothills and terai regions. The grey jungle fowl and the Ceylon jungle fowl are also found in India.

Habitat: Jungle fowl live in a wide variety of habitats such as dry scrubs, small woods and low altitude forests.

Physical feature: They have similarities in general appearance and anatomical features with their domestic counterparts. The mature body weight is 1.5 to 3.3 kg. The other salient features include a fleshy comb, lobbed wattles and a curved tail. The cock of red jungle fowl has fiery red and golden brown plumage. The hen is rusty brown.

Reproduction: The breeding season of red jungle fowl begins mainly from March to May. The nest which is a scrape on the ground lined with dry grasses and bamboo leaves, is found amongst thick undergrowth in the forest. She lays five or six pale buff to pale reddish brown eggs. The incubation period is 20-21 days.

Behaviour: Jungle fowl may be seen singly or in pairs or small parties of 5-6 birds. During breeding season, they are found in parties comprising of one adult male and several females. In winter months jungle fowl are also found in congregate. The Indian red jungle fowl are very shy birds and they come out in the open for food in the early morning and in the late afternoon.

Life span: The life span has been recorded 2-5 years.

Peafowl:

Zoological name: The common peafowl *(Pavo cristatus)*, the Burmese peafowl *(P. muticus)* and the Congo peafowl *(Afropavus congensis)* are well known species.

Distribution: Peafowl are distributed in India, Nepal, Myanmar, Congo etc.

Habitat: Wild peafowl occur in open dry forest country. The semi-feral peafowl, however, are found in the agricultural fields. The Congo peafowl occurs in the rain forest of east-central Congo basin.

Physical feature: Peafowl, the close relatives of pheasants, are admired for their exceptional beauties. Male (peacock) has spurs on his legs. Some peahens also bear spurs. The long ornamental feathers of males are actually not true tail feathers but these are elongated upper tail coverts. These feathers are shed after the breeding season. Peafowl have occipital crests of feathers.

Reproduction: The breeding season ranges from June to September in northern part of India, while it begins mainly in April and May in southern India. The female lays eggs in a scrape lined with grasses, leaves, sticks etc. on the ground. The semi-wild peafowl, however, may build the nests in buildings, forts etc. The clutch size varies from four to eight pale cream eggs. She alone incubates eggs for about a month. The young are pale buff.

Behaviour: Peafowl are found in small flocks consisting of a single cock and three to five hens. In the wild, they come out into the forest clearings or agricultural fields in the early morning and in late afternoon to obtain foods. They roost in tall trees in order to protect from their natural predators.

Life span: The life span of a peacock is about 20 years.

Common quail:

Zoological name: The common quail *(Coturnix coturnix)* is the widely known species.

Distribution: The common quail is found in India, Pakistan, etc.

Habitat: They are found in open country with standing crops and grassland.

Physical feature: The body weight varies from 120 g to 180 g. The males have rusty brown throat and breast feathers. The female lacks the black anchor mark on the throat.

Reproduction: The females mature at about two months of age. The common quails breed from March to July in India. The nest is built in the form of a scrape lined with or without grass and leaves on the ground. The clutch size is usually 6 to 8 eggs. The eggs are yellowish, buff to deep yellowish brown or reddish brown speckled and blotched with some shades of dark or chestnut brown. Young hatch after a period of 21 days incubation. The young are usually able to fly at seven days old.

Behaviour: They are usually found in pairs.

Life span: The common quail is reported to live up to five years.

Pigeon:

Zoological name: The common green pigeon *(Treron phoenicoptera)*, the large green pigeon *(T. capellei)*, the imperial fruit pigeon *(Ducula conciana)*, the wood pigeon *(Columba palumbus)* are well known species of pigeon.

Distribution: Pigeons are distributed in tropical and temperate regions of the world. The majority of species, however, are found in the tropical parts of Asia, Australia and the Pacific Island.

Habitat: The habitats include rocky hills, forests, near cultivated land etc.

Physical feature: Pigeons are stout- bodied, short-legged, small-headed and soft plumaged birds. Their necks are rather short. The rounded bill is more thickened towards the tip. There is a fleshy 'cere' at the base of the upper mandible in which the nostrils are lodged. It has a large crop. Most species have a large muscular gizzard.

Reproduction: The nest which is a flimsy platform of sticks is constructed in trees, in holes in trees, on building lodges, or on the ground. The clutch size is one to three pure glossy white eggs. Eggs may be yellow to buff in colour in some tropical members. Both sexes incubate eggs. Young are fed on 'pigeon milk' by regurgitation. The pigeon milk is secreted by special cell lining the crop of a pigeon. The cheesy milk is gradually replaced by the partially digested food materials. If eggs or young are lost, she will lay again. Two to three or more broods may occur annually. Young leave the nest when they are about 15-21 days old.

Behaviour: The majority members are gregarious. Some pigeons are migratory in nature. Fruits pigeons live almost in trees. Pigeons are used as pet and laboratory birds. A pigeon is trained to carry massages tied to it neck or leg.

Life span: Pigeons can live 22 to 25 years in the wild state.

Parrot:

Zoological name: The long-tailed parakeet (*Psittacula longicauda*), the rose-ringed parakeet (*P. krameri*) and the alexandrine parakeet (*P. eupatria*) are a few examples.

Distribution: True parrots are distributed in Asia, Africa, America and many other islands.

Habitat: They are found in forests, orchards, near cultivated land etc.

Physical feature: These brightly coloured birds usually have a solid red, green, yellow, white or black with contrasting shades of red, yellow or blue colours on the tails, wings, or head. The salient anatomical features include large head, short neck, short legs, and strongly down-curved hooked bills. The slightly movable upper mandible is loosely articulated with the skull. There is a bulging 'cere' at the base of the bill through which the nostril opens. All parrots have zygodactyls feet i.e. the first and fourth toes are backwardly pointed while the second and third toes are forwardly pointed. Toes are armed with powerful hooked claws. The crop is well developed. The tongue is thick and heavily muscled.

Reproduction: They build nests in cavities. However, parrots normally lay eggs in an unlined hole in a tree. The average number per clutch ranges from three to five round, glossy and white eggs. Both sexes share incubation although a few females are reported to incubate eggs alone. The incubation period is 17-20 days. The male and female feed the young by regurgitation like pigeon.

Behaviour: They are found in large flocks. Parrots are serious pests of agricultural crops. The rose-ringed parakeet is one of the most destructive bird pests in agricultural crop. They are also very popular cage birds.

Life span: Parrots are known to live up to 50 years in captivity.

Section 3

FEEDS AND FEEDING, RESTRAINT, HOUSING AND MANAGEMENT, AND HEALTH CARE

Chapter **15**

Feeds and Feeding of Wild and Zoo Mammals

The feeds and feeding of selected wild and zoo mammals are described in detail below:

Monotremes:

The monotremes include echidna and platypus. The diet of monotremes are given below:

Echidna: The common or short nosed echidna feeds preferably on ants and termites in wild state. In captivity, they are maintained on a diet of minced meat, evaporated milk, eggs and yogurt with vitamin and mineral additives.

The natural diet of long nosed echidna consists of earthworms and larvae of beetles. They can be maintained on a wet mixture of boiled oat meal, eggs, raw meat and earthworms with supplementation of minerals and vitamins. A little amount of soil may be added to the diet. The following dietary ingredients with quantity may be given to a captive adult echidna: canned feline diet (210 g), mixed baby cereal (60 g), egg yolk (one), soy-based infant diet (200 ml), mineral and vitamin supplements (45 g), caco$_3$ (1 g) and water (200 ml). All the ingredients should be mixed and make it a gruel.

Platypus: The dietary ingredients preferred by a platypus in the wild state depend on the season and the area where it lives. However, its diet generally consists of earthworms, mollusks, small crustaceans, insect larvae and probably frogs and tadpoles. Boiled eggs, mealworms, shrimp, earthworms can be provided to the platypus in captivity. A small amount of soil, sand and grit can be added to the captive diet in order to proper digestion of feeds. The following is the schedule of diet per animal per day in platypus: earth worms (450 g), live cray fish (2 dozen), leopard frogs (two), boiled eggs (two) and mealworms or cockroaches (one handful).

Marsupials:

There is a wide variation of feeding habits among marsupials. Some marsupials are herbivorous (e.g. kangaroo) animals, while others (e.g. Tasmanian devil) primarily subsist on carnivore diets. The herbivore diets primarily consist of leaves, various grasses, herbs, roots, bark, fruit and other plant materials. Carnivorous marsupials subsist on insects, small vertebrates, eggs, fish, frogs, mice, birds, carrion, snails, earthworms, seeds and fruit in the wild. However, some members are reported to be omnivores (e.g. bandicoot). They feed on both plant and animal materials.

Captive marsupials are maintained on a wide variety of diets. They include grain meal, hay, fresh green plant material, bread, fruit, chopped meat, mealworm, insect, small mammal, spider and bird. Some macropods (e.g. kangaroo) are offered rabbit pellets, green vegetables, banana, carrot and the fruits. However, the following text on the diets of kangaroo, wallaby and koala in their wild and captive states is given below:

Kangaroo: Kangaroo eats soft grasses, herbs and leaves in the wild. The captive diet of red kangaroo includes chopped vegetables, fruits, various greens (e.g. carrot tops), mixed grain, bread, powdered milk, fresh and pelleted alfa and dry commercial rabbit or rat pelleted feed.

Wallaby: They subsist on grasses, herbs, fruits, plant roots and bark.

Koala: Koala feeds in the wild on leaves and shoots of a variety of eucalyptus trees. In captivity, they are provided with mature leaves of eucalyptus trees and bread mixed with milk and honey. Some amount of earth is also given in order to meet the requirements of calcium and other minerals. About one kilogram leaves may be required per day for a single animal. Young eucalyptus leaves containing more amount of prussic acid may be lethal to the animal.

Insectivores:

Insectivores in their wild state feed on a wide variety of terrestrial and fresh water invertebrates. Carrion, small vertebrates, roots, seeds and plant materials are also eaten by insectivores. Cow milk, eggs, chicken, earthworms, mice, mealworms and cereal based mixtures are taken by many insectivores in captivity. Therefore, the general principle regarding the feeding habits of insectivores is that they have preferences for foods but consume what is available in lean periods like many other mammals. Young feed on glandular secretions (e.g. saliva), and parental and own faecal materials besides mother's milk. The feeds and feeding of hedgehog is given below:

Hedgehog: Hedgehogs are generalized omnivores. They consume small rodents, bird's eggs, frogs, toads, birds, snails, lizards, snakes, insects, earthworms, fish, crabs, fruits, small mammals etc. in the wild. However, their captive diets are chicken, mealworms, freshwater fish, milk, cheese, bread, boiled potato, apple, ripe banana, rice, oatmeal, insects etc. Eurasian hedgehog can be maintained on the following diet: meat mix (1/3 cup), dry dog food (1/4 cup), boiled egg (1/2) with mealworms supplement. Food is provided in a shallow earthenware pot. Water should be given in separate container.

Chiropterans:

Chiropterans or bats are voracious feeders. A small fruit bat may consume fruit more than its body weight at a single meal. It has been reported that 80 per cent microchiropterans feed solely on insects and the remaining subsists on various insects, fruits, small vertebrates and blood. The feeds and feeding of fruit bats and insectivorous bat are given below.

Fruit bat: The old world fruit bats primarily consume various types of fruits in the wild state. In captivity, they can be maintained by simply providing sweets and fruits (e.g. banana) with vitamin and mineral supplementations. However, other items like mice, ground meat, mealworm, cabbage, lettuce, sweet potato, boiled vegetable, milk, honey, papaya, citrus fruit, pineapple, melon, fig, dates, grapes, apple and cherry also form the dietary ingredients for captive bats. Fruit bats are very fond of water.

Insectivorous bat: In the wild, many insectivorous bats usually capture their preys while in flight. Captive diets for insectivorous bats include mealworms, crickets, fruit flies and blow fly larvae. Other items such as termites, cockroaches, spiders, woodlice, beetles, moths, grasshoppers, mosquitoes, caterpillars, raw meat, cottage cheese, chopped eggs, cray fish, veal liver, chopped vegetables and fruits (e.g. apple, melon and grape) are also given to the captive specimens. Commercial food like rat pellets and canned feline or dog food have also routinely been fed to the captive bats. Another important fact is that the insects fed to the insectivorous bats are typically low in calcium. Therefore, calcium-enriched diet with vitamins should be supplemented for captive specimens. The following diet is recommended for captive insectivorous and frugivorous bats: mixed baby cereal (20.7 %), wheat germ (4 %), nonfat dried milk powder (9 %), calcium caseinate (15.8 %), sugar (45.5 %), casein-based protein supplement (3 %) and mineral- vitamin mixture (2%). The ingredients should be mixed properly. Then 100 g of dry mix is mixed with 540 ml nectar, 260 ml water and 6 ml corn oil. Mixed liquid diet can be fed to the bats with banana. Food is offered in dishes. Bats may either be trained to feed from dishes or they may be hand fed. The remnants of the previous diet should be removed daily.

Vampire bat: Vampire bats are fed blood of cattle, sheep or chicken. Blood is defibrinated by stirring and is stored at 4^0C until it is given to the bats in bowls late in the afternoon. The amount of blood normally provided to the vampire bat is one half to two thirds of its body weight. This constitutes about 15 to 20 ml. They should be fed once daily. Frozen blood may also be given and this should be reconstituted with water. Vampire bats may also readily accept the blood of guinea pigs, donkeys and goats. Live chickens are sometimes placed in the enclosure so that vampire bats can take blood directly. When blood is not available, vampire bats should be given saline (0.09% NaCl) for one or two days. The supplementation of free water to vampire bats is highly debatable. It has been reported that many captive vampire bats may live without supplementation of free water.

Primates:

Nonhuman primates feed on a variety of foods in their natural states. It is reported that primates can eat up to 300 different types of foods in the wild. The wild diets of prosimians consist of leaves, fruits, insects and small animals. In captivity, most of the prosimians are omnivorous in feeding habits. They are adapted to feed a wide variety of food ingredients which include bread, eggs, leaves, small birds, fruits, milk, insects, lizards and frogs.

The apes prefer to eat vegetables, fruits, small sized vertebrates and insects. Most Old and New World monkeys are omnivorous in feeding habits. They feed on a large variety of foods in the wild. This includes fruits, leaves, bark, grass, seeds, ants, termites, young shoots, buds, flowers, small lizards, birds and other small animals. Langur and colobus monkeys, however, primarily subsist on vegetarian diet. Nevertheless, most primates in captivity can be maintained on commercially prepared diets. These diets are blended with a variety of natural foods. Moderate quantities of green vegetables, bananas, apples, carrots and sweet potatoes are also provided. Commercial diets given to captive primates are easy to prepare, relatively economical and have satisfactory results on growth and development. Commercial chows prepared from natural food ingredients have also successfully been used for captive nonhuman primates. The chow contains soybean meal, corn, skimmed milk, fish flour, cereal products, animal fat and vegetable oil. The ingredients are mixed, pressure cooked, blended and shaped into chow and fortified with proteins and minerals.

Primates spend up to 60 per cent of their waking time feeding. Therefore, food can be divided into 3-4 small portions and feed at different times of the day. The tree branches with leaves (e.g. jackfruit, fig, mango etc.) may also be provided.

Another important fact is that human race requires 50 to 60 nutrients and all these nutrients are also essential for nonhuman primates. Therefore, captive diets consisting of a variety of foods from both vegetable and animal sources should contain adequate amounts of essential nutrients. Adequate and balanced diet play an important role in controlling many diseases of captive nonhuman primates. In captivity, it is to be ensured that they are provided proper diet containing adequate amounts of energy, protein and other nutrients. Most new world primates, for example, require more protein than old world primates. It has been reported that Old World Primates need 15 per cent protein, while the protein requirement of the New World Primates is estimated at 25 per cent. The energy requirement is stated to be higher for smaller and younger nonhuman primates than those of larger ones. However, pregnant and lactating females need more energy. A young rhesus monkey, for example, requires 150-200 Kcal/kg, whereas an adult animal is needed 100 Kcal/kg. The pregnant and lactating females are required 175 Kcal/kg and 150 Kcal/kg respectively.

A dietary level of 3 to 5 per cent fat is said to be adequate for captive nonhuman primates. However, they are found to be well adjusted to total 10 per cent dietary fat. The excess amount of fat may cause diarrhoea. Vitamins and minerals are essential for captive specimens. Vitamin C in captive animals can be met through supplementation of green vegetables, orange and multivitamins. Calcium and phosphorus requirements are estimated at 0.6 to 0.8 per cent and 0.3 to 0.4 per cent respectively. Wild nonhuman primates mostly fulfill their water requirements through natural dietary ingredients. However, water should be supplemented for captive specimens. Food is offered in the morning and afternoon. The natural and captive diets of some nonhuman primates are given in Table 15.1.

Pangolin:

In the wild, the pangolin subsists mainly on ants and termites. They, however, also feed on beetles, cockroaches, worms, grubs, larvae of bees and wasps, and plant matter. Dry commercial dog food, minced meat, horsemeat, milk, eggs, carrots, boiled oatmeal, wheat germ and precooked cereals form the diets of captive pangolins. Cow's milk (2 parts) mixed with water (1 part) may be fed to the Indian pangolin. A feeding bottle with a rubber nipple can be used. Food is

Table 15.1: The feeds and feeding of selected primates in both wild and captive conditions:

Animal	Natural diet	Captive diet
Chimpanzee	Fruits, leaves, bark, roots, ants, termites, flowers, seeds, soft pith and resins are the natural food ingredients.	Ingredients for captive animals include ground wheat, lettuce, potato, carrot, cabbage, ground corn, wheat bran, bread slice, meat, egg, banana, orange, tomato, honey and sunflower seeds. A diet containing fruits (1 kg), milk (1 litre), bread slice (6 no), boiled egg (one), garlic and onion (100 g) and tea (2 cups) may be given to a captive chimpanzee. Water should be available regularly. However, a general ration containing ground wheat (10%), wheat bran (15%), ground corn (41%), dehydrated alfalfa (1%), dried skim milk (8%), soybean oil meal (6%), animal fat (2.5%), brewer's yeast (2%), sucrose (5%), meat and bone (2%), bran cellulose (3%) with mineral and vitamin supplementations (4.5%) can be given to captive chimpanzees.

Gibbon	Ripe fruits, young leaves, buds, flowers, young shoots, small ants, spiders, honey, insects and bird's eggs are the dietary components in natural state. The main item, however, is fruit.	Boiled rice, soaked gram, leaves, fruits, sunflower seeds, vegetables, eggs, biscuits, oatmeal cake and corn flour are the principal dietary items of captive gibbons. They thrive well on commercial monkey pellets with supplementation of vegetables and fruits. Grass hoppers (5 to 8 no.) may be given occasionally or daily. A typical daily food items for an adult gibbon may be as follows: apple (2 no.), orange (2 no.), banana (2 no.), bread slice (2 no.), hard-boiled egg (2 no.), cabbage (good amount), horse meat (small amount), monkey pellets (a handful) and fine bone meal and cod-liver oil (1 tsp each).Animals are fed twice daily i.e. morning and evening. A hoolock gibbon may be given the following ingredients: ripe banana (500 g), date (100 g), apple (100 g), grape (50 g), Bengal gram (20 g), ground nut (20 g) and egg (1 no).
Macaque	They thrive on a variety of plant and animal foods such as seeds, flowers, leaves, buds, bark, gum, roots, bulbs, insects, snails, crabs, fish, lizards, birds and mammals.	Captive diet is composed of commercial monkey pellets with supplementation of vegetables (e.g. pumpkin, cucumber, brinjal etc), insects and fruits (e.g. orange, apple, banana etc.). The following ingredients may be given to a captive macaque: boiled rice (125 g), vegetables (500 g), milk (250 ml), slice bread (2 no), groundnut (100 g), and roasted gram (50 g). The amounts provided to the animals, however, vary according to the body size, age, sex and individual needs. The feeding schedule of a lion-tailed macaque in the National Zoological Park, New Delhi is as follows: milk (100 ml), banana (2 no), cucumber (50 g), peas (100 g), boiled potato (100 g), onion (25 g), tomato (25 g), groundnut without shell (50 g), soaked gram (50 g), bread (100 g), green corn (100 g) and green leaves (100 g).
Langur	They feed on seeds, leaves, twigs, bark, buds, flowers, fruits and small vertebrates.	Boiled rice, fruits, vegetables, bread and leaves (e.g. pipal, bair, sajina and tamarind) are the dietary components for captive animals. Rabbit and monkey pellets are also provided. Ration should contain a good amount of crude fibre. An anaerobic bacterial fermentation takes place in the stomach of a langur. The following food items for a common or capped langur are as follows: bread (100 g/day), ground nut (20 g/day), mango (125 g/day), banana (750 g/day), egg (1 no/day), gram (50 g/day), vegetables (500 g/day) and pineapple (250 g/day).

served in a pan. Some amount of sand and pebbles may be added to the diet so as to enhance the digestion process. A young pangolin in captivity starts feeding on solid food when it is approximately 35 days old. The supply of drinking water should be ensured regularly.

A giant pangolin was maintained at the Antwerp Zoo, Belgium, on a blended mixture of 25 g corn flakes, 150 g chopped meat, half table spoon full wheat germ oil, 100 g warm boiled water and one drop formic acid. The above mixture was fed in morning and afternoon.

Lagomorphs:

Lagomorphs in their natural state feed on a variety of plant materials such as buds, twigs, grasses, leaves and flowers. Many wild species are known to be browsers. In captivity, lagomorphs are maintained on commercially prepared pellets with supplementation of greens, hay and other essential dietary ingredients. Legumes and hay are provided for captive specimens. The nutrient requirements for wild and captive lagomorphs depend on their body size, sex, age, environment and stage of productive functions. The feeds and feeding of hare and rabbit are given below:

Hare: In the wild state, hare consume twigs, bark, buds, leaves, grasses and seeds. Many wild hare also browse on flowers of shrubs. These animals swallow soft faecal material directly from their anus. During winter months, snowshoe hare changes its feeding habit and feeds on twigs, buds and bark of many soft woody and deciduous plants (e.g. maple, willow and oak). Captive hares are provided with commercial rabbit pellets containing oats, wheat grain, alfalfa hay, soybean meal, vitamins and minerals. Various types of green vegetables and fruits are also given to hare in captivity.

Rabbit: Rabbits, like other lagomorphs are non-ruminant herbivorous animals. Most types of vegetables, grains and hay are readily eaten by rabbits. They consume buds, tender twigs, leaves, grasses and flowers of shrubs in the wild. The diet offered to captive animals is based on the diet prescribed for the domestic rabbits. The dietary ingredients provided for captive rabbits include alfalfa hay, oat, wheat grain, barley, maize, soybean meal, turnip, carrot, sugar beet, sweet potato, grass, apple, orange and leafy materials. The wild and cultivated plants that are suitable for feeding of rabbits include groundnut (*Arachis hypogaea*), paragrass (*Brachiaria mutica*), guinea grass (*Panicum maximum*), elephant grass (*Pennisetum purpureum*), water hyacinth (*Eichhornia crassipes*), sweet potato (*Ipomoea batatus*), alfalfa (*Medicago sativa*), sugarcane (*Saccharum officinarum*), sorghum (*Sorghum vulgare*), Egyptian clover or berseem (*Trifolium alexandrinum*), wild vetches (*Vicia* spp), cow pea (*Vigna sinensis*) and maize (*Zea mays*).

A domestic rabbit on an average eats 4 per cent of its body weight. Commercially prepared pellet containing 12 to 20 per cent crude fibre, 12 to 17 per cent protein, 1.5 to 4 per cent fat and 50 to 75 per cent total digestible nutrient should be provided for various categories of captive rabbits. The total quantity of feed to be offered on daily basis to all types of rabbits is as follows:

Adults on a maintenance ration: 113-128 g.

Growing fattening rabbits (4-11 weeks): 100-130 g but usually feed *ad lib.*

Lactating does with litters: 340-400 g.

Stud bucks: 128-142 g.

Additives such as growth promoters and coccidiostats should be provided with the pelleted feed.

The following points regarding the feeding practices of rabbit are important:

- Rabbits are sometimes called pseudoruminants. They are known to practise a physiological phenomenon called coprophagy where rabbit consumes soft faecal pellets directly from its anus. These soft faecal pellets (also called night pellets or caecotrophes) contain appreciable quantities of high value proteins, water soluble vitamins and a small quantity of volatile fatty acids. Caecotrophy first begins to function in domestic or wild rabbits at about three weeks of age when young start consuming solid feed in addition to mother's milk. It is reported that coprophagy is normally practised at night by domestic rabbits, while the wild rabbits practise it during day time. Hard faecal pellets produced by rabbits largely contain lignocellulose.

- Although rabbits utilize crude fibre poorly, a fair quantity of fibre diet is required to promote gut motility as well as to minimize enteric diseases. Fibre also helps in absorbing toxins produced by pathogenic organisms. Diet containing high amount of fibre, however, may cause caecal impaction and mucoid enteritis.

- Rabbits need both water and fat soluble vitamins. Intestinal microorganisms are known to synthesize adequate amount of vitamins B group, C and K in rabbit. Therefore, dietary supplements of these vitamins are not required. The fast growing rabbits, however, positively respond to the supplements of thiamine (1-2 ppm), pyridoxine (1-2 ppm) and riboflavin (6 ppm). Other vitamins such as A, D and E are also required. Vitamins D and E can be given @ 900-1000 IU/kg and 50 ppm respectively. Vitamin A is usually not needed if the diet contains 30 per cent or more alfalfa meal. Ration containing 4 to 5 per cent minerals is adequate for maintenance purpose. Minerals may be given in the form of bone meal, finely ground chalk and salt.

- Care should be taken to avoid feeding of soiled, contaminated or mould feeds. Newly born lagomorphs may be raised by providing the following rabbit substitute milk formula: evaporated milk (240 ml), egg yolk (one), water (240 ml), honey (5 ml) and vitamins (5 ml). The quantity offered to the orphaned young is about 1 to 4 ml once or twice daily. The practice of excessive feeding of milk to the young should be avoided.

Rodents

Rodents may be herbivorous or omnivorous. Most members in the order Rodentia, however, are herbivorous animals. They subsist on a wide variety of plant and animal materials. Their preferences for dietary ingredients depend on many factors such as availability of feeds, physiological needs, seasonal changes, type and composition of feeds and in the areas where they live. Nevertheless, their basic diets in the wild state are comprised of seeds, vegetables, cereals, small vertebrates and invertebrates, and other plant materials. Many rodents store foods in burrows for future. The following descriptions (Table 15.2) cover the feeding habits of some selected rodents in their wild states.

Table 15.2 : The diets of selected rodents in the wild.

Animal	Natural diets	Remarks
Guinea pig	They consume a large quantity of vegetable diets such as grasses, legumes, bark, twigs, seeds, lichen and fruits. These dietary items are also eaten by their wild counterparts.	It is the domestic species of the genus *Cavia*. This herbivorous animal has a simple stomach like rat and mouse. But unlike rats and mice, the guinea pig's stomach is lined with glandular epithelium. A large semicircular caecum with numerous lateral pouches is present in guinea pig.
Chinchilla	They consume grasses and shrubs.	Chinchilla thrives almost on vegetable diets in the wild.
Hamster	Hamsters consume roots, green vegetation, seeds, peas, beans, lentils, grains, potatoes, fruits, frogs and insect larvae.	The stomach has two distinct chambers: a keratinized, nonglandular forestomach (cardiac) is separated from glandular region (pyloric) by sphincter like muscular marginal folds. The forestomach serves like the rumen of herbivores. The J-shaped caecum has many lateral sacculations. Wild species have a habitat of storing foods in underground burrows.
Rat	They thrive on a wide variety of seeds, grains and other plant matters. Rats also feed on invertebrates and small vertebrates.	Rats are omnivorous animals. The stomach has both glandular and non glandular regions. It has relatively a well developed caecum. The gall bladder is absent in the Norway rat.
Mouse	Mice feed on a wide variety of grains, seeds and other plant materials. They also consume foods of animal origins.	They are omnivorous in nature. The digestive tract of mouse resembles that of other omnivorous rodents.

In captivity, most rodent species are maintained on commercially prepared rodent pellets. They can be successfully kept in zoos by providing fresh vegetables, peanuts, soybean meal, carrots, apples, lettuce, biscuits and pellets. Many rodents do well on diets such as rolled oats, maize, sunflower seeds, millets and other plant materials. Soft diet should be avoided as far as practicable. Supply of clean water to captive rodents is very important. Newly captured rodents are gradually adapted to the commercially prepared feeds. The dietary ingredients provided for captive specimens are described in Table 15.3.

Table 15.3: Captive diets of some rodents.

Animal	Captive diets	Remarks
Guinea pig	Oats, ground wheat, ground yellow corn, soybean meal, dried skimmed milk, dehydrated alfalfa meal, animal fat, minerals and vitamins are usually given to captive guinea pigs. A captive ration for guinea pig is as follows: alfalfa meal (17% protein)-350.0 g/kg, soybean meal (40% protein)- 120.0 g/kg, ground whole oats-252.5 g/kg, ground whole wheat-236.0 g/kg, soybean oil-15.0 g/kg, dicalcium phosphate-5.0 g/kg, calcium carbonate10.0 g/kg, salt-7.5 g/kg and mineral and vitamin premixes-4.4 g/kg.	The commercially prepared guinea pig pellet contains a minimum of 18 % protein, 4% fat and maximum of 16% fibre. Captive diet contains more in energy and low in fibre as compared to the natural diet. A guinea pig eats several small meals during day time. The average food intake (g/day) is as follows: 35-45 (growing), 45-70 (adult), 70-80 (pregnant), and 100-130 (lactating). The average water intake is reported to be 100-250 ml/day for growing animal and 200-300 ml/day for adult animal. Vitamin C is essential for guinea pig. Pellets may be given as powdered form. A gel diet is also accepted by the guinea pig.
Hamster	In captivity hamsters are offered alfalfa meal, ground wheat, ground oats, ground maize, dried skim milk, fish meal, dried yeast, meat meal, bone meal, salt, vitamins and minerals. The following purified diet may be given to hamsters: wheat starch (635 g/kg), casein (200 g/kg), corn oil (50 g/kg), cellulose (50 g/kg), mineral mix (35 g/kg), vitamin mix (10 g/kg), $CaHPo_4$ (15 g/kg), DL-Methionine (3 g/kg), choline bitartrate (2 g/kg).	Hamsters may be given starch gel diets. Average food intake (g/day) by a captive hamster is as follows: 6-12 (growing), 10 -12 (adult), 12-15 (pregnant) and 20-25 (lactating).An adult hamster consumes 5 to 15 ml water per day.
Chinchilla	They can be well maintained on commercially pelleted diets. Ground grains, grain by products and alfalfa fiber, minerals, vitamins and antibiotics should be supplemented.	Chinchillas are commercially reared for pelts. Goat milk or evaporated cow milk may be given to young chinchilla. Milk should be diluted with water.

| Rat and mice | Both rats and mice can be maintained on the following dietary ingredients: ground wheat, soybean meal, ground maize, ground oats, dried skim milk, dried brewer's yeast, fish meal, dry molasses, soybean oil, alfalfa meal, maize oil, rice starch, egg albumen, salt free butter, minerals and vitamins. The following ration can be prepared for the captive rats and mice: ground wheat-230g/kg ,wheat middling-100 g/kg, dried skim milk-50 g/kg, fish meal (60% protein)-100 g/kg, soybean meal (48% protein)-120 g/kg, dehydrated alfalfa meal (17% protein)-40 g/kg, corn gluten meal (60% protein)-30 g/kg , ground yellow maize-245 g/kg, brewer's dried yeast-20 g/kg, dry molasses-15 g/kg, soybean oil-25 g/kg, salt-5 g/kg, Dicalcium phosphate-12.5 g/kg, ground limestone-5 g/kg, mineral premix-1.2 g/kg and vitamin premix-1.3 g/kg. | Average feed intake of a growing rat is 8-25 g/day. An adult rat consumes 25-30 g feed per day for maintenance. The feed requirements, however, are more during pregnancy (25-35 g/rat/day) and lactation periods (35-45 g/rat/day). A 10-15 per cent dietary protein is required for maximum growth. Adult rat consumes about 30 ml water per day. A growing mouse consumes 3-5 g feed daily. An adult requires 5-7 g feed per day. A lactating mouse, however, consumes 7-15 g feed daily. The diet should be free from microorganisms. The water requirements for growing and adult mouse are 3-10 ml/day and 5-10 ml/day respectively. |

Source: NRC: Nutrient Requirements of Laboratory Animals, 1995, USA.

Carnivores:

Carnivores are primarily meat eating animals. In the wild state, these animals subsist on a variety of preys. However, plant materials are also consumed by some carnivores with their regular animal diets. Captive animals thrive well on commercially prepared diets. Most prepared commercial diets are based on meat and meat-by products of various animals, fish meal, soybean, ground maize, beet pulp, minerals and vitamins. The requirements of food in captive animals depend on the physiological status of the animal such as growth stage, pregnancy period and lactation stage. Therefore, it is always important to provide feeds according to the individual requirements. This is essential particularly for captive animals. Based on the feeding habits studied both in wild and captive states, the following text highlights the feeds and feeding practices of various carnivores.

Felids:

Exotic felids are highly adapted to a carnivorous diet. They are intermittent feeders in the wild state. Muscle meat, kidneys, livers, hearts, bone meals, good quality fish, fish meals, chicken's eggs and chicken by-products are the principal dietary ingredients of carnivores.

Felids require high amounts of protein and fat in the diet. They need about 3 g protein per kg body weight per day as maintenance ration. Adult felids require a minimum of 21-22 per cent protein in their diet on dry matter basis. However, kittens are required a minimum of 33 per cent protein for growth purpose. Many essential amino acids are required in felids as they do not synthesize in sufficient quantities by conversion of other amino acids. They require relatively high levels of dietary methionine. The members usually do not eat voluntarily a protein-free diet. Moreover, diet in low protein content may not be acceptable by these animals.

Felids utilize animal fat very efficiently. The dietary supplement of cod liver oil containing poly unsaturated fatty acids may be toxic to the felids. This is especially important when diet lacks vitamin E. Deficiencies of fats of animal origins may result in poor coat condition and reproductive failures.

Felids do utilize soluble carbohydrates obtained from other sources. However, these animals derive most of their blood glucose from amino acid metabolism. They poorly utilize cellulose and unboiled starch. Sucrose is well tolerated by these animals. Felid lacks intestinal enzyme lactase resulting in less utilization of lactose. A diet may contain 3 to 4 per cent crude fibre

An adult small felid in captivity needs food about 4-8 per cent of its body weight daily, whereas young and growing animals may require more. The diet is provided once a day for captive animals. Young, however, should be given twice a day. Many zoos however, do not provide food to the animals one day in week. The frozen meat should be thawed before offering to animals. Plenty of water should be supplied when animals are given dry foods. Felids require vitamins A, D and E. They are unable to synthesize vitamin A from the plant derived precursor, beta-carotene. Therefore, animal tissue is the important source of vitamin A for felids. It occurs mostly in the viscera of prey, particularly the liver. Muscle meat contains very low calcium. So, fields should be given 100 mg of calcium per kg body weight per day. The feeds and feeding practices of some felids both in wild and captive conditions are presented in Table 15.4.

Young felids can be hand reared by providing powdered commercial milk replacer. It should be mixed with boiled water in the ratio of 1:1 or 1:2 in the first week to 9th week. The amount required for young felids of large cats is 60 to 90 ml in each feeding. However, the young felids of small cat require 30 to 60 ml in each feeding. Young may be fed five times at two and half hour interval. Initially young felids should be offered 5% glucose or sterile distilled water during the first 12 hour after birth.

Hyenas:

In the wild, hyena feeds on carrion. However, they subsist on a wide variety of foods such as small mammals, insects, reptiles, eggs, berries, crabs and fruits. The important feeding behaviour of hyena is that these animals regurgitate indigestible portion of consumed foods such as hairs, hooves, ligaments and horns

Table 15.4: Feeds and feeding practices of felids both in wild and captive conditions.

Animal	Wild diet	Captive diet	Remarks
Cheetah	In the wild, it preys on gazelle, wildebeest, impalas, and other hoofed animals weighing up to about 40 kg. Antelopes and guinea fowl are the main preys in their natural state. Hare, rabbit, newly born warthog, zebra and ostrich are also killed and eaten upon by cheetah.	The dietary ingredients for captive animals include fresh meat (e.g. horse, buffalo and cattle), liver, kidney, bones and processed canned food.	Cheetah usually hunts singly. The prey is suffocated by biting the underside of the neck. After killing the prey, cheetah usually prefers to eat the kidneys and heart first, then head, and finally rest of the carcass. Cheetah consumes about 2.8 kg or more meat daily. Drinking is said to occur once in every 4 to 10 days in the wild state.
Tiger	Cattle, wild pigs, deer, nilgai, porcupines, monkeys, bears, wild buffaloes, four-horned antelopes, leopards, goats, rabbits, hare, elephant calves, rhinoceros calves, gaur, fish and sea turtles are the preys of tigers. They usually prey cheetals, sambar, swamp deer, wild pigs, porcupines and nilgai. It has been estimated that on an average a tiger requires 3000-3200 kg of live prey or more annually in its natural state. So, a tiger, for instance, has to kill 40-50 adult deer in a year to meet the requirement. A female tiger with three juvenile cubs may, however, require 60 to 70 adult deer in the wild.	Dietary ingredients such as raw meat, kidneys and cod liver oil are provided to captive animals. A semi-adult and an adult tigers may be given 5 kg and 12 kg beef or buffalo meat per day respectively. The other dietary ingredients given to captive tiger include egg (1 no/day), orange (1 no/day), chicken (1 kg/week), liver (250g/day) and vitamin-mineral mixture (10 g/day). Some zoos supply bones with muscles for the purpose of exercising jaw muscles. Enough supply of water should be ensured. Deep frozen meat should be thawed before feeding.	In the wild, hunting is mostly done at night. When a tiger is disturbed by a prey, it usually leaves the same without protest. Hunting is usually done alone, although females are occasionally seen with cubs to kill langurs and other monkeys. Kills are dragged into a dense cover. They usually start feeding on hind quarters and then move towards the heart. It covers the remaining portion of the kill if it is not fed at one meal. On an average, it feeds 9-18 kg/day. However, it is not unusual for a tiger to eat 20-30 kg of meat in a single night. It may hunt once in every 3-5 days. They also feed on hair and skin. During scarcity of food, tigers even down to the level of insects. Freshly killed or whole animals with opened abdominal cavity can provide natural feeding stimulus.

Lion	In the wild state, lion feeds on cattle, zebra, impala, nilgai, sambar, wild boar, camel, goat, warthog, buffalo, wilde-beest and waterbuck. Lions occasionally con-sume hare, rodents, giraffes, young hippo-potamuses, baboons, reptiles, ostriches, guinea fowl and other small birds.	Horse meat, beef, buffen (buffalo meat), liver and kidneys form the basic dietary ingredients of captive lions. A semi adult and an adult lions may be given 5 kg and 10 kg beef or buffen per day respectively. The other food items for captive lions are egg (1 no/day), orange (1 no/day), chicken (1 kg/week), liver (250 g/week), and vitamin-mineral mixture (10 g/day).	Lions hunt mainly at dawn and at dusk. Hunting is normally done in packs. Prey is suffocated by a bite to the throat. They may drag their kills to cover. When a kill has been made, they usually first eat intestine, then hind quarters and progress towards the heart. However, their food habits are influenced by many factors such as size of prey, availability of prey and density.
Leopard	Leopard mainly feed on small mammals and birds. Langurs, deer, antelopes, dogs, porcupines, goats, sheep, mouse deer, hare, pigs, cattle, buffaloes, lizards, snakes and fish are the natural preys.	Dietary ingredients provided to the captive animals are similar to the diet given to a lion. A leopard may be given 4 kg buffalo meat per day.	In the wild state, leopard hunts singly usually at night. Leopards often kill the domestic dogs. Cheetal is the favourite prey of Indian leopard. African leopards also kill baboons. When a kill is made, leopard eats heart, liver, lungs, and flesh on the ribs first. Leopards often drag their kills up trees for feeding and storing purposes. Cubs have been known to be killed and eaten by other leopards living in the same den.
Snow leopard	In the natural state, it feeds on rodent, hare, ibex, markhor, wild sheep, musk deer, wild pig, chinkara and many birds.	Captive animals are provided raw horse meat and other meat items. Live fowl, rabbits or pigeons may be given two or three times weekly. 1-2 kg meat may be required daily for a captive animal.	In the wild state, snow leopards also move to different altitudes along with migrating preys such as wild sheep and goats. There may be one fast day per week in captivity.

| Small cat | Small cat prey on rodents, birds, fish, small mammals, small deer, lizards, insects, frogs and snakes. | The captive diet consists of chickens, pigeons, rabbits, rats, mice and sparrows. Raw meat fortified with fine bone meal and cod liver oil may also be provided. 1 kg buffalo meat may be given to leopard cat per day. | Small cats feed in a crouched position. |

in the form of pellets. This probably explains why hyenas do not practise in feeding regurgitated food for their young as seen in many other members of the order Carnivora. Group feeding is noticed among hyenas in the natural state. The feeds and feeding of striped hyaena is given below.

Striped hyena: They feed carrion and remains of kills of other predators e.g. tiger, lion, wolf and leopard. It also eats a wide variety of vertebrates, invertebrates, vegetables and fruits in the wild. The dietary ingredients provided to the captive animals include whole meat with bone, chopped raw meat, dog meal, bone meal, liver, eggs, milk and cod liver oil. The usual diet is occasionally supplemented with whole rabbits, rats, mice and fowl. This is practised for stimulating their normal appetite. Nevertheless, the striped hyena can be well maintained on commercially prepared carnivorous diets. An adult hyena may be given 3 kg buffalo meat daily.

Viverrids:

In general, the viverrids are omnivorous feeders. Their diet consists of small mammals, reptiles, fish, birds, bird's eggs, insects, crustaceans, roots, bulbs, nuts, fruits and other plant materials. The dietary ingredients that are offered to captive animals include minced meat, bone meat, boiled eggs, whole mice and chickens, frogs, lizards, live sparrows, snails, boiled rice, carrots, apples, potatoes, bananas, oranges, bread and biscuits. Most members can be well maintained on canned or frozen feline diets. Captive diets should be supplemented with vitamins and minerals. The feeds and feeding of mongoose is described below:

Mongoose: In the wild, mongooses are opportunists in feeding habits. They feed on small mammals, birds, eggs, reptiles, insects and occasionally fruits. The common Indian mongoose occasionally feeds on carrion. The rudy mongoose, on the other hand, subsists mainly on large snails. Fish is the common food item for the stripe-necked mongoose. Some mongooses (e.g. crab eating mongoose) largely consume frogs, fish and crustaceans.

Captive mongooses are provided a dietary mixture of raw meat, dog meal, bone meal and cod liver oil with small pieces of whole raw meat. Milk and eggs are also offered to captive animals along with minerals. The above diet is occasionally supplemented with fruit.

Bears:

Bears have diverse feeding habits in their natural state. They tend to be largely herbivorous in feeding habits. Polar bears, however, almost entirely feed on carnivorous diets such as fish and seals. Bears are usually fed at about three per cent of their body weight on dry matter basis. The amount of feed is increased prior to denning or during lactations. Most bears do well on frozen canine diets and fish. Green vegetables and fruits are also regularly supplied for captive animals. It has been observed that bears eat maximum amount of feed during summer months as compared to winter months. Bear tends to feed all the time.

The feeding regimen should be diverse as much as possible. When captive bears are fed ground muscle meat, then calcium carbonate should be supplemented @ 400 mg/100 g of meat in addition to multivitamin-trace mineral supplements. The milk substitute for young bear consists of low fat and low protein synthetic milk powdered and water. The natural and captive diet of Himalayan black bear and sloth bear is given below:

Himalayan black bear: They eat a variety of plant and animal matters. This includes nuts, fruits, ants, bark, leaves and insect larvae. The Himalayan bear tends to be omnivorous in feeding habits. These animals also attack domestic livestock or agricultural crops. It can be well maintained on commercially prepared omnivorous diet or dog food-based diets. The captive diet includes fish, bread, apples, carrots, dog biscuits and eggs. It is the most carnivorous in nature among Indian bears. The followings are the food items for captive bears: bread (250 g/day), ripe banana (2 kg/day), egg (1 no/day), vegetables (2 kg/day), rice (500 g/day) and milk (500 ml/day).

Sloth bear: The natural food of sloth bear consists of fruits, insects, termites, honey, sugarcane, grubs, eggs, carrion, flowers, ants, beetles, tubers, roots, and berries. The jamun, amaltas and ber are the favourite fruits in the forests. In the wild, sloth bear breaks the termite nests with powerful claws and draws out the termites by sucking action of the lips and protrusion of the long tongue. They are fond of sugarcane. It also likes mohua flower *(Maduca latifolia)*.

Sloth bear can be maintained on captive dietary ingredients that include raw horse meat, milk, honey, bread, banana, apple, boiled rice and a mixture of ground raw meat, dog meal, fine bone meal and cod-liver oil. Fresh green should be supplemented.

Mustelids:

Mustelids feed on a variety of plant and animal matters. Some members exclusively eat meat. Weasels, for example, feed rabbits, small rodents, birds, frogs, lizards, insects and earthworms. The feed and feeding of otter is given below:

Otter: The Oriental small clawed otter feeds on crabs, fish, snails and mussels. Most species, however, consume fish, crabs, crayfish and frogs. In general, their dietary ingredients consist of snails, crabs, fish, mussels, turtles, frogs, sea urchins, rabbits, insects, crayfish, lizards, birds, and aquatic invertebrates. The captive diet consists of chopped horse meat, chopped beef meat, bone meal, bran, ground dry cat food, carrots, tomatoes, rolled oats, orange juice and brewer's yeast supplemented with minerals and vitamins. Yogurt may be given to the captive otters.

Procyonids:

Procyonids are known to be omnivorous in feeding habits. Their natural diet is composed of a wide variety of plant and animal matters. Pandas, however, subsist mainly on plant materials. All the members can be maintained on canids diets. The information on feeds and feeding of pandas are presented in Table 15.5.

Canids:

Canids in their natural state prey deer, hare, wild pigs, porcupines, rabbits, beavers, mice, birds, crabs, fish and carrion. The captive animals do well on a mixture diet based on minced raw meat, dog meal, powdered bone meal and cod liver oil with supplementation of milk, vitamin D and calcium. The feeds and feeding of dhole and jackal are given below.

Dhole: The dhole mainly hunts cheetal, Nilgiri tahr, sambar, swamp deer, four-horned antelope, nilgai and wild pigs. They often attack on leopards, tigers and gaurs. Their diets in the wild also consist of insects, lizards, rodents and wild berries. Hunting is done by day, often in the morning. They usually hunt in packs.

Dholes are provided raw meat, a mixture of dog meal, bone meal, cod liver oil, apples, grapes, mice, sparrows and chickens in captivity. Like other carnivores, wild dogs are given a supplement of protein-mineral-vitamin concentrate to balance the nutrient requirements of the body. The composition for supplement is as follows: skimmed milk powder (42%), dried brewer' yeast (42%), steamed bone meat (8%), calcium carbonate (3%), salt (2%) and vitamins (3%). The diets of growing and adult dogs should contain 20-22 per cent protein. Young may be provided 50-60 g DM/kg body weight up to 6 months of age and 38-40 g DM/kg body weight from 6 months to one year of age.

Jackal: The jackals eat small animals (e.g. rabbits, mice and lizards), insects, plant materials and carrion in natural state. They are opportunistic feeders. Rodents and fruits are the main dietary ingredients for the silver-backed jackals. The captive diet consists of raw meat, dog biscuits, rats, chicken heads, boiled vegetables and fruit.

Table 15.5: The wild and captive diets of red and giant pandas:

Animal	Wild diet	Captive diet	Remarks
Red panda	In the wild, they mainly feed on bamboo leaves and fruits. Red pandas also eat roots, tubers, acorns, lichens, insects, birds and eggs.	The red or lesser panda does well on a gruel diet consisting of mixed baby cereal, evaporated milk, egg yolk, apple juice, grape, beet pulp, banana, honey, carrots, figs, dates, vitamins and minerals. Bamboo shoots and apples are also given. The following ingredients may be given to captive animal: bamboo leaves (500 g), vegetables and fruits (100 g), milk (500 ml), oat powder (25 g), wheat flour (250 g) and horlicks (25 g).	They are readily adaptable to zoos. Red pandas spend 10 to 12 hours for searching foods in the wild. In captivity, young are fed eight times during first 15 days of birth. They should be supplied fresh and good quality dietary ingredients. Adults are fed twice a day. The bamboo leaves comprise the major dietary ingredients.
Giant panda	In nature, their food mainly consists of stems and leaves of various bamboo species. They also feed on bulbs, grasses and occasionally rodents, small birds and insects. Giant pandas occasionally also forage on leaves, stems and bark of a variety of plants.	In captivity, they may be given a variety of dietary ingredients. These include bamboo leaves, green maize stalks, green soybean plants, spinach, raw carrots, apples, lettuce, beet tops, baked potato, sliced bread, milk, eggs, honey, orange juice and fish oil. An adult animal may be given the following diet: boiled rice (1.6 kg), canned feline diet (210 g), soybean oil (15 ml), honey (30 ml), powdered cottage cheese (100 g), water (1 litre), mineral mixture (50 g), and vitamin mixture (20 g). These ingredients are mixed into a gruel and offered along with carrots (430 g), apples (340 g), and sufficient amount of bamboo leaves and shoots. Animals may be fed twice daily.	A giant panda eats on an average between 25 and 30 cm of a stem in the natural state. An adult panda is estimated to eat 10-18 kg leaves and stem per day in the wild. The minimum protein requirement is 61.6g/day. In the wild, they are very choosy about their foods.

Elephants:

In the wild, elephants consume grasses, herbs, roots, bark and fruits. Their natural diets have low nutrients and high fibre contents. The amount of feed consumed by an elephant is equivalent to about four per cent of its body weight. The female elephant with suckling calf feeds about 6 per cent of her body weight. An elephant may spend 10-20 hours a day for feeding in the wild. African elephants enjoy the fruits of some palm trees, while their Asian counterparts are very fond of wild rice and the wood apple *(Feronia elephantorum)*. Elephants rip the bark off the trees and can even uproot them. This feeding behaviour is not uncommon in the wild state. A mature elephant may eat up to 300 kg of green fodder in its natural environment. However, an Indian elephant, as reported, may consume between 135 and 200 kg fodder in the wild. The requirements of large quantities of food for elephants are basically due to their poor utilisation of foods and large body size. One report indicates that 60-65 per cent of food intake leaves the body undigested. The poor digestibility of dry matter is due to consumption of large amounts of fibre diet as well as gravel and earth. The growing and lactating animals are required 13-14 per cent protein. However, captive animals may be provided 8-10 per cent protein for maintenance. The energy requirement for basic metabolic process is stated to be 12-13 Kcal per kg body weight per day. Pregnant and lactating animals need more energy over and above the basal metabolic requirements.

In the wild state, the elephant calf survives on mother's milk for first three months. During this period, it remains closely with the mother. They nibble at grasses between three and six months. An adult elephant may consume 140 to 230 litres of water in a single day. The water is sucked into the trunk and squirted down the throat. An adult African elephant drops about 100-150 kg dung per day.

In captivity, the basic diet of an elephant is hay. However, they are also given carrots, oats, leaves and twigs of many plants (e.g. fig, jack fruit, banana and banyan trees), lettuce, uncooked rice, bran and sugarcane. The captive elephants relish paddy plants in the wild state. Pelleted feeds mixed with molasses are provided to the captive animals in order to reduce the cost of feeding as well as to increase the appetite. The following diet is recommended for an elephant: green fodder/sugarcane-200 kg, tree leaves – 60 kg, dry fodder-15 kg, banana-24 (nos) and a boiled mixture of various ingredients (molasses -1 kg, pulse- 1 kg, rice- 2 kg, bajra- 1 kg, haldi- 100 g, salt- 250 g and mustard oil- 250 ml). However, the food items such as banana tree leaves (150 kg/day), maize plants (25 kg/day), banana (10 kg/day), banyan tree leaves (30 kg/day), vegetables (15 kg/day), gram (5 kg/day) and sugarcane (5 kg/day) are also recommended for a captive elephant. In south India, coconut is given in the diet of a pregnant mother.

Perissodactyls:

The feeds and feeding of perissodactyls are given below:

Rhinoceros: They are completely herbivorous animals. The black rhinos are primarily browsers, while the Indian and the white rhinos are grazers. The Sumatran and the Javan rhinos are solely browsers. In the wild state, the African rhinos mainly feed on woody plants, herbs, shrubs and grasses. Their Asian counterparts consume grasses and leafy vegetation in the wild. During scarcity, they also eat water hyacinth and tend to be semi-browsers. Rhinos consume submerged grasses during flood. On an average, an Indian rhino requires one and a half quintal fodder per day. Leaves of jack fruit and fig are also eaten by these animals. In the wild, the Indian rhinos graze early in the morning, late afternoon, and throughout the night.

The captive diet forms good quality berseem or lucerne hay and commercially prepared pelleted horse feed. The pelleted feed contains barley, oats, groundnut cake etc. The green fodder, carrot, potato, soaked gram, bran and fruits are also supplemented along with pelleted feed. Vegetables and fruits may be given one to two kg daily. An individual may drink as much as 200 litres of water daily. Black rhinos drink water primarily at night and spend most of the night near water bodies. However, all rhinos drink water almost daily at small pools or rivers when readily available. Excessive intake of protein causes abnormal growth of horns and hooves. Adequate amount of salt should be provided for captive animals.

Young can be hand reared with milk substitute. They may be given cow's milk, corn syrup and vitamins. A hand reared rhino calf is given the following ingredients: low fat cow's milk (50%), cow's milk without fat (50%), soluble vitamin powder (1ml/litre) and corn syrup (30 ml/litre). The young may be given 1.2 litres at single feeding for first week and the frequency of feeding is about seven times a day. The quantity may be increased up to 3 litres from day 50 to 180 days.

Zebra: In the wild, zebras consume tall and coarse grasses normally left by most antelopes. Their captive diet consists of crushed oats, maize, bran, carrots, apples, potatoes, fresh greens, vitamins and minerals. Grass hay should be given freely. Iodinated salt blocks should also be provided for captive animals.

Horse: They are maintained on a commercially prepared pellet diet. The amounts of diet to be fed to a captive horse, however, depend on many factors such as age, body weight, quality of diet and production status. A race horse requires more energy for maintenance. Gram, wheat, barley, hay, oats, maize, soybean meal, molasses etc. form the dietary ingredients. An adult animal requires 1 to 1.5 kg pelleted feed per 100 kg body weight. A good amount of hay should always be given to an adult animal. Vegetables and fruits may be supplied @ 0.5 to 1 kg daily. Captive Przewalski's horse may be given 500 g to one kg crushed oats and 1-1.5 kg pellets. Fresh lucerne crop and hay are also given.

Wild ass: The diet of wild ass consists of leaves, twigs, bark and grasses. The amount of captive diet required for an ass is reported to be similar to a horse.

Artiodactyls:

The artiodactyls feed on a variety of plant materials. Many members in this group are able to convert a wide variety of coarse roughage into usable nutrients. The nutrient requirements of wild members are reported to be almost similar to their domestic counterparts. In the wild, the majority of artiodactyls feed in the early morning and in the late afternoon. The wild ruminants spend about eight hours or more daily for feeding purpose. Wild animals normally tend to spread out during foraging and establish a regular pattern of spacing called individual distance. They sleep or loaf during mid day and late evening. Members of the order Artiodactyla prefer loafing under shade during peak heat in summer months.

Captive artiodactyls can be maintained on a variety of dietary ingredients. These include hay, pelleted diet, aquatic grasses, vegetables and fruits. The good quality grass hay is provided to many members in captivity. The commercially prepared pelleted diet containing proteins, minerals and vitamins adequately satisfy the needs of the captive animals. Another fact regarding the captive feeding of artiodactyls is that it is not always possible to provide pasture for captive ruminants. Therefore, vegetables or succulent grasses should be provided at the rate of 10 per cent of the total diet.

It has been reported that 60 to 70 per cent mortality among captive wild ruminants is due to poor nutritional management. Therefore, supplemental feeding is important to improve overall health and performance. The feeding of high level concentrate without roughage is uneconomical and it has also an adverse effect on proper rumen function and reproductive performance. Vitamin E is very important. The feeds and feeding of selected artiodactyls are described here.

Wild pig: The wild pigs are omnivorous in feeding habits. The wild diet consists of corns, grasses, leaves, buds, tubers, roots, ferns, fungi, small vertebrates (e.g. lizards, mice and frogs), insect larvae, earthworms, eggs, nuts and fruits. The warthogs are specialised herbivorous animals feeding on leaves, roots, bulbs and fruits of a wide variety of plants. The babirusa mainly thrives on grasses and fruits. The natural diet of bush pig consists of root, tuber, maize, groundnut, pea, fruit, egg, reptile, bird, bird's egg, insect and carrion.

Wild pigs normally forage in family groups. The pygmy hogs are omnivorous in feeding habits and thrive on leaves, grasses, insects, earthworms, eggs and carrion. During scarcity of food materials in summer months, wild pigs move to the salt marshes where they feed on grasses, tubers, roots and invertebrates.

Wild pigs do well in captivity on commercial diets along with greens, carrots, potatoes and apples mixed with bread, ground meat and hay. The following dietary ingredients may be given to the growing pigs maintaining in the zoo: ground shelled maize (68.79%), ground oats (10%), meat scraps (3%), dehydrated alfa meal (2.5%), soybean meal (14%), dicalcium phosphate (1%), and salt, vitamins and minerals (0.71%).

Hippopotamus: Hippopotamuses are nonruminants and thrive almost exclusively on vegetation. In the wild state, they forage on land at night. Aquatic plants and grasses form the principal dietary items. These animals are reported to spend about five to six hours for feeding purpose at night. Hippopotamus in its natural state may feed as much as 65 kg grasses in single night.

In captivity, the amount of diet required for hippopotamus depends on many factors such as age, sex, physiological status and individual preference. On an average an animal consumes one per cent of its body weight daily. A ration containing hay (40-50 kg), pelleted grain mixture (5-6 kg), bread (2-3 loaves) and vegetables (i.e. carrots, cabbage , potatoes, apples and onions) is said to be sufficient enough to maintain the requirements of an adult hippopotamus in captivity. The pygmy hippopotamus, on the other hand, may be given 7-8 kg of good quality hay and 2-3 kg of mixture containing pellets, bread and vegetables .Vitamins and minerals should be supplemented.

Mouse deer: Mouse deer or chevrotains are true ruminants. They are selective browsers and feed primarily on fallen fruits with some leaves in the wild. They feed at early in the morning and late in the evening. The captive diet of chevrotain consists of hay, pellets, rolled oats, gram, carrots, uncooked sweet potatoes, succulent grasses, bananas and apples. The hay should be of high quality. The mature hay should be properly chopped off for better utilisation. The concentrate mixture provided for captive animal should contain 18-25 per cent protein. In captivity, feed should be given early in the morning and late in the evening. The captive diet should be supplemented with vitamins and minerals mixture.

Giraffe: Giraffes are predominantly browsers. They browse on tall trees and crowns of small trees. These animals subsist on leaves and shoots of trees, shrubs, creepers and vines in their natural state. Giraffes also eat flowers, buds, twigs, bark, dicotyledonous herbs, seedpods and fruits. These animals occasionally eat grasses. In the wild state, their feeding preferences on particular plant species depend on the season , and nutritional content and digestibility of the herbaceous forage. For example, evergreen plants are chosen during dry season, while they select deciduous plants during wet season. However, the preferred forages eaten by giraffe are young shoots of *Acacia* spp., *Grewia* spp. etc. They are also very fond of beir (*Zizyphus jujuba*). Giraffe often eats salt or salty soil in the wild. The feeding occurs chiefly after dawn and dusk, though giraffe may feed during bright nights. They spend about 12 hours in browsing. The bull consumes about 66 kg of plant materials on fresh weight basis while the cow feeds about 58 kg daily. Giraffe ruminates the cud during hot mid day period. In the wild, they can survive for long periods without water like camel. However, giraffe drink readily if water is available. These animals drink by bending their front knees forward or by straddling their forelegs to either side.

Giraffes can be well maintained on artificial diets in captivity. They may be given about 10 kg of high quality alfa hay, 5 kg of commercial herbivore pellets and 2.5 kg of rolled oats mixed with carrots, potatoes, cabbage, apples and bread. Bananas are occasionally provided to the captive animals. Alfalfa hay may be replaced by good quality clover hay. The hay is provided in a hanging feeding rack which is placed about seven feet above the floor. Grains are placed in a removable manger which hangs on a wall of the same height from the floor as recommended for hayrack. Fresh and leafy tree branches should be provided during summer months. The captive diet given for giraffes normally contains 15-18 per cent protein, 2.5 percent fat and 7.0 per cent crude fibre. They drink frequently in captivity. A captive giraffe may drink 45 litres water in a day during hot weather. However, in cool weather condition it may drink about 12 litres per day.

The young giraffe can be hand raised. The colostrum is given for two days after birth. Young animals are needed calcium and vitamin D for proper bone growth. Young may nibble on leaves at about three weeks of age in the wild as well as in captivity.

Camelids: Camelids have a capacious stomach. The fundamental process of ruminant digestion including microbial fermentation, regurgitation and eructation takes place among all the members of this group. In their wild environment, they graze and browse on a wide range of herbage species. Camels prefer browsing on tree fodder and may continue to browse for 9-10 days without drinking water. They do not close their mouth while consuming the thorny plants. Camels are fond of babul and neem leaves.

Captive animals can be maintained on a similar concentrate mixture provided for other ruminants such as cattle, goat and sheep. Hay is readily accepted by these animals. A daily intake of 2 kg dry matter per 100 kg body weight is reported to be enough for guanaco, llama, alpaca and vicuna. Lactating, pregnant and growing animals, however, require more. Green grasses and tree leaves should be supplemented. Camelids are unable to lick. So, the supplement of minerals in the diet is important.

Deer: Deer are herbivorous animals. Some are known to be grazers (e.g. cheetal, and fallow deer), while others are browsers (e.g. muntjac and moose deer). However, the classification of deer based on the feeding habits is admittedly a very superficial one. This is due to the fact that their preferences of feeds in the wild may be determined by seasonal changes. Some authors are of opinion that they should be called as intermediate feeders. In their natural environment, deer feed on leaves, shoots, bark, flowers, grasses, twigs and fruits of trees, herbs and shrubs. They spend about four hours in the morning and an equal time in the evening for feeding purpose. Deer take shelter during mid-day and late evening.

Deer can be well maintained on a diet of good quality hay with supplementation of commercially prepared pelleted feeds. The common dietary ingredients provided for captive animals include ground oats, wheat bran, corn meal, grass, linseed-oil meal, soybean oil, beet pulp, wheat-germ meal, brewer's yeast, chopped alfalfa and molasses. They prefer to consume alfalfa or clover hay. Hay can be provided in mangers or in racks for captive animals. Grains are also placed in feeding boxes. Hayracks should be kept at eye level. The concentrate should be given at the rate of 0.5 to 1 per cent of their body weight. The concentrate ration provided to deer usually contains 13 to 19 per cent protein. Deer may be given 4 to 5 kg green fodder, 400 g to 1 kg and 500 g vegetables and fruits as a thumb rule basis. The important point to be remembered regarding the captive diet of deer is that any changes in the diet should be gradual. This is for the reason of avoiding upset of digestive system and scouring. The sudden supplement of a high concentrate ration may cause ruminal acidosis. This may result in death. The minerals and vitamins should be supplemented for captive deer. The following are the feeds and feeding practices of some deer species.

Cheetal: Cheetal thrives predominantly on grass in the wild state. They develop an unusual association with langurs in Central India. This phenomenon can be observed throughout the year except in monsoon season (July to October). Cheetals consume foliage fallen from the trees by feeding langurs. This association, however, declines rapidly during monsoon when foods are abundantly found. The captive diet of cheetals includes vegetables, hay and pellets.

Muntjac: The Indian muntjac preferably consumes leaf, herb, mushroom, bark and fruit in the wild. Muntjacs are generally browsers and are the most selective feeders in the wild. However, they readily consume grasses in absence of bark and leaves. Captive animals can be maintained on a diet consisting of hay, pellets and vegetables.

Swamp deer: In the wild, the preferred foods of swamp deer consist of grasses and herbs. In captivity, these animals can be maintained on a diet consisting of fifty per cent concentrate mixture (e.g. oat and carrot) and fifty per cent commercially prepared pellets with supplementation of fresh grasses and hay. The fresh grasses are given in summer months and hay is supplied during winter season.

Neonates are raised on commercial milk replacer. Young may be given undiluted cow's milk @ 50-100 ml per feed for 4-5 times per day for about two weeks in addition to water, hay etc. The amount may be raised up to 750 ml after two weeks. Concentrate may be started after three weeks.

Antelopes: Their preferences for various foods can be influenced by many factors such as area, season, age of the animal, stocking density and presence of other competitors. Antelopes can be classified into grazers, browsers and mixed feeders. Water buck, wildebeest, reedbuck, roan antelope, sable etc. are known to be grazers, whereas eland, kudu, bushbuck etc. are called browsers. Impala and springbok are mixed feeders. Nevertheless, antelopes generally feed on a wide variety of plant materials. Some antelopes consume bark, roots, seed pods, flowers, tubers, sap and fruit.

In captivity, antelopes do well on a diet containing good quality hay and concentrate. Alfalfa hay is good for captive antelopes. A concentrate ration containing wheat bran, yellow maize meal and sunflower cake meal may be given to antelopes. A supplement of vitamins and minerals is important. The feeds of selected antelopes are given below:

Nilgai: Nilgai or blue bull mainly eats grasses. They also consume a considerable amount of leaves (e.g. *Acacia* spp and other trees), cereal crops and wild fruits. In captivity, nilgai may be given green fodder (6 kg), gram (1 kg) and vegetables and fruits (500 g) daily.

Blackbuck: Blackbucks are predominantly grazers. They occasionally consume leaves. The diet of four-horned antelopes consists of buds, grass, etc.

Chinkara: They feed on grasses, crops, leaves, buds and fruits.

Wild sheep and goat: Wild goats graze on very short grasses and they browse on foliage which are not generally accessible to sheep. However, goat is reported to be more browser and less grazer than sheep. These animals move to the lower slopes during winter months. Captive animal can be maintained on hay and concentrate ration. Fresh vegetables (e.g. carrots and potatoes) and greens (e.g. lettuce and cabbage) should be given.

The mountain goats browse on leaves of various plants. However, they also eat mountain grasses, lichens and mosses in the wild. The captive animals are provided good quality hay and rolled oats supplemented with carrots, potatoes and apples.

Goat antelopes: Foods consumed by takins in their natural environment consist of grasses and foliage. In captivity, they do well on a diet containing lucerne hay, pellets, rolled oats, cabbage, sweet potatoes, carrots, bananas, apples and fresh foliage.

The Nilgiri tahr can be maintained on a diet containing commercially prepared calf chow and alfalfa hay with supplementation of vitamins. Grass hay and mineral blocks should be available *ad libitum*.

Wild cattle: The wild cattle feed on grasses, leaves and other plant matters. The captive diets of wild cattle are almost similar with the diets consumed by domestic cattle. For this, readers can consult any standard textbook on animal husbandry.

Yak: The yak gets its nutritional supply through grazing on alpine pastures and other herbages. They prefer grazing in morning and in evening. They have great craving for salts. Yak spends more than 7-8 hours for feeding on the pasture.

Mithun: They are mainly browsers. Like yak, mithun has a strong craving for salt. They are reported to be selective feeders and often travel long distances in the deep forest in order to search of palatable feeds.

Wild buffalo: Wild buffaloes rarely browse. They graze in early morning and late in evening. Wild buffaloes feed on large varieties of grasses along the bank of rivers. The prescribed dietary ingredients given to domestic buffalo may be provided to wild buffalo in captive condition.

Chapter **16**

Feeds and Feeding of Wild and Zoo Reptiles

In the wild, most reptiles feed on carnivorous diets which include fish, rats, mice, small birds, small rabbits, guinea pigs and amphibians. Some are omnivorous reptiles (e.g. desert iguana) while others are herbivorous (e.g. larger iguanas). However, any change in environmental temperature causes the change of feeding behaviour of the animal. Higher temperatures increase digestion and metabolic rates. Other factors such as photoperiod, humidity, type of food and population density also play a vital role on the feeding behaviour of reptiles. Red and yellow coloured foods, for example, are preferred by turtles and some lizards. Some members are known to eat other reptiles. The vivid example is the king cobra that eats other snakes. Some insectivorous reptiles (e.g. many lizards) feed earthworms, mealworms and crickets in the captivity.

Captive reptiles suffer from many nutrient deficiencies. Therefore, balanced diet is very important for animals kept in the zoos. A deficient in proteins, minerals and vitamins may cause various diseases to the reptiles. The diet should contain 11 to 12 per cent protein for reptiles which feed on mostly plant materials and about 18 to 20 per cent protein for carnivorous reptile. The diets containing inadequate protein will cause muscle wasting, loss of body weight, reproductive disorders and increased susceptibility to infections. The protein deficiency is usually found in those reptiles whose diets are based on plant materials. Roughage diet is also given for the normal functioning of the digestive tract. Vitamins and minerals are essential for captive reptiles. Hypovitaminosis commonly occurs in chelonians. The feeds and feeding of chelonians, crocodilians and snakes are described here:-

Chelonians: Chelonians may be carnivorous, herbivorous or omnivorous in feeding habits. Aquatic turtles are chiefly carnivorous reptiles feeding on snails, fish, mice, slugs, worms and insects. Lettuce or other plant material is occasionally given to captive aquatic turtles. The captive diet is also supplemented with liver oil or vitamins containing A, D and E. Aquatic turtles swallow foods with their heads submerged in water. The marine turtles subsist on fish, invertebrates and sea weed. Many adult sea turtles are virtually herbivorous. The box turtle is omnivorous in feeding habit and can be offered worms, snails, lettuce, tomatoes, mouse pups, melons, fruits and cottage cheese. The olive ridley turtle is also omnivorous in feeding habits. They feed on dead fish, crabs and other crustaceans, and soft parts of mollusks. The green turtle is herbivorous and feeds on sea algae (e.g. *Gracillaria* sp and *Gelidiella acerosa*) and sea grass (e.g. *Cymodacea* sp). The hawksbill turtle is largely carnivorous animal. They feed on sponges and other invertebrates, and fish. The leathery turtle primarily feeds on jelly fish.

Indian pond terrapin lives mainly on vegetarian diets. They forage on land at night. The Indian mud turtle is omnivorous in feeding habit and feeds on water plants and small animals such as snails, crustaceans, frogs and fish. The starred tortoise feed on grass, fallen fruits, snails, animal excreta etc.

In captivity, tortoises eat flowers and leaves mixed with succulent grasses, carrots, green peas, fruits, pumpkins, tomatoes and melons. A small amount of bread or dog food may be given. Food materials should be placed on dry land for ground living tortoises. Larger aquatic turtles may be fed thrice or less a week. However, some authors are of the opinion that as most aquatic turtles do not tend towards obesity, food may be provided daily. Young turtles, however, should be fed daily. Young can grow well on a commercially prepared feline diet. Wild turtles often swallow stones, rocks and shells while foraging.

Crocodilians: Crocodilians are carnivorous reptiles. The diets of crocodilians consist of rats, mice, pigeons, chickens, fairly large mammals (e.g. deer, dogs and monkeys), turtles and fish.

The gharials predominantly eat fish. However, they occasionally feed on small mammals, turtles, birds and corpses. Stones are also swallowed by these animals.

The marsh crocodile feeds on fish and other animal that is with in its capacity to kill (e.g. cheetal, sambar, four-horned antelope, blue bull fawn, wild dog, barking deer, monkey, goat and hyena). They also live on soft shell turtles, snakes, wild birds and pigs. Young marsh crocodiles feed on snails, frogs, water insects and fish.

The estuarine crocodiles feed mainly on fish occurring in the estuarine habitat. They also prey other animals. The estuarine crocodile is reported to attack on human beings.

In captivity, crocodiles should be offered intact preys such as whole fish or mice. The viscera along with strips of meat may also be given. Diet should be provided in water. Young are fed *ad libitum* on chopped baby mice and fish with vitamin and mineral supplements every second day. The captive gharial should be fed more often. It is reported that they feed better when the dietary ingredients are tossed to them just lateral to the mouths. Like lizards, the forced feeding is difficult in crocodilians. Some animals may be stimulated by teasing them with a piece of food dangled from a pole. Large crocodiles can be chemically immobilized prior to forced feeding. Normally feeds are provided two times a week. An adult crocodiles may consume 30 kg beef in each feeding.

Snakes: All snakes are carnivorous animals. Their natural diet consists of mammals, birds, amphibians, fish, reptiles and invertebrates. Some snakes, however, feed solely on eggs, snakes, slugs, crayfish and termites. The king cobra, for instance, is almost strictly snake eater. Boas and pythons prefer to feed on warm-blooded preys such as rodents and birds. Earthworms and frogs are the natural food items of garter snakes. Therefore, it is always better to feed snakes after considering their natural feeding habits.

In captivity, reptiles are fed whole animals. The snake always swallows its prey intact. Poisonous snakes kill or paralyze their preys by envenomization before swallowing. Snakes often eat dead food or even strips of meat. Though, the feeding of live preys stimulates their natural feeding behaviour, it is said that the live prey for captive snakes, if possible, should be avoided. This is due to the fact that the live prey may bite and injure the animals. Snakes can be fed once or twice a week. This is because snakes tend to become obese in captivity. The general rule is that the larger reptile is to be fed less frequently. However, some highly active snakes such as cobras, vine snakes and garter snakes should be fed more than once weekly. Pythons should be fed once every four to six weeks. Another important fact regarding the feeding behaviour is that when two or more snakes are kept in the same enclosure, it is always advisable to feed individually by holding the food animal in suitable device (e.g. tongs and spoons).The reason is that when two snakes begin swallowing the same prey from opposite ends, the larger snake will continue to eat until it has swallowed the smaller snake. Therefore, it is very important to avoid cannibalism while group feeding is practised. Some snakes can survive for several months without food. Sometimes, forced feeding is necessary for captive snakes. Some nutritious fluid (e.g. fish broth and meat broth) may be given by stomach tube. The fluid should be fed very quickly in order to avoid regurgitation. The meat broth may be mixed with raw egg, milk, glucose, vitamins and minerals.

Pythons feed on a wide variety of mammals, birds and reptiles in the wild. They include cheetal, sambar, muntjac, chinkara, leopard, langur, jackal, mouse deer, hog deer, hare, rat, porcupine, peafowl, poultry, wild duck, monitor lizard, frog and toad.

In captivity, python accepts any dead or live mammal, bird or reptile. The duration between two successive feedings depends on the size of the prey, time of the year and the condition of the animal. Rats and crows, for example, are digested in about eight days and the digestion of a goat requires about three weeks in summer. It has been reported that the captive pythons often refuse foods for many months.

The common Indian kraits mainly eat snakes including other kraits. However, they occasionally feed on small mammals, lizards and frogs.

Toads, frogs and rats are the main foods of Indian cobras. They also eat lizards, birds and other snakes including cobras. The Indian cobra is famous for stealing eggs from poultry house. It swallows the whole egg and the digestion takes place in about two days. The staple diet of king cobra is snakes. All snakes including other poisonous snakes and monitor lizards are taken by the king cobra. Horse meat and dead rats are acceptable in captivity.

In the natural state, Russell's viper feed primarily on rats and mice. In captivity, they may be provided small birds, rats, mice, kittens, squirrels, shrews, frogs and lizards. The young are reported to be cannibalistic. Many adult Russell's viper snakes may live without food for longer periods.

Chapter 17

Feeds and Feeding of Wild and Zoo Birds

Many wild birds are very popular exhibits in the zoo. They do well on captive diets. The dietary ingredients required for birds depend on the type of birds kept in the captivity. Some birds are omnivorous (e.g. hornbills and kingfishers) while others are carnivorous birds (e.g. herons and storks). Protein, fat, and carbohydrate are the basic organic food components required for birds. Vitamins, minerals and water are also important for captive birds. The diet required for both wild and captive birds are presented in Table 17.1:-

Table 17.1: The wild and captive diets of selected birds.

Bird	Wild diet	Captive diet	Remarks
Ostrich, emu, rhea kiwi and other ratites	In the wild, their diet consists of a variety of fruits, flowers, seeds, leaves, shoots of shrubs, creepers and succulent plants. Lizards, caterpillars, beetles, grass hoppers and mollusks are also eaten by these birds. A large quantity of stones and grit are also taken by ratites in the wild.	All large ratites do well on commercially prepared dog foods, and rabbit and poultry pelleted feeds. A captive ostrich may be offered gram and maize (500 g), wheat flour bread (250 g), fruits and vegetables (500 g) and onion and garlic (500 g). A captive kiwi may be provided the following dietary ingredients: beef heart (200 g), cooked rolled oats (20 g), vegetable oil (2.5 ml), wheat-germ flakes (2 g), water (160 ml) and vitamin-mineral mixture (2 g).	The diet containing 20-24 per cent crude protein and 12-19 per cent crude fibre is suggested for growth and maintenance of ratites. Kiwis are nocturnal feeders. Small amount of banana and apple may be offered to captive cassowaries. Ostriches are omnivorous ratites.

Crane	Cranes are omnivorous birds and thrive on insects, mollusks, mice, amphibians and plant matter.	The captive diet consists of various dietary ingredients such as ground oats, wheat, maize, vegetable oil, brewer's dried yeast, fish, meat and bone meal. Vitamins and minerals should be supplemented. The following dietary ingredients may be offered daily to a captive crane: gram and maize (150 g), wheat flour (50 g), mash (50 g) and green vegetables (150 g).	A starter ration containing 24 per cent protein may be given to young cranes for two months. The maintenance ration containing 15 per cent protein is suitable from two months to one year old birds. Mouldy food should not be given to captive birds.
Flamingo	They subsist on insects, larvae, seeds, crustaceans, marine shrimp and algae. Some flamingos (e.g. lesser flamingo and Andean flamingo) are surface feeders whereas others (e.g. greater flamingo and American flamingo) are bottom feeders.	Flamingos can be given commercially prepared diets. The dietary ingredients offered to captive flamingos include maize, wheat, oat, rice, barley, carrot, lettuce, sprouted grain, bread, soybean meal, boiled egg, fish, dried skim milk, vitamins and mineral mixture. A captive flamingo can be offered 200 g millet and 50 g green vegetable per day. The diet used for the Caribbean flamingo (1-21 days chick) at San Anterio Zoo is as follows: smelt (56.7 g), peeled shrimp (56.7 g), hard boiled egg yolk (56.7 g), gerber high protein baby cereal (28.4 g), vionate (1.1 g) and bone meal (1.1 g). Chicks may be given 4.5 ml on first day and then the amount should be increased. Chicks may be fed five times a day for the first ten days.	The captive diet should contain 4-5 per cent fibre, 20 per cent protein and 3 per cent fat.

Pelican, stork, heron, spoon bill and other aquatic birds	The diet consists of insects, fish, frogs, rodents, young birds, mollusks, tadpoles and plant material. Some storks such as adjutant stork of India and marabou stork of Africa feed on carrion. However, stork primarily feeds on locust, cricket and larvae of various worms. They feed occasionally on mice, frogs, small fish, lizards, mollusks and crabs.	In captivity, they are given chopped meat, insects, earthworms and a variety of fish. Commercially prepared poultry and trout diets may also be given. A captive adjutant or painted stork may be given fish (250 g) and ground meat (250 g) daily. Pelican may be offered 250 g fish per day per bird.	In the wild, herons normally follow wait and watch method of hunting. All are primarily aquatic birds.
Duck, goose, swan and other waterfowl	In their natural state, water fowl eat animal and plant matters. These include grasses, tubers, seeds, leaves, fish, frogs, snails, worms, insects, reptiles and tender shoots of wheat, gram and other crops.	Commercially prepared duck or game bird pellets may be offered to most waterfowl. The ingredients used in various combinations for waterfowl diets are as follows: grass, grain, seed, biscuit, cabbage, apple, fish, seafood, sprouted wheat, shrimp and crushed grit. A pintail or teal may be offered 100 g millet and 15 g green vegetables per day. The following ingredients may be given to a captive goose per day: millet (150 g), wheat flour bread (50 g), and onion and garlic (25 g).	Protein requirement for young (6-8 weeks) duck and geese is estimated at about 17 per cent. An average water fowl may be given grain @ 130 g per bird per day and biscuit @ 10 g per bird per day. Captive birds are usually received two meals a day. The frequency of feeding should be more for plant eater water fowl.

Parrot, macaw, cockatoo and other psittacine birds.	The dietary ingredients taken by psittacine birds in the natural state include nuts, buds, fruits, flowers, seeds, roots, berries, nectar, leaves, larvae and insects. Pygmy parrots feed on slime like fungi. Lories mainly eat nectar.	Large seed eating captive psittacine birds feed on seeds (e.g. sunflower, oat and safflower), vegetables (e.g. carrot, sweet potato and green vegetables), peanuts, monkey biscuits, dry dog food, fruits (e.g. banana and apple) and mealworms. Dried egg yolk, meat, cheese, alfa leaf meal, yeast, wheat germ, vitamins and minerals should be supplemented. The dietary ingredients may be given in the following proportions: seeds and nuts (20%), yellow vegetables (25%), greens (15%), fruit (25%), dry dog food (15%) and adequate minerals and vitamins. Smaller seed-eating psittacine birds may be provided with seed mixtures, bread, fruit and chopped greens. A large Indian parrot may be offered paddy (10 g), grain mixture (20 g), groundnut (10 g), cooked rice (10 g), fruits and vegetables (25 g) and green chilies (10 g) per day.	The daily intake of food is estimated at about 10-15 per cent of the body weight. Water and oyster shell grit should be supplied round the clock. Food and water may be placed in earthenware bowls. Seeds should be soaked in water. These birds relish sprouts. Vitamins and minerals can be mixed with water, fruits and vegetables.

Pigeon and dove	The wild diets consist of seeds, nuts, berries and fruits. Some birds in this group, however, eat worms, caterpillars and snails.	Yellow maize, cowpea, wheat, pea, banana, apple, tomato, papaya and grape are the dietary ingredients of captive pigeons. Seed-eating birds can be provided commercially prepared pigeon pellets. Vitamins and minerals should be supplemented. A captive ring dove may be offered millet (50 g), and green vegetables (25 g) per day. The amount of millet may be increased up to 100 g for pigeons.	The ration prepared for domestic pigeon contains 13-15 per cent protein, 3-6 per cent fat, 60-70 per cent carbohydrate and less than 5 per cent fibre. Food should be given as small and bite-sized morsels.
Quail, peafowl, jungle fowl, turkey, partridge and guinea fowl	They feed on nuts, seeds, shoots, buds, worms and insects. Peafowl are omnivorous birds and feed on grains, seeds, tender shoots, berries, insects, worms, lizards and small snakes. Guinea fowl thrive on seeds, leaves, bulbs, insects and amphibians.	Commercially prepared game bird diets may be given. The dietary ingredients such as insect, chopped lettuce, succulent grasses and other green vegetation should also be supplemented. Oyster shell grit should be supplemented. A peafowl may be offered 250g grain mixture, 125 g green vegetable, 50 g mash and 50 g onion and garlic per day. The following diet can be given to a pheasant and jungle fowl per day: vegetable (25 g), grain mixture (50 g), mash (25 g), and onion and garlic (25 g).	Crickets and mealworms can be given to young birds.

241

Owl, hawk and falcon	They eat mice, lizards, snakes, insects, fish, frogs and birds.	The captive diet consists of whole mice, rats, birds and fish.	The diet recommended for raptors may contain 45-50 per cent crude protein, 18-20 per cent ether extract and 2.2-2.5 per cent crude fibre.

Chapter 18

Restraint, Housing and Management of Wild and Zoo Mammals

The following text describes the restraining, housing and management of selected wild and zoo mammals.

Monotremes:

Restraint: Monotremes can be restrained by both physical and chemical means. The echidna is controlled by placing the gloved hands underneath the back of the animal to grasp both the hind legs. The platypus can be controlled by a hoop net. However, care should be taken while restraining the male platypus. This is due to the fact that platypus bears a venomous spur on the medial aspect of the tarsal joint of hind legs. As monotremes are extremely fragile and delicate animals, special care is required in transporting these animals.

Housing and management: The echidna can be housed in a single open-topped enclosure measuring 4.5 m x 5.8 m x 0.9 m high. They are kept singly, in pairs, or in small groups in a cemented floor house. The cemented floor is recommended for the captive animals with the aim at preventing animals from digging out. Blue or red fluorescent light can be provided so that animals are visible during day time in captivity. Nest boxes, rocks, hollow logs and tree stumps are provided in the enclosure.

Platypus can be maintained singly or in pairs in captivity. They need a non slippery floor and a den. Wood shavings and leaves are kept in the den. This is mainly important for rearing the young. Due to its semi aquatic nature, a pond measuring 1.8 m x 1.2 m x 0.6 m with tunnel is essential for better management.

Marsupials:

Restraint: Most marsupials scratch the intruders with their strong claws. Moreover, the larger marsupials can injure the restrainers by kicking with their hind legs (Fig.18.1). Therefore, restraining of marsupials by physical and chemical means should be performed very cautiously. Small marsupials can be handled with towels, gloves, tubes and other special devices. They can be physically restrained by placing one hand behind the head and extending the hind limb with the other hand. Larger marsupials are difficult to control. They are efficient jumpers and their hind legs are used as weapons. A net with a very fine mesh can be used for capturing a small kangaroo. They can be gripped by placing one hand at the base of the tail and the other hand on the neck region. During stress kangaroo can stand upright balanced up on the tail and slash down with front or hind legs. The animal raising up on its toes and tail indicates its attacking mood. In that situation, a forked stick can be used to control the animal. When an aggressive kangaroo attacks in unprotected intruders, it is always safe by striking the animal in neck region to drive it off.

1   2

Fig.18.1: Warning (1) and slashing kick (2) by kangaroos.

Another important point is that before capturing kangaroo, it is desirable to confine the animal in a solid walled enclosure. Then a mattress can be used to press the animal against the wall for controlling.

Housing and management: The housing for marsupials depends on the size and nature of the animals to be kept in the zoos. Small marsupials can be kept in an enclosure measuring 4 ft (1.21 m) x 4 ft (1.21m). A single animal is usually maintained in the enclosure. Larger marsupials may be kept in barns with access to outdoor enclosures. Kangaroos are traditionally housed singly or in pairs. A female with her newly born young may be kept in a single cage. The concrete floor is suitable for marsupials. The cage should be provided with proper shades of trees. This is particularly important for kangaroos and wallabies because these animals have the less ability to withstand high temperature and humidity. The following management should be given due importance for captive animals.

- The floor of the sleeping area should be kept dry and clean. The floor should be provided with bedding materials such as straw and wood shaving.

- The provision for hollow logs or rock crevices or nest boxes for shelter in cold weather is very important. Marsupials need protection from bright light. Blue fluorescent tubes may be provided.

- Most members are notoriously quarrelsome animals. Two males can not be kept together. There should be a sufficient space so that over crowding can be avoided. This is especially true for kangaroos in order to reduce the risk of infections like necrobacillosis. Moreover, overcrowding may cause severe fighting among cage mates.

- Wire mesh fences or moated yards can be used.

Hedgehog:

Restraint: Insectivores are delicate animals. They need careful handlings. Moreover, some insectivores may bite during restraining of animals. Therefore, these small mammals should be handled carefully. Light gloves and plastic tubes should be used for handling insectivores. Hedgehogs become docile if they are kept for long periods in captivity. However, it is always desirable to handle this animal with gloved hands.

Housing and management: Hedgehogs are reported to be hibernating during colder months. Therefore, these animals may be kept in a quiet place so that they can spend colder months in natural torpor. Hedgehogs may be housed in groups comprising one breeding male and three to ten breeding females in cage measuring 5.5 m x 3.7 m x 1.5 m. However, they seem to do well when kept individually. Pregnant animal should be separated. A nest box should be provided.

Chiropterans:

Restraint: Nets, gloves, forceps and towels are the basic tools for manual restraining of bats. One needs to be careful during handling bats because of their biting habits. Moreover, some species including vampire bats are the potential reservoirs of rabies virus. A caged bat may be caught with gloved hands. A towel or cloth can be dropped over the bat in order to capture. Hoof nets are very useful in retraining bats. A forceps with long handle may be used for carrying bat from one cage to another. Bats may be transported in cardboard boxes. Sometimes, bright light is useful for grasping bats. The vampire bats can be physically controlled by wearing soft leather gloves.

Housing and management: Bats can be kept in wooden or metal cages. A single cage measuring 1 m x 1 m x 1 m can accommodate 15-20 long-tongued and short-tailed bats. The most important factor to be given priority for management of captive bats is the proper maintenance of temperature and humidity. Though captive specimens may survive in a wide range of temperatures, the ideal cage temperature should be maintained at 21 -26^{0}C. A relative humidity of 60 to 70 per cent should be maintained in order to prevent desiccation. The cage height may be increased for flight purpose. The Indian flying fox can be exhibited with other zoo animals such as pangolin and slow loris. The giant flying fox can be housed with parakeet, barking deer, tree shrew and mynah bird. The following practices are important for captive bats in order to keep them healthy.

- Water should be provided in a small jar. Blood should be kept in a flat bottom vessel for vampire bats. Dishes and other utensils are to be cleaned daily. The regular cleaning of cage floor is very important. This is especially true for vampire bats which produce copious urine and solid waste.

- Sawdust and plastic backed absorbent floor paper may be used as bedding materials. Dead bodies causing smell should not be left in the cages. The other necessary tools such as roosting boxes and perches should be provided in the cage. A dark roosting area for hiding during daytime should be provided.

- Claws should be trimmed at an interval of two or three months. The newly captured wild bats should be kept in quarantine for at least six months in order to observe whether they carry rabies virus or not.
- Workers handling captive bats should have some protective measures including wearing soft leather gloves and immunization against rabies. Adequate sanitary measures and application of some selective insecticides for controlling ectoparasites are also important for keeping bats in healthy conditions.

Primates:

Restraint: Primates are controlled manually for various purposes. It defends by scratching and biting. It is particularly true for larger primates. Restrainers should always be cautious during handling of primates. Larger primates especially apes are very powerful animals and can crush hands and dislocate joints of the handlers. Primates have the habit of throwing objects such as stones, hypodermic darts and faeces at the restrainers. Another important point is that handler should not use full length neck tie and clothe with pockets while handling of primates. Gloves, hoop nets and squeeze cages are the important tools for working with these animals. Primates should not be handled with bare hands. Heavy leather gloves should be used in order to prevent injuries. However, some zoo animal specialists are of the opinion that bare hands are sometimes desirable for properly gripping the animals. A bath towel may be used for controlling small primates. A net or bag net can also be used for controlling the animal. A medium sized primate can be physically restrained by holding its arms behind its back. Then the restrainer's fingers can be inserted between the arms for safeguard. A small to medium sized primate may be grasped by holding its forelimbs. Plywood sheets can be used for transporting large apes.

Housing and management: Like many other mammals, primates are popular zoo exhibits. Primates such as monkeys and chimpanzees are widely attracted by the visitors in the zoological gardens. The proper designing of a house for primates depends on climate, kind of animals and type of display. The followings are the important principles for housing and management of captive primates.

Captive chimpanzees are the best known primates for their highly social and adaptable nature. A cage measuring 17 ft (5.18 m) x 16 ft (4.87 m) x 17 ft (5.18 m) may provide adequate space for a single animal. A water moat with a proper barrier can be provided. The chimpanzee may enjoy walking through water. Other necessary items such as pillars, sleeping branches and bars should be provided in the cage. This animal is known for its nest building activity. Temperature, humidity and ventilation should be maintained properly.

Macaques are relatively easy to maintain in captivity. They may be kept in pairs or in family groups. A cage measuring 2.0 m x 1.0 m x 1.5 m may be adequate for keeping a pair of animals. The minimum size of outdoor per enclosure per pair is 500 sq m. When monkeys are kept in adjacent cages, it is essential that partitions should be made of solid materials. The wire netting partitions may cause injuries during fighting. Another important point is that the adult male will be violent in the presence of its potential rival in the same quarter. Therefore, two mature males should not be kept in the same enclosure.

Langurs or hanuman monkeys are very difficult to maintain in captivity. However, they may be kept in a cage with sufficient space. Branches and elevated shelves should be provided for langurs.

New World monkeys can thrive well in captivity. Most species in this group can be housed in a glass-fronted indoor cage measuring 3.4 m x 2.5 m x 2.2 m. Rocks, logs, branches, gravel, sand and growing plants should be provided in the enclosure.

The general management principles required for captive non-human primates in healthy condition are as follows.

- The provision for adequate space for the animals is important. Inadequate space may cause stress in animals. Therefore, sufficient space with several feeding stations should be provided. Stressed animals may be aggressive and may throw objects at the keepers. It is always better to house a pregnant animal in a separate enclosure before delivery. The canine teeth of a dominant male may be removed in order to prevent injuries during fighting.

- Regular cleaning of enclosures is very essential for maintaining hygienic conditions. Bedding materials should be removed from the enclosure daily. Floors should be cleaned at least once daily with proper disinfectants. The whole cage should be completely emptied and thoroughly cleaned every few months. However, care should be taken to prevent infection from one cage to another.

- Most nonhuman primates are wasteful feeders. It is particularly true for monkeys. Therefore, moderate amounts of feed should be supplied at different time intervals. Adequate number of food dishes and water pots should be available. Rhesus monkey requires up to one litre of water per day.

- Captive primates need tree branches, swinging branches, nest boxes, pet toys and other necessary furniture in order to stimulate their natural behaviours. Proper sheds are required for the protection of the animals during hot and rainy days.

- Paints used for cage materials, furniture and utensils must be free from toxic metals such as lead. Lead toxicity results in morbidity and mortality in many captive primates.

- The newly arrived animal must be kept in a quarantine shed for at least two to three months. During this period, the animals should be physically examined. Tuberculin test is routinely performed for newly arrived animals. It is especially true for monkeys that are imported from the countries where tuberculosis is more prevalent.

- Infants of nonhuman primates need proper care. Human milk replacer can be used to rear the young. They may be fed at every two hours interval.

- Newly received animals have been found to have a high incidence of morbidity and mortality. They are very much susceptible to gastroenteritis and infectious diseases. Moreover, these animals often do not eat commercially prepared food. Therefore, a natural diet with adequate amount of water should be freely available. Water mixed with electrolyte should be given for one week or until the animals start taking food regularly. A large amount of fruits and vegetables should be provided.

- Attendants working with primates should be protected against some dreadful diseases such as rabies and hepatitis. Tetanus toxoid should be taken as a precautionary measure. The use of protective clothing, goggles, face masks and gloves are very important. Zookeepers should take care when using needles, scalpels etc. that causes injuries.

Pangolins:

Restraint: Pangolin can be physically handled by grasping the tail with the help of light gloves. They curl up into a ball when pangolins are touched, as mentioned earlier. So they should be handled carefully. Care, however, should be taken to avoid any injury caused by its heavy claws. As a defensive measure, pangolin may urinate.

Housing and management: Their adaptability in captivity is not encouraging. However, many zoos keep pangolins as popular exhibits. Bedding materials such as sand, straw etc. should be avoided. This is because these animals tend to ingest all these materials causing gastrointestinal disturbance. The newly born young may be hand reared. The young should be placed in an incubator with a temperature of 29^0C and 70 per cent relative humidity. The young are weaned at about four months of age. A Malayan pangolin in Calcutta zoo had been found to swim across a lake of 110 ft (33.52 m) wide. Pangolins rarely breed in captivity.

Lagomorphs:

Restraint: Lagomorphs need gentle care during handling. Although, they are not aggressive animals, caution should be exercised during handlings to avoid severe scratches caused by their powerful hind legs and sharp toe nails. Therefore, one should wear light gloves during physical restraining of these animals. The best way to control lagomorphs is to grasp the animal with one hand at the nape of the neck and support the abdomen by the other hand. Animal's feet should be directed away from the restrainer's body in order to avoid scratches. The animal may be picked up by holding the loose skin over the shoulders (Fig.18.2).

This method is usually applied for controlling heavy breeds. A proper method by which a young lagomorph can be controlled is to grasp the animal across the loin. The most important point to be remembered during handing of lagomorphs is that they must never be picked up by the ears or legs alone. If the animal is required to be lifted by the ears, the restrainer should place one hand under the animal to take the full weight (Fig.18.3). Restraint box and cat bag can be used for restraining lagomorphs. Improper handling of these animals may cause fracture. The technique of carrying rabbits is given in Fig.18.4.

Fig.18.2: Control of a rabbit by holding its loose skin over shoulder.

Housing and management: Wild lagomorphs other than domestic rabbits are rarely kept in captivity. This is due to the fact that wild hare, pikas and cottontails are difficult to adapt to the conventional animal husbandry housing and management practices. However, wild lagomorphs are popular exhibits in many zoos, though they are rarely visible to the public primarily because of their nocturnal habits.

Fig.18.3: Placing of one hand under the

Nevertheless, the domestic rabbits are thrived well in confined housing and management systems. An indoor rabbitry has several advantages. It protects animals from harsh environment as well as from their natural enemies. Moreover, they can be well maintained in the conventional housing. The design of a rabbit house is determined by the climate and other factors. In moderate climatic conditions, animals need a minimal housing whereas an environment-controlled rabbitry is advocated in area where there is extreme heat or cold. The size of the cage or hutch may vary according to the size of the animal and space availability. A cage measuring 3 ft (0.91 m) in length x 2 ft (0.60 m) in width and 15 inches in height is generally suggested for Indian condition. Modern rabbitry specialists, however, have suggested that a cage where animal is kept for breeding purpose should have the size of 2.5 ft (0.76 m) x 2 ft (0.60 m) x 16 in for small breed, 3.5 ft (1.06 m) x 2 ft (0.60 m) x 20 in for medium breed and 4 ft (1.21 m) x 2 ft (0.60 m) x 20 in for large breed. Cages may be wooden (Fig.18.5) or metal. Wooden cage is usually not advisable for keeping animal because of their chewing behaviour. Moreover, wooden cage tends to soak urine.

Fig.18.4: Technique of

The following management principles are important to maintain lagomorphs healthy and productive.

Fig. 18.5: Outdoor wooden cage for rabbit.

- Cages must be made water proof and floors should be maintained as dry as possible. This is because the wet environment creates a favourable condition for coccidiosis.

- The cages should be provided with feed hoppers, water bottles, forage racks and nest boxes. Feed hoppers are used for keeping grains or pelleted feed. It is made up of metal sheet with holes or a screen under the bottom for removal of small broken feed articles. Sometimes, small feeding troughs are kept for feed mash. Watering of each cage may be done by automatic watering system having a series of connecting pipes with individual watering nipple. Watering is done by providing an inverted water bottle drinker. Earthenware pot, glass or used old cans may also be used for watering purpose. However, these may cause hygienic problems as rabbits tend to soil the water. Generally a rabbit requires 350 to 600 ml of water daily. Wooden or welded wire nest box measuring 16 in length x 10 in width x 8 in height should be kept in the cage. The nest box serves in protecting young from inclement environment. Wood-shavings, straw, saw dust or shredded sugar cane may be used as nesting materials. The nest box is provided into the enclosure on 28th or 29th day following mating.

- Environmental factors such as temperature, humidity and ventilation play a crucial role in maintaining the rabbits in healthy and productive conditions. Rabbits become comfortable in temperatures ranging between 10 and 30^0 C.

- Wild rabbits are less sensitive to high environmental humidity. This is due to the fact that much of their lives spend in under ground burrows. A relative humidity of 60-65 per cent may be ideal for captive rabbits. Lagomorphs become sensitive when humidity is below 55 per cent. However, high temperature with high humid condition also detrimental to the survival of rabbits. Proper ventilation is also important for rabbit production.

- Feeding is one of the important aspects for keeping animals in productive conditions. Wild lagomorphs generally graze for feeding in the early hours of day, late evening and during the night. Therefore, the condition in captivity should be such a way that it stimulates the animal for natural grazing habits.

- The study of reproductive parameters in rabbit is important like any other domestic livestock. For commercial rabbitry, a ratio of 1:10 (male: female) is a common practice. However, the ratio may be one buck to 20-25 does depending on how intensively they are managed. In intensive system of production, does are bred immediately after kindling, while they breed 10-20 days after kindling in semi-intensive system of production. In extensive system of rabbit production, females are bred soon after weaning (5-6 weeks of birth). The other important parameters like servicing date, kindling date, litter weight etc. should be recorded properly. It is always essential to transfer the does into the male's cage for mating purpose.

• Sanitation is vital to successful rabbit production. Cages, water troughs and feeding racks should be cleaned regularly with proper disinfectants. The area where rabbits are kept should be free from their natural predators like rats, mice, civets etc. Sick animals are to be separated in a quarantine hutch. Moreover, wild-caught animal should also be kept in quarantine. Rabbit usually produces 250-400 g of faeces daily. So, the regular removal of manure from the cage is important from the hygienic point of view.

• The other salient husbandry practices that are important for commercial rabbitry farm include identification of young (e.g. tattooing and ear rings), weaning, selection of young for future breeding stocks, culling of sick and unproductive animals and maintenance of records (e.g. mating date, litter number and average birth weight), control of natural predators, isolation of sick animal and proper health care including treatment with prophylactic measures.

• The behavioural study is also important for proper management of rabbits. Although rabbits have been domesticated for the last 200-300 years, the behaviours of domestic rabbits still remain the same with their wild counterparts. Wild rabbits live in groups and in each group the females are out numbered by the males. The female becomes aggressive towards the offspring of other female. Although, wild rabbits are primarily nocturnal, they however, drink and feed at any time during day and night. Therefore, water and feed should be available in the cage for several hours. In the wild state, doe makes nest in burrow for kindling. A nesting box should, therefore, be provided in the cage. Sometimes doe eats her young. This behaviour is usually noticed when does are exposed to under severe stress conditions such as noise, over crowding and absence of drinking water. So, the stress should be removed.

Rodents:

Restraint: Many rodents bite during restraining causing injuries to the restrainer. Moreover, their incisor teeth are used for chewing and gnawing. Therefore, one needs to follow some precautions during handling of rodents. It is especially important for controlling wild members. Rodents can be manually controlled by many protective devices such as leather gloves, nets, traps, cages, boxes and plastic tubes. One should not exert excessive pressure on animal during handling. Nevertheless, guinea pig can be manually handled by placing the right hand around the neck while the hindlegs are extended by the left hand (Fig.18.6). Guinea pigs may also be captured by bare hands.

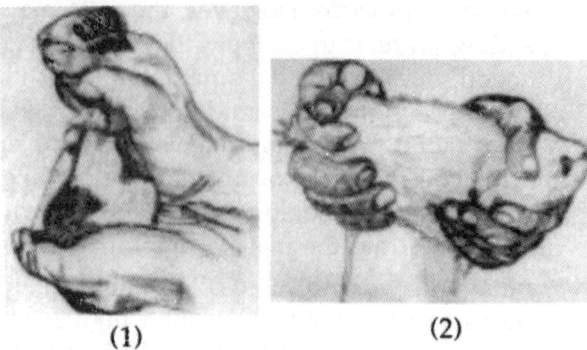

(1)   (2)

Fig.18.6: Manual restraint of a guineapig (1 & 2).

Rat can be picked up very gently by the palm of the hand over the back with the fingers supporting underside of the body. Rats should not be lifted by the tail alone (Fig.18.7). It can be restrained by holding the end of the tail with one hand and place the other hand over the neck and wither. The important principle in controlling rats and mice is that their movements should be restricted slowly, gently and without sudden jerks. Care should be taken during capturing of spiny rats. This is due to the fact that the tail may easily break off during handling.

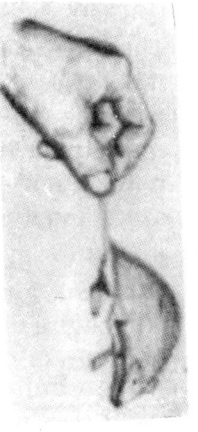

Chinchillas can also be restrained manually. However, one needs care during handling because of its delicate body fur. Improper handling may result in loss of some fur. This phenomenon is called "fur-slip". Therefore, a chinchilla can be controlled by grasping the animal at the base of the tail with one hand and at the same time the ears can be grasped with thumb and fingers with the other hand.

Fig.18.7: Rat should not be lifted by the tail alone

The controlling of porcupine is a matter of patience. When these animals get frightened, the quills may be released. So, they can be controlled with a piece of heavy plywood or metal. Porcupines can be captured by placing the hand low and grasp hairs at the tip of the tail. Snares can also be used for controlling porcupines.

Housing and management: Many rodent species are maintained in zoos for public exhibits. Some members of the order Rodentia are also kept in laboratories for biomedical research. Rodents have a diverse habit. Some members such as porcupines and chinchillas are nocturnal in habits. So, each group of animals needs separate housing and management practices. Moreover, the application of management principles also depends on their biological characteristics. The housing and management of some selected rodents are described here.

The guinea pig or domestic cavy is kept in metal cage with wire mesh floor. They may be housed singly or in small groups comprising two to four individuals. A cage measuring 14 in x 9 in x 8 in is adequate for keeping two animals. Six animals can be kept in a cage measuring 20 in x 14 in x 12 in. Pregnant animals, however, require more space. Guinea pigs thrive best at temperatures between 65 and 80^0F with 45 to 55 per cent relative humidity. However, they will adapt to temperatures well below the freezing, if guinea pigs are gradually habituated in low temperatures with dry housing.

The cages should be washed and sterilized at least once a week. Cleaning and washing should be done more frequently if animals are infected with diseases. Clean water should be made available daily, though animals fed with green greases do not require much water. Guinea pigs tend to eat and drink throughout the day and night. If they are habituated to consume a particular diet, any change in the composition of food may result a sharp decline in food consumption. Wood savings and peat moss are satisfactorily used as bedding materials. Hay is used by the guinea pig for preparation of nest. Bedding materials should be protected from any contamination. Guinea pig house should be free from wild rats and mice.

Chinchillas can be well bred in captivity. They are generally housed in pairs in pens with wood or metal frames. However, several females can be accommodated with a single male, if sufficient space is provided. They are best thrived at temperatures between 40 and 70^0 F. Females may be kept in separate pens due to their aggressive behaviour. A retiring box should be provided. Another box containing earth and dust should be provided with in the cage for a dust bath. Chinchillas are very sensitive to heat. Water and food should be made available *ad libitum*. A light and dark ratio of 14:10 hours respectively is adequate.

Hamsters are widely used for scientific studies. They may be housed in simple cages made for rats. An individual or paired hamster can be accommodated in a metal type cage measuring 15 in x 12 in x 6 in. A dark chamber is useful for the purpose of breeding. The environmental temperature may be maintained between 20 and 22^0C with 40-60 per cent humidity. The cage should be well ventilated. A large number of animals with similar sexes can be housed in a large cage. It is always important to remove all the existing animals from the cage when newly arrived animals are to be mixed in the same cage. After putting in sufficient food in the cage, all individuals including newly arrived animals should be mixed and returned into the cage. Hamsters secrete highly concentrated urine containing large quantities of mineral salts. Sometimes, it is very difficult to remove salt deposits. So, the dilute acids may be required to remove the deposits.

Old World porcupines are largely nocturnal and ground living animals. They must be kept in secured enclosures. A dark area may be provided for these animals. Old world porcupines are sensitive to cold. Therefore, heated quarters must be provided.

Squirrels are popular exhibits in many zoos. In captivity, tree squirrels can be kept in pairs or in small groups with a single male. Cage should be provided with tooth resistant materials. A sleeping chamber may be provided for tree squirrels. Other necessary items such as rocks, live plants and nest boxes should also be provided in order to stimulate their natural behaviour. However, the above mentioned necessary items should be replaced at regular intervals, as they get damaged by chewing. Dry leaves are provided as bedding materials. The floor should always be kept dry. A cage measuring 11 in x 8 in x 8 in may be useful in laboratory for individual ground squirrel. A pair of Malayan giant squirrel can be maintained in a glass-fronted exhibition cage measuring 3 m x 3 m x 2 m high. The enclosure should be furnished with natural tree branches, dried leaves and nest boxes.

Rats are adapted readily in captivity. They are widely used in laboratories for experimental purposes. Usually, there are three types of rat cages. The large cages are made for mating and housing does. The second type of cage is the smaller breeding cages in which individual doe with young is kept. The third type of cage is usually kept for maintenance of rats during scientific studies. Smaller species of rats are normally housed in wire mesh or glass sided cages with removable wire tops. Water is provided with drip bottles. Small hollow logs or nest boxes are provided for hiding purpose. One to three animals can be kept in a single cage measuring 12 in x 8 in x 8 in. A group of rats comprising four to ten individuals may be housed in a cage measuring 20 in x 14 in x 8 in. Soiled cages should be removed immediately. Regular cleaning of cages is an important part of rat management. Sterilization of cages is done once a week or more depending on the nature of experiments to be conducted. Rats can be ear-tattooed for identification purpose. Ears are also satisfactorily marked with v-shaped snips in rats. The optimal environment temperature is ranged from 65 to 70^0 F with a relative humidity between 40 and 70 percent. The cage bottom may be covered with a sheet of newspaper. A thin layer of powder may be spread over the sheet in order to absorb urine and excreta.

Like rats, mice can be well maintained in captivity. Generally an individual mouse requires 0.05 to 0.07 sq ft area. Mice are kept in small or in large groups. A single cage measuring 12 in x 8 in x 5 in may be sufficient for keeping five to ten individuals. The large group cage measuring 18 in x 12 in x 5 in can maintain ten to twenty mice. The ideal temperature for maintaining mice in captivity is 60 to 70^0 F. The room should be well lighted. Saw dust can be satisfactorily used for bedding purpose. Paper, hay or straw may be served as nesting materials.

General management guidelines of rodents:

The proper housing is one of the basic components that can influence the well being of the rodents. Space provided for these laboratory animals should be adequate. The size of individual animal-holding room should be sufficient to accommodate standard cage and equipment and to allow adequate space for servicing animals and equipment. They can be kept either individually or in groups. However, the space required to maintain these animals depends on many factors such as age, body size, weight, sexual maturity, breeding activities and behavioural characteristics. The cage should be free from sharp edges. The bottom of the cage can be either of wire or solid metal. The floors of solid bottom cages are normally covered with bedding materials such as woods and shredded newspaper. These bedding materials absorb urine and moisture from faeces thus improving the hygienic condition of cages. Proper temperature, relative humidity, air flow and illumination should be maintained. Rodents should be kept in less noisy areas. It is particularly important with in the area of a zoo where sound may produce by noisy animals such as non-human primates, tigers, lions and dogs. The other management practices are as follows:

- Animals should have easy access to food and water. Rodents should be provided nutritionally balanced diet. Laboratory rodents should be supplied fresh, potable and uncontaminated drinking water *ad libitum*. Water can be provided through water bottles and drinking tubes.

- Adequate sanitation is an important aspect of day to day management. All components of animal rooms should be cleaned with appropriate disinfectant solutions so as to remove accumulated dirt and debris. Soiled bedding materials should be removed and replaced with clean and dry bedding. Cages, cage racks and accessory equipment such as feeders and watering devices should be cleaned and sanitized regularly in order to keep them free from contamination. As a general rule, rodent cages and cage accessories are required to be washed at least once every two weeks. Solid bottom cages and water bottles are needed weekly cleaning.

- A pest control programme is essential for proper maintenance of rodents. It includes prevention and controlling or elimination of pests, insects and wild rodents. Pests can be prevented by sealing potential points of entry. Relative nontoxic substances e.g. boric acid and amorphous silica gel can be used. The use of toxic compounds should be avoided as far as possible.

- Individual identification and record are always important for proper management of rodents. Identification can be accomplished by ear punching, tattooing (normally on the tail), and ear tagging or implanting electromagnetic transponders. Tags should be light enough. The temporary identification of fur skin or tail is done by dyes. Cage identification cards can also be used for identification of individual animals or group animals. The identification card should have sufficient information about species, strain, sex, body weight, number of animals in the cage and source of animals. Detail records such as feeding, breeding, disease and treatment records on individual animals are important.

- Wild-caught rodents normally carry pathogens and parasites. So, wild rodents should be kept in quarantine until their health status is known. Serological tests should be conducted during this period. During quarantine, effective sanitization programmes should be conducted. Daily animal care is important.

- The most important aspect of rodent management is the control of infectious diseases. Wild rodents are the carriers of many zoonotic diseases like leptospirosis and lymphocytic choriomeningitis. One should observe the appearance and behaviour of the animals daily. Abnormal signs such as weight loss, ruffled hair coat, dry skin, abnormal gait, lethargy, swellings, seizures, lacerations, diarrhoea, dyspnoea and discharges from natural orifices should be observed properly. Therefore, good management practices like purchasing disease-free animals, using protective clothing, using protected cages, staff for proper clinical observations and controlling the movement of personnel and visitors within the premises should be properly accomplished.

- Rodents infected with bacterial diseases can be treated with antibiotics. Antibiotics can, however, adversely affect animals especially hamsters and guinea pigs. It may cause an imbalance of intestinal microflora. Therefore, caution should be exercised in administration of antibiotics.

- Monitoring for control of ectoparasites should be done regularly by examining the skin. The skin scrapings from infected rodents can be examined microscopically. Frequent washing of floors, walls and cages is important for controlling parasitic infections.

- As mentioned earlier that rodents are sensitive to noise like lagomorphs. So, external disturbances if any should by kept minimum. Workers handling rodents should be offered immunization programme against various zoonotic diseases.

Carnivores:

Restraint: The restraining of carnivores is essential for performing various activities such as physical examination, administration of drug and collection of blood. Carnivores can be controlled both manually and chemically. Special care and precaution, however, are two aspects that need to be given importance while handling carnivores. Some carnivores have powerful claws that may cause severe injuries to the restrainers during handling. Moreover, their enlarged canine teeth are also very dangerous. The habit of regurgitation is very common in case of carnivores. Therefore, handling of these animals after a meal should not be performed in order to avoid aspiration pneumonia. Squeeze cages, snares, tongs, nets and gloves are the common devices used in controlling wild and captive carnivores.

Smaller felids are controlled manually. Heavy leather gloves are very important tools in controlling young felids. Smaller animal can be grasped by placing the hands on the nape of the neck, Cubs can be wrapped in a towel for restraining. Snares, nets and squeeze cages are also routinely used for handling of felids. Larger felids should be controlled by the squeeze cages. Smaller felids are squeezed from top to bottom, while larger felids are squeezed from side to side in the squeeze cages. A cheetah can be controlled by using a snare. Small and medium sized felids can be controlled by nets.

Wild canids can be controlled by snares and squeeze cages. Small canids should be handled with gloved hands. Although, heavy leather gloves do not offer full protection against biting, the wearing of leather gloves, however, is very important especially for controlling larger canids. Snares poles can be used in controlling of small canids. Squeeze cages made especially for larger animals should be used for both physical and chemical means of restraining of larger canids.

It is really a difficult task to control an adult bear by physical means because of its tremendous strength and compact body. A large bear is capable enough to kill handler even with a single swat of paw. Therefore, caution must be exercised while handling bears. Immature animals can be manually controlled with the help of snares or nets. Squeeze cages designed especially for bears should be used for restraint purpose. When bear is required for transportation from one place to another, it is always necessary to chain or rope the crate to the door.

Procyonids are small to medium sized animals and can be easily controlled by various devices. However, these animals can also inflict severe injures. Squeeze cages, nets and snares are the usual devices for controlling pandas. Leather gloves are useful in handling these animals.

Mustelids can be restrained by snares, squeeze cages or nets. Sea otter can be controlled by throwing net over the body. Leather gloves are also very important devices for controlling mustelids. The metal cone of suitable size can be used for controlling of animals in order to collection of blood or other purpose. Mustelids can be transferred in a wire cage.

Hyenids can be controlled by squeeze cages. Their jaws and teeth are so strong that hyena can even tear a net. These animals are generally controlled by chemical immobilization and it is better to avoid handling manually particularly for larger animals.

Small viverrids can be controlled by the similar techniques and devices applied, for mustelids and canids. Small viverrids can be handled with leather gloves and nets.

Housing and management:

Felids: Felids are very popular exhibits in the zoos. Wild felids can be kept in large outdoor cages. They are required spacious outdoor quarters with sleeping dens. They can be housed in an enclosure with heavy wire mesh. Other requirements such as logs, rocks and sleeping quarters should be provided. The followings are the essential housing and management practices for captive felids.

- The minimum space required for each animal is as follows: 8 m x 4 m x 4 m (large sized felid), 4 m x 2 m x 2.5 m (medium sized felid) and 2 m x 2 m x 2.5 m (small sized filed). Minimum size of outdoor per enclosure per pair animals is about 1000 sq m for tiger and lions. An extra of 250 sq.m area is required for an additional animal.

- The enclosure can be surrounded by dry or water-filled moats. The water moat, for example, may be 25 ft (7.62 m) wide and 16 ft (4.87 m) deep for tigers. For lion, it may be 21 ft (6.40 m) wide and 12 ft (3.65 m) deep. The land area is gradually sloped downward from front to back so that the moat wall on the visitor's side is higher than the inner one. The other necessary structures such as fencing should be constructed according to need of the animals. Leopards are good climbers. So, they can be safely exhibited in an enclosure with properly constructed roof.

- Felids can be kept in pairs or in small groups. A pair of lions with their cubs can be maintained in a single enclosure. However, it is a normal practice to keep male in a separate enclosure except in breeding periods. Nevertheless, there is no definite rule in this regard. It depends on many factors such as compatibility, adequate space and management facilities.

- Enclosures should be provided with separate cubbing dens, logs, shelves, tree branches, rocks and sleeping quarters. The pregnant and sick animals should be kept into a separate quarter. The enclosure should have proper shades in order to comfort animals from hot summer months.

- The floor may be constructed with concrete or wooden materials. Some zoos have grassy areas and rocky ledges. There should be a proper drainage system. The floor should be cleaned daily. Some felids such as cheetahs and jaguars require a small heated quarter during winter months.

Bear: In the natural state, bear moves around its range for searching vegetation. Therefore, it is to be ensured that bears get regular supply of foods with in the range in different seasons. The area should have many fruits bearing and other trees so that wild bears do not face any shortage of seeds, nuts and different kinds of fruits. The protection and preservation of dens are important. A den spot may be created. Dispersed log piles should be kept near den area. Moreover, the development of water spots near den area is very essential.

Bears are very popular exhibits in the zoos. They are readily adapted to the captive condition. These animals should be kept in a large enclosure. The size of a night shelter or feeding cubicle is 2.5 m x 1.8 m x 2.0 m. A minimum of 1000 sq. m area is required per pair of animal as outdoor area. Physical barriers like wall, moat and fence are essential. The top of the enclosure may be of metal sheets or electrical wires. Enclosures must be sturdy and reinforced. An ordinary net fence should have a height of 3.5 m to 4.0 m in order to prevent from escaping through climbing. The enclosure needs two important aspects- optimal visibility for the public and easy access to the enclosure for observations and services. Bears are known for their digging habits. The floor, therefore, should be cemented. They can be well exhibited in a large moated enclosure. The other essential items such as bathing pools and retiring dens should be provided. The enclosure should be provided with logs or trees for stimulating their natural behaviours like scratching. The rocks are provided for climbing purpose. The pregnant female should be separated into a maternity pen. Disturbances should be avoided during pregnancy and delivery periods. This is due to the fact that bears are known to kill their young if disturbances occur immediately after delivery.

Another important fact is that a period of dormancy in a sheltered nook during the colder months is found among bears. This is more important for northern species. Therefore, animals should not be disturbed during this period. The other management practices for captive bears are as follows:

- Bears are known for their unpredictable temperaments. So, keepers should avoid in entering into the enclosure when animals are present.

- The animal should receive its main feed early in the morning. In addition to two main feeds, bears should be provided with two additional feeds during the day so that they can spend much time in foraging and feeding. Food should not be left in the vicinity of the enclosure. The feeding regime should be diverse.

- Bears should be provided with a variety of objects for manipulation and play. They should be provided nesting materials in the form of leaves, straw and tree branches for stimulating natural behaviours. The enclosures should incorporate natural substrates such as soil, pebbles, bark and pits of sand. Daily cleaning of den is very important.

Canid: Wolves can be housed in small groups of similar sex. These animals can be kept in a wire mesh enclosure with concrete flooring. The dimension of a cubicle is 3 m for keeping single animal. Minimum size of outdoor per enclosure per pair of animal is 400 sq m. Canids have the habits of escaping by burrowing. If it is not possible to provide a concrete floor, the wire should be run deeply in the ground at least 2 ft (0.60 m). Keepers should enter into an enclosure with some precautions. Animals may be shifted into different cages for sanitary reasons. It is important for proper management of forest habitat where wolves are naturally occurred.

Jackal may be kept in an indoor quarter with open shelter. Animals are well accommodated in cold environment. However, there should be a provision for heated quarter during colder months.

Foxes are housed in individual pens. The enclosure should be provided with a shelter box. Saw dust, wooden shavings and straw may be used as bedding materials. The floor should be dry. There is a need for an underground well drained den. In their wild state, foxes are required brushy habitats and scattered trees.

Asiatic wild dogs may be kept in captivity. However, captive animals do not thrive well.

Mustelids: Otters are popular zoo exhibits. These animals can be kept in an enclosure with outer walls. The size of the cubicle is 2.5 m x 1.5 m x 1.0 m. Otters have the habits of swimming and diving. Therefore, a pool with clean fresh water should be provided. Each animal should have a separate den. Breeding male and female, however, can be kept together. Marine otter needs sea water in the pool. Unpolluted stream and proper vegetation are very important for otters in their wild state.

Procyonids: An enclosure of two to three metres height with tree or branches, hollow logs and other essential furniture are the ideal environment for keeping procyonids.

The giant panda needs a strongly built housing. A moated enclosure with rocks, trees and sleeping platforms is very important. This curious creature enjoys cold weather. The indoor quarter with air conditioning is ideal for keeping giant panda. Hot weather causes stress to these animals. The giant panda, however, can withstand temperature up to 28^0C with 50 per cent humidity without visible discomfort.

The lesser or red panda is housed in an enclosure of mesh, walled, moated and fenced, glass screened or any combination built in the zoo. A pair of red panda can be kept in an enclosure measuring 3.0 m x 1.5 m x 1.0 m. with 300 sq m area of outdoor enclosure. The height may be increased up to three metres. The enclosures should have natural grasses with a variety of elevations. A varied environment including trees, logs, pools, rocks and clumps of vegetation is essential. There should be a minimum of three nest boxes within the enclosure. In extreme weather conditions, indoor access should also be provided. The enclosure of red panda should not be located at the adjacent to the quarters having aggressive animals. They should be maintained in monogamous pairs.

Hyenids: Hyenas may be kept in pairs or in family groups in a moated enclosure. However, adequate space should be provided when they are kept in together. Generally, hyenas are maintained in a warm enclosure. As these animals are shy and nocturnal in habits, the enclosure should be provided with proper dens or nest boxes. Nest boxes should have sufficient space for loafing as well as delivering of litters. Hyenas are known for their digging habits. Therefore, enclosure should be built in such a way so that they can not escape.

Viverrids: Most members of the family Viverridae inhabit in tropical parts of the world. Civet does well in a glass or wire fronted cage measuring 2.0 m length x 1.5 m breadth and 1.0 m height. There is a need for sleeping dens and climbing branches like many other carnivores. A box with straw should be kept in a sheltered corner for pregnant animal.

Mongoose may be placed in a cage measuring 2.4 m x 1.2 m x 1.0 m. Sleeping platforms, rock crevices, sleeping boxes and stout branches should be provided. Other objects such as ping-pong balls, golf balls, cardboard boxes, newspaper, hollow logs and cloth are provided in the enclosures in order to stimulate their natural instincts. The floor should be constructed in such a way that it satisfies digging habits as well as to keep their nails worn down. Mongooses can be kept singly or in small groups depending on species. The African dwarf mongooses, for example, live in small family groups, while the small Indian mongoose is solitary animal. The captive small Indian mongoose can be best exhibited in a natural house with reversed lighting system. A pair of male and female may be kept in a single enclosure. Additional females, however, can be added in the same enclosure but not males.

General management principles:

Although, the management practices vary according to the kind of animals kept in zoos, a general principle of management, however, can be applied to all types of captive carnivores. They are as follows:

- Carnivores with soft paws (e.g. lion) require non abrasive floor surface to avoid bleeding. Cleaning is an important part of day to day management. Feed and water containers and food preparation utensils should be cleaned and disinfected daily. All the dens should be cleaned regularly with proper disinfectants. Food and animal waste should be removed daily. A strongly built cage is needed for transporting animals. This is particularly true for larger cats. Moreover, the cage without projections should be used. Rigorous pest control measures in and around the enclosures should be taken.

- Wooden flat utensils are usually used to provide foods for the tigers and lions.

- The maintenance of various records of each animal is very essential. Important records maintained for captive animals include growth rate, place of origin, sex, feeding, breeding and health aspects.

- The pregnant animal should be kept in separate den and she should be given adequate balanced ration.

- The enclosures should be designed in such a way that they resemble natural habitats. Moreover, it should also solve the odour problems of felid species maintained in closed environment. The provision of natural day light in the enclosure is important for cat species.

- Other management practices such as immunization against dreaded diseases, routine examinations, isolation and quarantine measures and highest standards of personal hygiene should be followed. Newly arrived felids must be kept in quarantine for at least 30 days. However, it may be kept for long duration. During this period, the animals should be carefully observed every day.

- The feeding of wild birds and mammals (e.g. pigeon and rabbit) to some carnivores is not suggested. They may carry a variety of transmissible diseases. The intestines from chicken carcasses should be removed so that the risk of salmonella infection is reduced.

- It is important that captive animals should be free from parasitic infestations and infectious diseases. The programme should be implemented in such a way that it reduces parasitic infestations to as low a level as possible. Here, some anthelmintics used against different endoparasites and ectoparasites are given in Table 18.1.

Table 18.1: Common anthelmintics used in carnivores.

Parasite	Drug	dose	Route	Remark
Roundworms (nematodes)	Ivermectin	0.2mg/kg body wt.	Oral or subcutaneous or intramuscular	Severe parasitic infestations result in diarrhoea, vomiting, loss of appetite, emaciation, loss of hair, poor coat condition and even death depending on the types of parasites found in the animal. Faecal samples should be examined at least twice a year. Flotation and concentration methods are used in detection of parasites and their eggs. In chronic parasitic infestations, animals should be treated with the anthelmintics at frequent intervals (e.g. every three months). Drugs are often effective when given for several days (normally 3) than recommended for a single day. Faecal samples should be re-examined after the completion of any course of treatment.
	Piperazine adipate	80-100 mg/kg body wt.; repeated after 2-4 weeks.	Oral	
	Pyrantel parnoate	3-5 mg/ kg body wt. for 3-5 days	Oral	
	Fenbendazole	20 mg/kg body wt. daily for 5 days	Oral	
	Mebendazole	15 mg/kg body wt. for 3-5 days	Oral	
Tapeworms (cestodes)	Fenbendazole	20 mg/kg body wt. daily for 5 days	Oral	
	Praziquantel	5 mg/kg body wt.	Oral	
Coccidia (protozoans)	Sulphadimethason	50 mg/kg body wt.	Oral	
Mites (ectoparasites)	Ivermectin	0.2-0.4 mg/kg body wt.	Oral or subcutaneous or intramuscular	

Captive carnivores are routinely vaccinated against many infectious diseases. Effective vaccines are commercially available for controlling the many infectious diseases in domestic carnivores. However, it is to be noted that the most appropriate vaccine types, their effectiveness and the duration of immunity produced have not fully studied for the majority of exotic species. Normally live vaccines for immunity are avoided in exotic carnivores. A wide variety of vaccines used in exotic carnivores are presented in Table 18.2.

Table 18.2: The schedule of vaccinations recommended for exotic carnivores.

Animal group	Disease	Vaccine type: Killed(K)/ Modified Live Vaccine(MLV)	Vaccination frequency
Exotic felids	Feline panleukopenia	K/MLV	Annual
	Feline rhinotracheitis	K/MLV	Annual
	Feline caliciviruses	K/MLV	Annual
Canids	Canine distemper	MLV(Avian origin is preferred)	Annual
	Canine adenovirus-2	MLV	Annual
	Canine parvovirus	K	Annual
	Canine parainfluenza	MLV	Annual
Hyenids	Canine distemper	K/MLV	Annual
	Feline panleukopenia	K/MLV	Annual
Bears	Canine adenovirus-2	K	Annual
	Leptospirosis	K	Annual
Mustelids, viverrids and procyonids	Canine distemper	K/MLV	Annual
	Feline panleukopenia	K/MLV	Annual
	Canine adenovirus-2	K/MLV	Annual

Source: The Merck Veterinary Manual, 6th edn.

Management of captive neonatal: Maternal care is very important for a new born young. They are nearly deaf with their eyes closed and helpless. Ideally, they should be reared by mother. A phenomenon called "maternal neglect" has been observed in carnivores. Moreover, cannibalism is also common in these animals. In that situation, new born cubs are required special care. Therefore, hand rearing of neonates is an essential part for zoo keepers. After initial care (e.g. removal of mucus from mouth and pharynx and physiological examination), the following management practices are essential for rearing of young.

(1) The young may be bathed antiseptic soap water to reduce external contamination. The young should be kept in an incubator for at least five days.

(2) The young should be fed properly. Colostrum from goat, bitch or supply of human milk replacers (Lactogen I & II) is essential. It is important to note that hypothermic cubs should not be fed till its body temperature becomes

normal. Initially neonates are given a warmed oral electrolyte-glucose solution through feeding bottle. Another important fact is that young should be fed in sternal recumbency with its head elevated. The young's head should be supported by the feeder's fingers. Neonatal carnivores are normally fed 8 to 10 times daily. The stomach capacity of young is about 50 ml per kg body weight. Therefore, 20-30 ml liquid diet per kg body weight should be fed at single feeding. They can be successfully raised on milk replacers and supplements. Adult diet (e.g. meat or chicken soup and meat along with milk) may be slowly introduced after three month of age. Solid diet may be placed in a pan.

(3) Anogenital stimulation is important for urination and defaecation. Therefore, after each feeding a warm and moist cloth may be placed on the anogenital region to stimulate urination and defaecation. Their nails should be trimmed regularly. Some playing objects (e.g. balls) are kept for stimulating their natural instincts. However, care should be exercised so that cubs do not swallow these objects. This may cause gastrointestinal disturbances.

(4) Neonates are susceptible to many diseases. Pneumonia, gastroenteritis and naval ill are very common. The affected animal may be treated with suitable medicines. Young should be vaccinated against dreaded diseases as prophylactic measure. The vaccination is done when young carnivores attain at three to four weeks of age. Initially they are given killed vaccines. Modified live vaccine, however, may be given later in life. Sanitary measures should be given utmost importance for keeping the animals in healthy conditions.

Elephants:

Restraint: Like other animals, restraining of elephants is done for various purposes such as examination of blood, introduction of medicine, trimming of nails and physical examination. Elephants are very sensitive and intelligent animals. Therefore, they should be handled with great precautions. It is especially true when elephants become excited or annoyed. Elephant hooks, chains, ropes, squeeze chutes etc. are used for physical restraint of the animal. Young elephant can be controlled by a hook if it is properly trained. However, care should be taken so that hook does not cause any physical injury to the animal. Chains of various sizes and strength are indispensable tools for controlling these animals. Elephant can be controlled by leg chains. Captive animals should be trained to accept chains on all four legs. Training to accept chains should be done at early stage so that they become habituated to standing quietly when chained. The chain should not be too long for controlling elephant. The rope is sometimes used for restraining the animal temporarily. A squeeze chute may be used for collection urine, blood and musth fluid in bull elephant. Young elephants can be cast by using chains. However, a number of precautions should be taken during restraint of elephants. They are as follows:

- It is always desirable to observe the mood and behaviour of the animal first. Elephant can inflict fatal injury to the restrainer when annoyed. Therefore, it is advisable to handle the animal by the known keeper. This is due to the fact that elephants are highly responsive to the known trainers or keepers.

- During musth the elephants become aggressive and the behaviour is extremely unpredictable. Elephants even kill their known keepers during musth. So, the animal must be chained in order to avoid damages.

- Elephant can grasp an intruder by its trunk. Trunk serves as a mode of defensive or offensive weapon. Tusks are dangerous weapons. Therefore, care should be taken when handling elephants.

- Chemical sedation and immobilization of elephants may cause radial nerve paralysis, myositis, heart failure, traumatic injury etc. So, extreme care should be taken during chemical restraint of elephants.

Housing and management: The enclosure made for sheltering elephant needs sufficient space with an indoor stall. The size of a feeding cubicle or night shelter is 8 m x 6 m x 5.5 m. Elephants may be kept singly or in small groups. However, the provision for separate enclosures for males, females and sick animals should be the ideal practice in captivity. Floors should be solid and dry in order to avoid foot problems. An outdoor enclosure with sand pit and water pool is important for wallowing and other activities. If it is not possible to provide a water pool, there should be a provision for water shower. A healthy elephant should be bathed at least once a day. In summer season elephant, however, may be bathed for two times or more. For drinking purpose, a water trough of at least 240 litres capacity should be provided. A permanent water container should not be kept in the night barn.

Elephant can tolerate cold better than heat. Outdoor shades are required for shelters during summer months. The rubbing posts should be provided for captive animals. All electrical fixtures, water pipes and other loose objects should not be kept near elephant's trunk distance.

Captive elephants are chained at night. They may be chained to alternatively anterior or posterior legs each night so as to avoid any injury to a particular leg. Chains should be shorter in length. In this regard, it is important to note that when many elephants are required to be kept in a single stall, it is desirable that all the animals are to be chained both in the anterior and posterior legs in order to avoid them from getting tangled. Other day to day management practices such as grooming, cleaning of water feeder, trimming of nails, health care and growth records should be properly followed. During summer months, pool should be drained and bleached at least twice a week. Faecal materials should be removed daily.

Care of new born elephant: Sometimes new born calf is rejected by mother for unknown reasons. So, the practice of hand rearing of elephant calf is important in captive condition. The calf should get colostrum immediately after birth or at least within 24 hours of birth. Colostrum is important for developing immunity in the body against diseases. A new born calf consumes about 2-10 litres of colostrum within two days.

Baby elephants can be raised by providing rice-based diet with cow's milk. However, young are intolerant to cows milk. This is due to variation in fat composition and higher concentration of capric and lauric acids. The following diet may be given to a baby elephant: 500 g boiled rice, 500 g dried whole milk, 200 g sucrose and 8.5 litres water. The above diet should be supplemented with vitamins (e.g. vit C and D) and minerals (e.g. calcium and phosphorus). Initially young calf may be fed every two to three hours interval. The frequency of feeding may be reduced to about four times in a day when the calf is about nine months old. Overfeeding should be avoided as practicable as possible. The dried milk formulas prepared for human babies are more suitable for elephant claves. Coconut milk, cerelac, rice gruel or ragi gruel may be added to the reconstituted bay food.

Perissodactyls:

Restraint: The physical control of an adult rhinoceros is really a difficult task and thereby the restraint of these animals is a matter of patience and experience. Although, rhinoceros do not kick or strike, they may crush or trample the restrainer against a wall. Rhinoceros often bites. A heavy built crate may be used for transportation. Chemical restraint therefore is an important practice for controlling of rhinoceros.

Equids are dangerous animals. So, utmost care should be taken while controlling of equids. Horse, zebras and asses are potential kickers, biters and strikers. Wild horses are more violent kickers than domestic horses. Many devices such as lip chains, rope halters etc. are normally used for controlling of domestic horses.

Housing and management: Rhinoceros requires a spacious yard with an indoor stall. The outer yard is floored by hard-packed earth and gravel and is surrounded by concrete wall. The provision for indoor housing is important especially in cold area. An enclosure measuring 5.0 m in length x 3.0 m in width x 2.5 m height is suitable to keep one rhinoceros. A minimum size of outdoor per enclosure per pair is 2000 sq m. Compatible animals may be kept together in captivity. However, a single stall is best suited for an individual animal. The enclosure may be surrounded by a strong brick wall (about three feet height) with strong iron fencing (about two feet). Rhinos can be exhibited in the moat.

The size of a cubicle for wild ass is 4.0 m x 2.0 m x 2.5 m. Another 1500 sq m area is required as outdoor enclosure.

The other management aspects that need to be given more emphasis for captive rhinos are as follows:

- The floor surface should be cleaned regularly. The dry ground is important for captive rhinos.

- In captivity, rhinoceros require grasses, rubbing posts and boulders. Rubbing posts and boulders are needed for stimulating their natural behaviour. Trees are also very useful for shades and rubbing purpose. Rhinoceros are very fond of wallowing. Therefore, they should be provided a water bath or mud wallow. The African species prefer mud wallows. The Indian rhinoceros, on the other hand, prefers clean water. Wallowing is essential for removal of external parasites. Sufficient grazing space should be provided for captive animals.

- A separate house is required for sick or pregnant animal. Pregnant animal should be isolated at least one week before parturition.

Artiodactyls:

Restraint: Artiodactyls should be handled cautiously. Many members of the order Artiodactyla are vulnerable to any physical handling. They may be injured during physical restraint. Crates, squeeze chutes, opaque plastic sheets, ropes and nets are used for controlling of artiodactyls. Many members such as giraffes, deer, antelopes and camelids can be captured by placing crates in the yards or stalls and then wait for the animals to walk into the crates. This method of capturing is primarily used for translocation of animals. Smaller artiodactyls can be restrained by hand. However, restrainer should be careful to avoid any injury. The abrasions caused by horns, antlers and hooves can be prevented by wearing long-sleeved shirts and jackets. Animals can be driven along the lane by using opaque plastic sheets. Nets are occasionally used in controlling these animals. A net may be thrown over the animal for capturing. Some members in this order can also be restrained by ropes. Squeeze chutes and catching chutes may be used in controlling larger bovids.

Care should be taken during physical restraint of camels. They are known to kick, strike and bite in any direction. Llamas and other South American camelids normally do not kick. Circumstances may, however, lead animal to strike, bite and kick. A long rope may be used for manually restraint of camelids. South American camelids are known for their spitting behaviour during restraining.

Wild pigs can bite. The tusks are dangerous weapons of pigs. Therefore, care should be taken during handling of these animals. Nets, snares and squeeze chutes are the usual devices used for physical control of wild pigs. Gloves may be used to prevent injury caused by their sharp teeth.

Hippopotamus use its tusks both for defensive and offensive weapons when restrainers approach for controlling. Therefore, one should be cautious in restraining hippopotamus. Heavy crates can be used for transportation of these animals.

The danger potential in handling deer is their antlers. Males become dangerous during rutting season. For that reason, handling of male deer needs a greater caution. Smaller deer may bite. Moreover, deer may strike with their front feet. Nevertheless, small deer can be controlled by hands. Ropes and nets can be used for restraining purpose. Crates are used for transportation of deer. However, physical restraining sometimes may cause damage to the animals as well as the restrainers.

Although, giraffes are not known for their biting habits, they can kick and strike in any direction. Therefore, extreme caution should be exercised during handling of these agile animals. The specially designed squeeze cages are used for controlling giraffes. A plywood shield may be used for physical restraint of newly born calf.

Horns are important weapons of wild cattle. Many members in the family Bovidae also kick. Ropes, nets, plywood shields, rubber balls, transfer crates and squeeze cages have been used for controlling the wild bovids. Rubber balls are sometimes fixed on the tips of the horns in order to prevent injury caused by horns. Young animals can be physically restrained with ropes or nets.

Chemical restraint is necessary to control artiodactyls. Animals with full stomach should be avoided for chemical immobilization.

Housing and management: Artiodactyls usually remain under stress for few days after capturing. Therefore, they need utmost care during this period. Initially, the animal should be provided with limited space followed by actual space required by an individual animal. They should be given enough water, hay, grasses and green vegetations. Food should be provided according to the feeding habits of the animals. The environmental stress should be reduced as much as possible. The newly caught wild animals exhibit the symptoms of salivation, piloerection, shivering, diarrhoea, tachycardia etc. However, animals will be back to normalcy after a few days.

Wild pigs can be maintained in moated or fenced enclosures. The indoor floor should be made of concrete. This is because of their digging and rooting habits. Proper bedding materials such as straw should be provided. Since pigs have a habit of wallowing, a water filled moat on the outdoor enclosure is important in order to stimulate their natural instincts. They can be kept in an enclosure as a group consisting of one male and a number of females. However, the pregnant females should be separated at furrowing (act of parturition) time. The quarters should be warm during winter months. This is particularly important for piglets.

Hippopotamuses are amphibious animals. They spend much of the day times in sleeping on the banks or in shallow water. A single Nile hippo can be kept in an enclosure measuring six hundred square feet floor space. For additional animal, there is needed about 150 sq ft floor space. The height of the wall should be 5 to 6 ft. They are usually kept in concrete structures. The usual practice is to keep male and female separately except for mating period. If both male and female hippos are maintained in a single enclosure, it is always better to separate the female at the time of parturition as well as in early lactation periods. There may be steel bars or moat in front of the house. Hippopotamus frequently defaecates in water. Therefore, cleaning and draining of water are very essential in order to keep the animals in healthy condition.

Camels are housed within enclosures surrounded by light wire netting or moats. There may be dry or water filled moat. A camel, according to American Association of Zoological Parks and Aquariums, is required 600 sq ft floor space. South American camelids can be kept in an open shelter with wire netting of 6 to 7 ft high. They may be kept in groups comprising one adult male, several females and their young. However, when the young males become older, they must be separated from the group.

The giraffe is primarily timid animal. They are readily adapted to the zoos. A single animal can be kept in a large enclosure measuring 750 sq ft floor space. The height of the wall is to be kept at about 18 ft (5.48 m). A few females can be housed in a single indoor quarter. However, separate enclosure should be provided for male giraffe. The breeding bull is maintained with the female herd. The following management are important for captive giraffes.

(1) The floor should be slightly abrasive. It should not be too soft or too smooth. The floor may be sprinkled with sharp sand. The abrasive floor helps in preventing the overgrowth of hoofs. The ideal flooring for giraffe is the brushed concrete with proper drainage.

(2) The indoor temperature should be maintained at about 70^{0}F. The giraffe should be kept in indoor during rainy days.

(3) The provision for proper shades during summer months is very important for captive animals.

Deer are very popular exhibits in the zoos. They are readily adapted to the captive condition. Deer can be accommodated as a group in single enclosure. The group may consist of one adult male, several females and their young. According to the American Association of Zoological Parks and Aquariums, a fallow deer is required 500 sq ft floor space with 125 sq ft area for additional animal. The fence height should be about 7 ft. However, according to the Central Zoo Authority, the size of a cubicle for swamp deer is 3.0 m x 2.0 m x 2.5 m. Another 1500 sq m area is required for a pair of animal as outdoor enclosure. The floor should be well grassed with proper drainage system. The other management guidelines that are important for captive deer are as follows:

• Deer are known for their digging habits. Therefore, the fences should run well into the ground.

• Properly constructed shelters with roofs should be provided in order to protect animals form cold. There should be a concrete water pool for drinking purpose. Proper shades are essential. It can be maintained by providing scattered trees. The tree trunks must be covered with wires so that trees can be protected from nibbling as well as rubbing from antlers. Moreover, trees may also be damaged due to their bark stripping behaviour. Wooden stands, tree trunks etc. should be provided for stimulation of their natural instincts. Some deer are more sensitive to cold. So, they are required mildly heated shelters during the colder months. This is especially true for thamine deer.

- Some deer have the habit of mud bathing. So, a small marshy area should be provided for these animals. Wallowing also helps in protecting animals from biting flies and mosquitoes.

- Males become dangerous during rutting season. Females may be injured or even killed by the male during this period. Therefore, the practice of the removal of antlers in rutting season is suggested by some zoo authorities. It is always better not to capture deer during summer months in order to avoid stress to the animal. The stress, however, can be minimized by providing proper food and fresh water. The deer may be provided with artificial salt licks in the enclosure. Overcrowding should be avoided.

- The young can be accommodated in a penned grassy area with shrubs and trees or they can be kept in an artificial shelter. They should be provided adequate shades, fresh water and small paddocks. Young start to browse or graze within 15 days after birth. Colostrum should be provided @ 40-60 ml per feeding for three days. Fawns are fed evaporated bovine milk and may be given @ 60-100 ml per feeding for the first week. The amount may be increased up to 300 ml per feeding in subsequent weeks. The young may die due to digestive disorders. The nutritional diarrhoea can be checked by controlled feeding.

The Nilgiri tahr can be maintained in dry moated enclosure. The size of an indoor enclosure for a Nilgiri tahr is 2.5 m x 1.5 m x 2.0 m with an additional 350 sq m area for outdoor enclosure. Too soft earth surface, however, causes an excessive growth of hoofs. Large rocks should be provided in the enclosure. The Nilgiri tahr is reported to be tolerant or amiable to other species (e.g. squirrel) kept in the same enclosure.

Antelopes can be kept in large outdoor areas. Chinkara is required an indoor cubicle space of 2.5 m x 1.5 m x 2.0 m with an area of 350 sq m for outdoor enclosure. Both females and males of smaller species of antelope can be put together in the same enclosure. However, the males of larger species should be separated from the females. The blue bull is well adapted to captivity. They can be maintained in a similar house constructed for other bigger species of wild ruminants. Black bucks are commonly found in zoological gardens. They can be maintained in open and unheated shelters. Blackbucks may be kept in small groups numbering one adult male, six to eight females and their young. The young male with sufficient size and weight should be separated. Hay racks and grain boxes should be provided. An enclosed barn may be constructed for occasional shelter.

Yaks are maintained in zoos. They can be confined in open enclosures with open-fronted shelters. The general management practices that are important for captive ruminants are given below:

- Animal should be housed in adequate space. They require a barn with individual stall. Secondary stalls may be constructed for sheltering purpose. The water and feeding troughs should be provided.

- The out door enclosure should be covered with permanent grasses. They can be kept in fenced enclosures or in water-filled moats. The moats may have steep or gradual slopes.

- Adequate tree cover is important. Logs, boulders and pile of soils can be kept within the enclosures so as to stimulate the natural behaviour of many wild ruminants. Young seek refuge near all these places.

- Good sanitary measures are very essential. Proper cleaning of faecal and bedding materials is vital. The barn should be cleaned and disinfected regularly. The water should be available *ad libitum*. The potable water should be cleaned and germ free.

- The care of health is very important. Sick animals should be housed in a separate barn. Young animals should be vaccinated against dreadful diseases. The faecal samples should be examined periodically.

Management of newborn ungulate: The newly born ungulate needs special care and management. It is especially important for injured, sick or weak calf. Sometimes, calf is rejected by the mother. Therefore, it necessitates the hand rearing of infants. The following managemental practices are important for rearing of a newly born ungulate.

- The infant should be kept in a separate stall in order to reduce stress and contamination. The young should be examined thoroughly. The body temperature, respiration rate and heart rate should be recorded. The naval cord is cut by sterilized scissors about ½ in from the body. The tincture iodine (3%) should be painted on the exposed part of the umbilicus in order to check infection.

- The infants can be hand reared successfully through feeding of colostrum. Normally, bovine colostrum is fed to the neonates. The colostrum contains a much higher proportion of immunoglobulins that protect the young from many diseases. Moreover, colostrum is the rich source of minerals and vitamins. It also serves as a laxative. The colostrum should be fed as soon as after birth. The delayed in feeding of colostrum impedes the absorption of immunoglobulins through the gut. Normally colostrum is given 10-20 per cent of infant's body weigh and it should be provided for 3-5 days. The rubbing of perineal region with a warm, moist cloth for urination and defecation is very important.

- The diet given to neonates is mainly based on evaporated cow's milk. Normally, neonates are given milk @ 10 to 20 per cent of their body weight. Infants should be fed four to six times per day.

- The other managemental practices such as identification of animal by different methods (e.g. tattooing, tagging, ear notching etc.), recording of body weight, proper exercising, , proper cleaning, vaccination programme and proper medication should be carried out regularly.

Chapter 19

Restraint, Housing and Management of Wild and Zoo Reptiles

Restraint: Reptiles are required restraining for physical examination as well as other purposes. Most lizards can be controlled manually. Animals should be grasped near the shoulder girdle and the body for restraint. Larger lizards can be restrained by grasping the animal behind the head and at the pelvic girdle. Lizards should not be caught by their tails. Nets, snares, ropes, plastic tubes and gloves are the usual devices for physical controlling of lizards. A long-sleeved shirt or gloves should be used to prevent scratches. Some lizards are known for their biting habits. So, proper care should be taken during controlling of the animal.

Snakes are captured and restrained by various tools such as pinning sticks, tongs, hooks, nooses, wire screens and snake squeeze boxes. Nonpoisonous snakes will not bite unless they are severely tortured. They can be controlled by grasping the neck immediately behind the head and supporting the body with the other hand. However, the body can be placed on the floor if the snake is very large. Snakes can suffer spinal cord damages if they are dangled by the necks for longer periods. Therefore, snake should not be restrained by holding it with one hand behind its head while the rest of the body hangs (Fig.19.1). Snake hooks are essential tools for restraint purpose. It can be used to remove snake from the cage. A small plastic shield is used to capture a large nonpoisonous snake. The plastic wedge is used for opening the mouth of a snake.

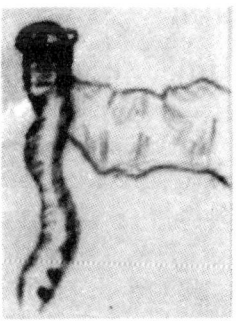

Fig.19.1: Snake is improperly lifted by hand.

Poisonous snakes are always dangerous to control without extreme cautions. Therefore, they should not be handled without experience. Transparent plastic tube with appropriate caliber can be used to physical control of poisonous snake. When half of the snake has crawled into the tube, the opening end of the tube and the snake are grasped together and held as a unit. The restrainer should not hold the tube with one hand and the snake with the other. This is due to the fact that the snake might back out of the tube and bite.

Squeeze boxes are important tools for controlling more aggressive poisonous snakes such as king cobras. Another important fact while controlling of poisonous snakes is that restrainer must use a plastic shield or goggles to protect the eyes. This is especially true when spitting cobras are handled.

Small crocodilians can be manually controlled. The gloves should be used in order to prevent scratches. These animals can deliver hard blows with their tails. Therefore, extreme precautions should be taken while handling crocodilians. It is always difficult and hazardous in physical controlling of large crocodilians. So, chemical immobilization is very important for restraining large animals. Smaller animals can be controlled with a snare. Small crocodile is manually controlled by placing one hand in the neck region with the other hand on the caudal region (Fig.19.2). Large crocodilians can be restrained by heavy cargo nets.

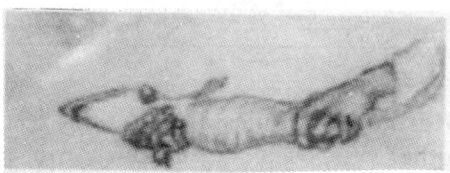

Fig.19.2: Controlling of a small crocodile.

Chelonian can be restrained by grasping the shell. Snapping turtle can be lifted by the tail with its belly. These turtles are aggressive bitters. Other turtles such as soft shelled turtle may occasionally bite. Forceps are used for extraction of heads of tortoises and turtles. The mouth of a chelonian can be opened by wooden wedge.

Housing and management:

Small reptiles can be housed in wooden boxes, screen cages and glass jars. Crocodilians and chelonians can be maintained in fenced or moated yards. The enclosure with sufficient space to permit normal movement is important. Cages should be properly constructed without openings so as to prevent them from escaping. There should not be any projections in the enclosures. This may cause injury to the animal. A number of substances such as sand, stone, wood shaving, peat moss, and newsprint may be used as floor coverings. The substrate used for flooring purpose should be nontoxic, non abrasive, inexpensive and easily disposable. The cage should be provided with logs, rocks, tree limbs, small boxes and other furniture which stimulate their natural behaviour such as climbing, hidings etc.

The enclosure constructed for crocodilians should be large enough for both bathing and other purposes. Artificial rocks, trees and plants should be present in the enclosure. The depth of water is maintained at about two feet. A basking area is essential for crocodilians. More areas for basking purpose may be created if sufficient space is available. The floor of land surface may be of concrete made so that it can be easily cleaned. The exhibit is illuminated by fluorescent tubes.

A suitable tank is needed for keeping freshwater turtles. Three to four small turtles can be kept in a single tank. Like crocodilians, the cage should have a land area for basking and a water area for bathing and feeding purposes. The water level may be just deep enough to cover the animal's carapace. Rough stones should be put in to the land area. Terrestrial turtles such as tortoises and box turtles are provided a larger space as compared to aquatic turtles. The wire fence should be run into the ground several inches so that turtles can not escape by digging out. Sand, soil and leaf litter may be provided as cage substrates.

Tortoises have the habits to make long treks in search of food and water in the wild. Sometimes, they climb over difficult terrains. Therefore, the enclosure should have several sizeable climbing mounds. The smooth concrete pool with a maximum water depth of 1.3 m should be constructed. Tortoises are known to wallow in mud during moderate heat or cold periods. So, provision for a mud wallow is essential. A heated winter shelter may be constructed. Winter quarter is bedded with hay.

Snakes and lizards require some basic and natural elements such as logs, rocks, tree branches, water bowls and artificial plants. Arboreal species are needed more tree limbs. The enclosure of large and heavy snakes (e.g. python) should have large and heavy logs and tree limbs. The floor may be covered with sand. However, small amounts of sand may be consumed during normal feeding. As a result, snakes may suffer from intestinal impaction.

The general care and management of captive reptiles are summarized below.
- Most reptiles are exothermic animals. They become thermostable by way of behavioural adaptations like climbing to cool or warm surfaces, seeking out sun or shade, or burrowing in sand and in mud. Therefore, a proper temperature should be maintained for the captive reptiles. The preferred body temperature of most snakes, turtles and nocturnal lizards ranges from 25 to 32^0C. Crocodilians have a preferred body temperature of 32 to 37^0 C. They hibernate at lower temperature. The tropical species generally do not hibernate. Therefore, reptiles should be kept warm and active all year long. A relative humidity in the range of 35-70 per cent is recommended for most species of reptiles. Forest animal requires more humid environment. Desert specimens normally tolerate lower levels of humidity. High humidity is essential for normal ecdysis in snakes. Very low humidity will cause dry skin and impedes removal of skin. Excessive moisture, on the other hand, may cause infection to the skin.
- Reptiles thrive well on artificial light. An alternative period of light and dark should be provided for captive reptiles. The sufficient lighting improves the appetite and reproduction of the animals. Incandescent bulbs may be provided for light and heat purposes. However, care should be taken so that animals are not directly in contact with these lighting devices. Direct contact with the bulbs may cause thermal burns. A lamp may be placed about 5 ft (1.52 m) above the floor for crocodilians.
- Adequate sanitation is important for keeping captive reptiles in healthy. Sani-

tation helps to prevent the spread of various diseases. Cages should be kept free from waste matter and uneaten food. The snake's skin left after ecdysis may shelter mites. So, it should be removed as early as possible. Cages and water dishes should be disinfected with proper disinfectants. Phenol, cresol and coal tar derivatives should not be used. These disinfectants may be toxic to the animals. The water tank should be kept clean regularly. The water should be changed twice a week for chelonians.

- The incomplete shedding of skin is commonly found in snakes. So, unshed skin can be removed with forceps after soaking the animal in lukewarm water. The turtle's shell should be regularly examined for any abnormal curvature of traumatic lesion. The substrates provided in the enclosure should be changed regularly.

- Overcrowding should be avoided. This may cause stress to the animals. Aggressive animals should be separated from the other cage mates. The regular supply of water is vital. Several factors such as stress, environmental temperature, seasonal variation and nutritional status affect the immune system of reptiles. So, these factors should be corrected. Crocodiles being translocated should be kept warm and the eye should be lightly bandaged.

- The young reptiles should be kept individually or in small groups. The cage should be provided with rocks, branches and other objects. The young should be moistened with a light mist of water. This will help in proper shedding of skin. Earthworms, raw fish, mealworms, cricket, mice etc. are the dietary ingredients for juveniles. One baby mice is given to a young reptile weekly. Captive reptiles may vomit due to over feeding. Fruits and vegetables may be given to turtles and lizards.

- The newly born tortoise hatchlings are kept in a plastic box. When tortoises become one year old, they may be placed in outdoor enclosures. Most young feed on plant materials readily. After one to two weeks of birth, tortoises may be provided chopped carrots, oranges and bananas. Dog food and raw eggs may be given in juveniles to promote growth. However, diet with high protein content causes irregular shell growth especially in smaller specimens. Excessive feeding of banana may cause diarrhoea. Excessive feeding also causes obesity in the captive tortoise particularly males. The supplementation of calcium is important for the proper growth of carapace. The ingestion of gravel, sand and other related materials by young inside the enclosure may cause intestinal impaction. So the enclosure should free from these substrates.

- There are different marking techniques used in identifying reptiles. They include scale-clipping, toe-clipping, branding, tattooing, notching the shell and hand drawn or photographic records. The initial period of quarantine is the suitable period for marking the reptiles.

- Quarantine is important for both wild and captive reptiles. Newly arrived reptiles should be quarantined for three months.

Chapter 20

Restraint, Housing and Management of Wild and Zoo Birds

Restraint: Improper handling results in stress on caged birds. Therefore, care must be exercised when restraining of birds. Bird can be physically controlled by grasping its head and body. Nets, crates, cages, plywood shields, gloves and towels are the common tools used in capturing and restraining of birds. Flightless birds have a sharp beak that serves as a weapon for defense. An ostrich bears clawed feet. It can inflict severe injury to the handler. The restrainers should never stand in front of these birds during restraining. Moreover, one should take precaution while entering into an enclosure containing ratites such as ostrich. A plywood shield can be used in order to protect from these birds.

Nets are the common tools for capturing waterfowl. Some waterfowl have sharp winged spurs. So, care should be exercised. The method of handling waterfowl is given in Fig. 20.1. Raptors have sharp tearing beaks and claws. Some members in this group use their beaks as a weapon of defense. Leather gloves should be used while handling raptors. Raptor can manually be controlled by throwing a towel over it and then grab the bird around its body. A hood (leather covering for a hawk's head) is an important tool for restraint. Raptors should be approached from the behind and then grasp the wings, body and legs together.

Fig. 20.1: Proper method of restraint of a water fowl.

Galliforms have claws. It may scratch during restraint. Males bear large tarsal spurs that may inflict injury. Galliforms should not be grabbed by their tail feathers. Long-legged birds such as cranes and storks have sharp and pointed bills. They may inflict serious injuries to the eyes and faces of handlers. The head should be controlled first by grasping either the neck or the bill while capturing. However, care should be exercised because their legs and wings are liable to be injured during handling. The restrainer should not enter into the crane's enclosure without protection (e.g. a stick in hand). A flamingo can be manually controlled by grasping the neck just below the head with one hand and the other hand can be placed at the base of the legs.

Most parakeets can be controlled by naked hands. However, they should be handled with care. Psittacine birds have large heavy bills and strong jaws. Therefore, care is needed during restraint. A cardboard can be used in controlling a macaw (Fig. 20.2). Pigeons and doves should be handled gently (Fig. 20.3).

Fig. 20.2: A cardboard is used in controlling a macaw

Housing and management:

Ratites in general are maintained in small groups within a large enclosure. The male ostrich is kept in a single small enclosure. Emus can be kept in flocks if adequate space is available. Normally, ratites become aggressive during breeding time. Therefore, birds should be confined properly.

Raptors can be maintained in a wire cage measuring 10 m x 5 m x 3.5 m. Sand, earth, straw and wood chips may be used as floor materials. A wooden shelter should be kept in each pen.

Fig. 20.3: Dove should be handled gently.

Many birds such as flamingos, cranes and water fowl are maintained in yards with low fences or on ponds and lakes surrounded by low fencing. Pelicans require an open water area measuring at least two to three times the land surface. Most aquatic birds can be kept in fresh water bodies. Marine species requires saline water. The maintenance of water quality is important for aquatic birds. The water should be changed periodically (e.g. once a week). This will help in maintaining the optimal water quality. The minimum environmental temperature should be maintained between 35 and 40^{0}F. Aquatic birds should be provided a minimum of 12-14 hours of photoperiod during breeding. The area for grazing and loafing is essential for waterfowl. A large brooder pen is required for flamingo chicks. A shallow water pan is needed in order to stimulate drinking and bathing habits of young chicks. Jungle fowl can be kept in both indoor and outdoor houses.

A variety of cages are recommended for pigeons and doves. Nesting boxes should be provided in the enclosures. The pigeon house should be dry and well ventilated. The separation of aggressive cage mate is very important. Newly arrived birds should be kept in quarantine shed. Parrots can be maintained in a wide variety of commercially prepared cages. Nest boxes, leaves and paper stripes should be provided. The fresh and nontoxic tree branches should be kept within the enclosure. Toys, ladders and bells are sometimes provided in the cage. Some food items such as bones, sprouts, tree branches, buds and leaves may be provided for "occupational therapy". A breeding pair can be housed in a minimum area measuring 7.2 m x 1.4 m x 2.4 m.

In captivity, cranes are required considerably more space than most birds. In the wild state cranes usually rear one chick per year. However, they may produce as many as five fertile clutches per pair in captive breeding programme. Crane chicks grow rapidly during first two months. Special care needs to be given for proper bone development. The condition of weakened leg caused by too rapid a weight gain is a common problem in the zoo. A high protein diet may cause higher incidence of leg twisting and wing abnormalities as compared to a diet containing lower protein.

Storks are gregarious birds and primarily terrestrial. Storks can be kept in a small aviary. The outdoor area may be provided with nest boxes which are fixed on a wall. During winter months the temperature of indoor quarter should be maintained at 20 to 25^0C. Newly born chicks are placed in an incubator. Hand reared chicks should be provided adequate calcium and vitamin D_3 to avoid nutritional bone diseases.

The identification of birds is important for various purposes. Loose plastic bands or metal leg bands are tagged in caged birds. Care should be taken so as to prevent any injury caused by metal bands. Nail trimming is a routine management practice. Care, however, is needed to avoid bleeding during nail trimming. Debeaking is also periodically done. The upper beak is normally trimmed in some birds. The primary and secondary feathers are sometimes clipped. However, clipped wing birds are able to flutter on the floors without loosing balance. There should not be any bleeding during clipping of wing feathers. The other management practices routinely followed in captivity are described below.

- Food should be provided at same time, same person and same container as far as possible.
- Inadequate diet causes nutritional stress for captive birds. Therefore, diets should be scientifically evaluated. The body weight should be measured regularly.
- Stress should be avoided during capture and physical examination of birds as far as possible.
- Reduced appetite, abnormal breathing patterns, soiled vent, loss of feathers, bleeding, loss of body weight and reduced vocalization are the clinical characteristics of sick birds. Sick birds should be isolated. They should also be properly examined.
- Regular cleaning of floors, utensils and other necessary items is important.

Chapter 21

Health Care of Wild and Zoo Mammals

Monotremes:

Physiological parameters: Monotremes have a lower metabolic rate as compared to placental mammals. The cloacal temperature varies from 28.7 to 31.5^{0}C in echidna. The nonnucleated erythrocytes are found in echidna. The heart rate of platypus varies from 69 to 98 per minute. Blood samples can be collected from the femoral or the ventral coccygeal veins. The milk of echidna is reported to have 19.6 per cent fat.

Diseases: There is paucity of information on various diseases of monotremes. However, pneumonia and enteritis have been reported in platypus. In captivity, monotremes are likely to be infected by other common pathogens such as *Streptococcus* and *Staphylococcus* spp during stress condition. The platypus is known to be infected with the trematodes *Moreauia mirabilis* and *Maritreme ornithorhynchi*.

Salmonella infections are known to occur in captive echidna. It causes acute enteritis. They are mainly associated with the stress.

Marsupials:

Physiological parameters: During high environmental temperature, marsupials exhibit some physiological and behavioural responses, viz. sweating, salivation, panting and licking. The cloacal temperature of marsupials varies from 32 to 37^{0}C depending on the species. The heart rate ranges from 125 to 150 per minute in kangaroo. Blood samples are collected for various studies and the collection sites are the lateral caudal veins or the cephalic veins. The femoral and the ventral coccygeal veins are also the sites for blood collection in small marsupials. Although, marsupials are characterized in having nonnucleated erythrocytes, koalas, however, are reported to have 4-40 per cent nucleated erythrocytes.

Diseases: Captive marsupials are susceptible to a wide variety of infectious diseases. The common bacterial diseases that are reported in marsupials include tuberculosis, salmonellosis, pasteurellosis, chlamydiosis and pneumonia. Diseases like brucellosis and vibriosis have also been reported in some marsupials.

Lumpy jaw has been reported in captive marsupials. The clinical signs of lumpy jaw include severe inflammation, swelling of soft and bony tissues, salivation and pus formation. The affected animal is unable to masticate. Death occurs due to starvation, toxaemia or septicaemia. Treatment of lumpy jaw includes local debridement of abscess, long term antibiotic therapy and hygienic measures. Overcrowding should be avoided.

Tuberculosis caused by acid fast organisms (*Mycobacterium tuberculosis*, *M. bovis* and *M. avium*) has been reported in marsupials. Animal suffered from tuberculosis shows the clinical signs of coughing, dullness, emaciation, diarrhoea and ataxia. Animals infected with tuberculosis should be culled.

Salmonellosis has been reported in many members of the order Marsupialia. Young animals are more vulnerable than those of other age groups. It is caused by *Salmonella* spp. Diarrhoea, septicaemia, enteritis, emaciation, dehydration and interstitial nephritis are the clinical signs of salmonellosis. Strict hygienic management and antibiotic therapy are the two important measures for controlling of salmonellosis.

Pasteurellosis caused by *Pasteurella multocida* has been reported in kangaroo, wallaby, opossum etc. Overcrowding, inclement weather, transportation stress and shortage of food are some predisposing factors causing pasteurellosis in marsupials. Animal may die without showing any clinical sign in acute case.

Viral diseases (e.g. rabies) have been reported in marsupials. Lymphoid neoplasia is the most common form of neoplasia found in captive as well as in wild koalas. It is characterized by loss of body weight, lethargy, abdominal pain, diarrhoea, dyspnoea, lymphadenomegaly and neurological signs.

Common parasitic diseases found in marsupials are coccidiosis, babesiosis, toxoplasmosis, trypanosomiasis and leishmaniasis. Captive kangaroos are more prone to coccidiosis. Marsupials are very susceptible to toxoplasmosis. It is characterized by anorexia, lethargy, diarrhoea, dyspnoea, weight loss, jaundice and nervous dysfunction.

External parasites such as ticks, mites and fleas occur in marsupials.

Photosensitization occurs in marsupials. Erythema and oedema are the two early lesions. Unpigmented and sparsely haired areas are mostly affected.

Insectivores:

Physiological parameters: The hibernation during low environmental temperatures has been reported in insectivores. The normal body temperature of hedgehog is reported to be 95.2°F. The femoral vein is the site for collection of blood in insectivores. Hedgehog milk contains about 10.1 per cent fat.

Diseases: Insectivores are susceptible to many infectious diseases. They include leptospirosis, salmonellosis, pasteurellosis and foot and mouth disease. Fleas (e.g. *Doratopsylla dasycnema* and *Palaeopsylla sorius*) are commonly occurred in shrews. Fur mites and larval ticks are very common in these animals.

Chiropterans:

Physiological parameters: Bats may be homoiothermic or heterothermic in part of their lives. Flying foxes are reported to be obligate homoiothermic mammals. The rectal temperature in fruit bats varies form 37 to 38.5^{0}C. Vampire bats have been reported to produce copious quantities of urine. The heart rates in bats tend to be varied from 100 to 970 beats per minute. This, however, depends on the size of animal, body weight, climatic conditions etc. The haemogram of Malaysian flying fox is as follows: RBC (x10^6/mm$^{3)}$) :7.4-9.8; Hb (g/dl):12.9-15.7; PCV (%):42-47; WBC (x10^3/mm^3):7.8-18.6; Segmented neutrophils (%):11-56; Eosinophils (%):0; Basophils (%):0; Lymphocytes (%):43-88 and Monocytes (%):0.0-2.0.

Blood may be collected by clipping a toenail or by heart puncture. Venipuncture for collection of blood in bats is also done by stretching the wing and locating the vein on the anterior edge between the carpus and shoulder. The milk of fringed bat *(Myotes thysanodes)* contains 17.9% fat, 12.1% protein and 3.4% lactose.

Diseases: Bats are known to be infected with some infectious diseases. They include rabies, leptospirosis, tuberculosis, haemorrhagic septicaemia, salmonellosis, spirochaetosis, brucellosis and shigellosis. Bats are infected by at least 28 different viruses. Yellow fever and vesicular stomatitis are known to occur in bats.

Rabies, the most important viral disease, is found in these mammals. Both furious and dumb forms have been reported in vampire bats. Most bats may recover from rabies infection. Some bats, however, become carriers of this disease and may shed virus in the saliva for up to 16 months. The important clinical signs of rabies in bats include aggressiveness, restlessness, resting on the ground, biting of inanimate objects, difficulty in swallowing food or water and unusual daytime flying. Captive bats should be vaccinated with inactivated tissue culture vaccines.

Salmonellosis has been reported in bats. Infected animal shows the clinical symptoms of high rectal temperature, anorexia, depression, enteritis and septicaemia.

Bats may be the carriers of parasites (e.g. *Trypanosoma cruzi* and *T. equinum*) causing trypanosomiasis. Toxoplasmosis has been reported in bats. Infected animal shows chronic febrile syndrome. Bats may infest with a great variety of flukes (e.g. *Ophiosacculus mehelyi and Prosthodendrium dinanatum*). Nasal mites are found in bats. A large number of external parasites are also found in these animals. Bats infested with external parasites may be treated with malathion and DDT. Free filarid worms are found in the peritoneal cavity of many long-tongued and short tailed fruit bats.

Toxocariasis caused by *Toxocara pteropodis* has been reported in island flying fox (*Pteropus hypomelanus*). Dilated cardiomyopathy caused by hypovitaminosis E has been reported in flying fox. The clinical signs include lethargy, anorexia, hypothermia, reluctance to fly and cranial oedema.

Primates:

Physiological parameters: Primates are widely used for the purpose of bio-medical research. Among all the primates used for research, monkeys have been received the most attention worldwide by the scientific community. Many primates are caught frequently in the wild state for laboratory use.

The normal rectal temperature is reported to vary from 95.5 to 101.8^{0}F. The normal respiratory rate in large primates ranges from 12 to 20 breaths per minute, whereas, in small primates it varies from 20 to 50 breaths per minute. The normal heart rate ranges from 95 to 112 beats per minute in medium and large primates. The milk of rhesus monkey contains 12.2 per cent solids.

Blood may be collected from the radial, jugular or femoral veins in primates. Sometimes, blood is obtained from the marginal ear veins. The blood values of rhesus monkey are as follows: RBC (x10^6/μl):3.1-8.1; Hb (g/dl):8.7-14.7; PCV (%):41-47; WBC (x10^3/μl):10-16; Neutrophils (%):35-50; Eosinophils (%):1-6; Basophils (%):0-1; Lymphocytes (%): 36-75 and Monocytes (%): 0-6.4.

Diseases: Many infectious diseases found in human race have been known to occur in nonhuman primates. This is due to the close phylogenetic relationship between nonhuman primate and man. Many diseases reported in nonhuman primates are known to have zoonotic importance. Therefore, caution must be exercised during handling of nonhuman primates. Primates may also be the carriers of numerous infective agents. Some important diseases reported in nonhuman primates are given in Tables 21.1, 21.2 and 21.3.

Mycotic Disease: Several mycotic diseases such as cryptococcosis, nocardiosis, phycomycosis and candidiasis have been reported in nonhuman primates. Ringworm is very commonly found in these animals. The causal agents are *Microsporum* spp. and *Trichophyton* spp. The affected animal shows alopecia and scaliness. It can be treated with griseofulvin (25 mg/kg body weight) orally for 3-4 weeks. Cryptococcosis causes by *Cryptococcus neoformans* has been reported in macaques and marmosets. It is characterized by coughing, anorexia and ocular discharges. It can be treated by fluconazole in combination with flucytosine. Candidiasis caused by *Candida* spp. is also a common mycotic ailment in nonhuman primates. Dermatophilosis has been reported in orangutans. It is caused by *Dermatophilus congolensis*.

Table 21.1: Some reported bacterial diseases in nonhuman primates:

Disease	Etiological agent/susceptibility	Clinical signs	Diagnosis	Treatment and control
Tuberculosis	*M. tuberculosis var hominis* (human type), *M. tuberculosis var bovis* (bovine type), and *M. tuberculosis var avium* (avian type); all primates.	Clinical signs appear in advanced cases. Symptoms include anorexia, cough, emaciation, lethargy, diarrhoea and enlarged lymph nodes.	Skin test (administration of 0.1 ml of mammalian tuberculin intradermally into the upper eyelid or abdominal skin; positive case shows eye swelling, erythema and skin necrosis; radiography of thorax and culture of sputum and faeces	Administration of isoniazid (5-10 mg/kg body weight) once a day for continuous three months followed by three months rest and repeat for another three months ; euthanasia for all positive cases to prevent public health hazards; tuberculin test of newly arrived animals; and proper sanitation.
Salmonellosis	*Salmonella* spp.	Diarrhoea, rapid dehydration and cramps.	Clinical signs and culture of blood and faeces.	Administration of antibiotics with fluid therapy.
Shigellosis	*Shigella* spp.; *S. fxnleeri* is commonly found in nonhuman primates.	Diarrhoea with blood and mucus in the faeces, anorexia, lethargy, abdominal pain, dehydration and facial oedema.	Clinical sings and culture of blood and faeces.	Antibiotic with fluid therapy. It is associated with poor hygiene and sanitation.
Tyzzer's disease	*Bacillus piliformis*; monkeys.	Elevated body temperature, diarrhoea and jaundice.	Clinical signs, postmortem lesions and bacteriological culture.	Administration of antibiotics.
Leptospirosis	*Leptospira icterohemorrhagiae* and *L. ballum*; macaques and baboons.	Icterus, abortion, still birth, mucosal haemorrhages and convulsion.	Clinical signs, urine culture and serological tests.	Administration of antibiotics.

Table 21.2: Some reported viral diseases in nonhuman primates:

Kyasanur forest disease	Arbovirus; langurs and bonnet macaques; found in Kyasanur forests of Karnataka in India.	Fever, anorexia, diarrhoea, encephalitis, weakness and epistaxis.	Isolation of virus.	Symptomatic treatment and control of tick vectors (*Ixodes* and *Haemaphysalis*).
Yellow fever	Arbovirus; large apes, monkeys etc.; occur in Africa, Central and South Americas.	Jaundice, fever, haemorrhages, albuminuria, prostrating illness and death.	Isolation of virus from blood and liver.	Yellow fever vaccine and control of mosquitoes (e.g. *Aedes*).
Rabies	Rhadovirus; rhesus monkeys, chimpanzees etc.	Fever, diarrhoea, vomiting, conjunctivitis, hyperirritability and paralysis.	Clinical signs and serological tests.	Vaccination (killed vaccine) with quarantine measures.
Monkey pox	Pox virus; chimpanzees, rhesus monkeys, squirrel monkeys etc.	Fever, facial oedema, maculoppapular rash and various pustules.	Serological tests.	Vaccination.
Herpes B virus infection	*Herpesvirus simiae*; macaques.	Lip and tongue ulcers, nasal discharge, anorexia, diarrhoea ad conjunctivitis.	Clinical signs and antibody titer.	Symptomatic treatment, care in handling and protective clothing.

Table 21.3 : Some reported parasitic diseases in nonhuman primates:

Amoebiasis	*Entamoeba histolytica, E. coli, Balantidium coli* etc.; all primates.	Profuse diarrhoea, dehydration, enteritis and severe depression.	Faecal examinations.	Administration of metronidazole (30-60 mg/kg body weight daily for 5-10 days) and supportive electrolyte therapy.
Malaria	*Plasmodium* spp.; monkeys apes etc.	Anorexia, headache, and joint and muscle pain.	Blood smear examinations.	Administration of quinine and control of mosquito vectors.
Trypanosomiasis	*Trypanosoma cruzi*; rhesus monkeys, gibbons and liontailed macaques.	Anorexia, dehydration and loss of body weight.	Blood smear examinations.	Administration of specific parasitic drugs.
Toxoplasmosis	*Toxoplasma gondii*; primates such as chimpanzees and macaques.	Anorexia, lethargy, diarrhoea, emaciation and abdominal pain.	Serological tests, histological findings and blood examination in acute case.	Administration of sulphadiazine and control of rodents.
Coccidiosis	*Isospora* spp.; baboons, marmosets etc.	Diarrhoea and rectal prolapse in chronic cases.	Faecal examinations.	Administration of sulfamethaxine with supportive oral electrolytic therapy.

Nutritional Diseases: Diseases caused by deficiencies of various vitamins have been reported in primates. Vitamin A is required for proper vision. The deficiency of vitamin A causes retarded growth, xeropthalmia and dry hair coat. Captive primates are commonly associated with vitamin C deficiency causing scurvy. The deficiency symptoms of vitamin C include haemorrhages of gums, swelling of bones at the epiphyses and diarrhoea. A daily intake of vitamin C @ 4 mg/kg body weight will meet the requirement of the animal. Citrus fruits should be regularly supplemented for captive primates. Vitamin B_1 deficiency causes muscular weakness, loss of body weight and ataxia. A dietary level of 0.03 mg/kg body weight is essential for maintenance. Vitamin B_6 deficiency causes anaemia, dermatitis, dental caries and arteriosclerosis. Primates require a dietary level of 1.5- 2 mg/day of vitamin B_6.

Other diseases: Leaf eating langurs consuming excessive indigestible *Acacia* can suffer from gastrointestinal linear foreign body syndrome. It is characterized by abdominal pain. Animals may exhibit a hunched posture with nonspecific vomiting, anorexia and depression symptoms.

Acute gastric dilatation or bloat has been found in many nonhuman primates. It is characterized by depression, subnormal temperature and distended abdomen. The evacuation of gas from the stomach is the immediate treatment to save the animal. Administration of surface reducing antacids through stomach tube and parenteral administration of fluids containing adequate amounts of sodium, potassium and chloride are advocated.

Simian immunodeficiency virus, a member of lentivirus subgroup of retroviruses has been found in nonhuman primates. It is related to HIV virus.

Pangolins:

Physiological parameters: The normal body temperature of pangolins ranges from 32^0C to 35.2^0C.

Diseases: There is scanty information on bacterial and viral diseases of pangolins. Parasitic diseases caused by tape worms and ticks have been reported. Drugs like piperazine and ivermectin are used in controlling the cestode infections. Ticks can be controlled by applying pesticides like DDT. However, precautions should be taken to avoid accidental poisoning of the animals with pesticides. It has been reported that death of captive specimens in most cases occurs due to stress and malnutrition. Lacerated wound in pangolin can be treated with povidine-iodine (Betadine 5%) ointment with intramuscular injection of cefotaxime sodium (250 mg twice a day for five days).

Lagomorphs:

Physiological parameters: Lagomorphs, as mentioned earlier, thrive well in different habitats ranging from semi-arid deserts to tropical and temperate conditions. The usual rectal temperature of rabbit is recorded 102.5^0 F, though the body temperature of the New Zealand white rabbit may range from 99.1 to 103^0F. The respiration rate is reported to be 39 breaths per minute (in New Zealand white rabbit: 32-60 breaths/ minute) in rabbits. The heart rate is reported to be 205 (range 123-304) beats per minute. The total solids in rabbit milk (4th to 21st day of lactation) has been reported as 26.1 to 26.4 per cent (protein: 13.2-13.7 %; fat:

9.2-9.97%; lactose: 0.86-0.87% and ash: 2.4-2.5%). Doe produces about 30-50 g milk in the first two days of lactation. However, milk production increases up to 250 g towards the end of third week of lactation. After third week, the milk production is reported to decrease rapidly.

Blood may be collected from the jugular vein and the marginal ear vein. The auricular artery may also be the site for collection of large amounts of blood in lagomorphs. The salient feature of rabbit blood is that it contains pseudo-eosinophils (or called amphophils) which are comparable to the neutrophils of large domestic animals. Another important haematological feature among these common laboratory animals is that the basophils are regularly found in small to modest numbers in the circulation (frequently 8-10%). The normal blood values for a rabbit are as follows: RBC (x 10^6/μl): 4.6-6.9; Hb (g/dl):12.1; PCV (%): 36-48; TLC (x 10^3/μl):5-21.5; Neutrophils (%):32-59; Eosinophils (%):1.0-4.0; Basophils (%):2.4-9.0); Lymphocytes (%):20-68 and Monocytes (%):1.5-16.0.

The total urine excreted is reported to be about 130 ml/kg body weight daily in rabbits. It is important to note that the information given on physiological parameters in this text is based on the study conducted on domestic rabbits.

Diseases: A wide variety of diseases have been reported in lagomorphs. The various diseases recorded among the members of the order Lagomorpha are mainly based on findings in domestic rabbits. Diseases may occur as a result of infections or simply may be due to poor husbandry practices. Overcrowding causes many diseases in domestic rabbits. Besides microbial and parasitic infections, the other categories of lagomorph's diseases such as nutritional deficiency, metabolic disorders, physical injuries and other non specific ailments have also been reported. Therefore, an early diagnosis of diseases is vital from the production as well as mortality point of view. Nevertheless, some widely known diseases occurred in rabbits are discussed in the following text.

Bacterial diseases: Bacterial diseases reported in lagomorphs include pasteurellosis, pseudotuberculosis, spirochaetosis, listeriosis, salmonellosis, conjunctivitis, mastitis, anthrax, brucellosis, necrobacillosis, Q fever, tularemia, tuberculosis and enteric disease. A short description of selected diseases is given below:

Pasteurellosis: Pasteurellosis is commonly found in rabbits like rodents. It is a highly contagious disease and is caused by *Pasteurella multocida*. Some may be the healthy carriers. However, the occurrence of this disease in rabbits depends on the immunity power of the animal. The disease is characterized by snuffles, abscesses, nasal rash, pneumonia, conjunctivitis and septicaemia followed by death. The testicles may also be affected resulting sterility in males. The purulent discharges are noticed in eyes and nose. Pasteurellosis can be diagnosed by the clinical symptoms and the serological tests. The response to treatment of affected animal is poor. However, drugs like oxytetracycline and penicillin may be used with limited success. Tetracycline at a dose rate of 50 mg/kg body weight may be useful for treatment of pasteurellosis. Caution must be exercised in application of antibiotics. Antibiotics may cause the disruption of natural environment of gut flora resulting in diarrhoea. The animal may even die.

Pseudotuberculosis: Pseudotuberculosis is also known as yersinosis. Wild rabbits and hare are commonly encountered with this disease. The causative organism of pseudotuberculosis is *Pasteurella (Yersinia) pseudotuberculosis*. The affected animal shows the clinical signs of anorexia, emaciation, dyspnoea and lassitude. On postmortem examination, whitish-yellow nodules are found in lymph nodes, intestine, lungs, liver and spleen. Joint lesions have been reported in very advanced cases. It can be diagnosed by clinical signs as well as postmortem lesions. Practically there is no treatment of this zoonotic disease. The incidence of pseudotuberculosis can be reduced by prevention of taking contaminated food materials. The incidence of the disease can also be reduced by adopting general hygienic measures. Rodents may be the carriers of pseudotuberculosis. Therefore, control of rats and mice within the farm premises is important.

Spirochaetosis: It is also known as treponematosis, rabbit syphilis or vent disease. Spirochaetosis is caused by a spirally twisted organism called *Treponema cuniculi*. This venereal disease is transmitted by coitus in both sexes. However, infection may transmit to offspring through mother. The clinical signs include infertility, formation of small vesicles or ulcers, scabs, and hair loss from the surrounding of penis or vulva. Other body parts such as lips, eyelids, ears and nose may also be affected. Diagnosis of spirochaetosis is based on genital lesion and microscopic examination of the organism. The animal may be treated with antibiotics (e.g. penicillin 42000 IU/kg body weight for three weeks at seven days interval). As a control measure, infected animal should not be used for mating.

Listeriosis: Listeriosis is sporadically found in rabbit farms. The causal agent of this disease is *Listeria monocytogenes*. Diagnosis of listeriosis based on clinical signs is very difficult. Affected animal, however, may show anorexia, loss of body weight, depression and torticolis. Abortion may occur in does. Small white focal lesions are found in liver tissues. Listeriosis shows poor response to medication.

Salmonellosis: Salmonellosis in rabbits is commonly caused by *Salmonella typhimurium* and *S. enteritidis*. It is characterized by rise of temperature, anorexia, depression and sometimes diarrhoea. Abortion may take place in the pregnant animal. Septicaemia is also reported in per acute cases. Necrotic lesions are found in liver. As salmonellosis is largely spread through contamination of water, food and litter, strict hygienic condition, therefore, is the only corrective measure to check this disease. However, administration of antibiotics and sulphonamides is also suggested to combat salmonellosis. Some lagomorphs may be the carriers of this bacterial disease.

Mastitis: Mastitis is also called blue breast. It is commonly occurred in rabbits maintained at farm with mesh floor. The disease is usually caused by *Staphylococci* and *Streptococci* organisms. Mammary glands become hot, redden and hard in the congestive state. The affected area may be cyanotic (hence the name "blue breast"). Treatment includes bathing of mammary glands with warm water containing an antiseptic. The affected animal may be treated with penicillin intramuscularly (10000-15000 units/0.5 kg body weight daily). It is uneconomical to treat animals with purulent mastitis. Lagomorphs with fibrosed udders should be culled. Mastitis can be prevented by taking adequate sanitary measures.

Mucoid enteritis: Mucoid enteritis is also known as bloat or scour. This disease has a worldwide distribution. Mostly growing rabbits are susceptible to mucoid enteritis. The incidence of mortality in adult animals is reported to be less. The etiology is largely unknown. However, some causative organisms such as *Clostridium perfringens* type E, *C. difficile, Escherichia coli, Salmonella typhimurium, Bacillus piliformis and Pasteurella pseudotuberculosis* have been found in the animals affected with mucoid enteritis. It is characterized by anorexia, dehydration, subnormal body temperature, rough hair coat, dullness and humped posture. Affected animals may also exhibit profuse diarrhoea or constipation. Animals often show a bloated abdomen. The colon and rectum are filled with a considerable amount of mucus which are translucent and gelatinous in nature. Inadequate supply of water and roughage to the animal are likely to be the predisposing factors for mucoid enteritis. It can be diagnosed by clinical signs and necropsy findings. Animals may be treated with intramuscular injection of antibiotics e.g. streptomycin at a dose rate of 50-100 mg/animal. Electrolytes therapy is also useful. Strict hygienic measures, restricted feeding, supplementation of adequate amount of hay and drinking water with vitamins are some important husbandry measures to be practised for controlling of mucoid enteritis.

Viral diseases: The viral diseases reported in lagomorphs include myxomatosis, rabbit pox, rabbit viral haemorrhagic diarrhoea, European brown hare syndrome and papillomatosis. A short description of each disease is given below:

Myxomatosis: Myxomatosis is also called mosquito disease. It is a highly fatal infectious disease of wild and domestic rabbits. The cottontails are reported to be the dangerous carriers. Myxomatosis is caused by Myxovirus. This virus is very resistant to disinfectants, weather and physical changes (e.g. heat and cold). Mosquitoes and fleas are the main vectors through which the transmission of this disease is taken place. Animals may also be infected through contaminated equipment or contact. The symptoms of myxomatosis are conjunctivitis, anorexia, high rise of temperature (up to 42.2^0C), and oedematous swelling in eyelids, lips, ears, nose and external genitalia. In acute cases, some rabbits may die within 48 hours of infection. Fibrotic wartlike lesions are found on ears, nose, eyes, vent and forefeet of animals. These nodules, however, disappear within three to four weeks after infection. The spleen of the affected animal becomes enlarged. There is no treatment for myxomatosis. Good hygiene and control of insects are two essential steps for prevention of myxomatosis.

Rabbit pox: Domestic rabbits are susceptible to rabbit pox. It is clinically characterized by nasal discharge, fever and skin rash. This virus is closely related to the vaccinia virus. The vaccination is effective against rabbit pox.

Rabbit viral haemorrhagic disease: It has been reported in free living rabbits. This is caused by Calicivirus. The mortality is very high. It can be spread by oral and nasal transmission.

European brown hare syndrome: This has been reported in free ranging hare. It is caused by Calicivirus. The salient pathological features of European brown hare syndrome are rapid progression, mild nervous symptoms, diffuse or petechial haemorrhages on serosa and mucosa, occasional jaundice ,degeneration and congestion of liver, spleen and kidney. There is extremely high morbidity and mortality among the affected animals.

Papillomatosis: Papillomatosis has been reported in domestic rabbits. The affected animal shows papillomatus lesions around the mouth. Animal can be protected by vaccination.

Fungal or mycotic diseases: Ring worm is commonly found in rabbits. It is caused by various species of fungi in the genus *Trichophyton* and *Microsporum*. This contagious disease can occasionally transmit to human race. It is characterized by the formation of circular, reddened and greyish or yellowish crusts on the ears, face, nose, head and forepaws. The skin becomes inflamed and irritated. It generally begins with circular bald patches. Treatment includes the topical application of fungicides. Griseofulvin may be used in the feed @ 825 mg/kg of feed for about 10-14 days. Separation of infected animals and good hygienic measures can rapidly eliminate ring worm infections.

Parasitic diseases: Several species of parasites have been found in lagomorphs. Information on parasitic diseases in literature is primarily based on the findings reported in domestic rabbits. The parasitic infections in rabbits occur mainly due to poor husbandry and unhygienic conditions. Some important parasitic diseases of lagomorphs are described below:

Coccidiosis: This protozoan disease is very common in rabbits. It is caused by several species of *Eimeria*. There are two forms of coccidiosis: hepatic coccidiosis and intestinal coccidiosis. The hepatic form is mainly due to *Eimeria stiedae*. Growing animals are highly susceptible to this form. Clinical signs of hepatic coccidiosis include anorexia, rough hair coat, diarrhoea and emaciation. Yellowish white nodules are found in the liver. Hepatic coccidiosis can be diagnosed by the presence of oocysts in the liver tissue and in the bile ducts.

The main symptoms of intestinal coccidiosis are diarrhoea, weight loss, rough hair coat, potbelly and low intake of food and water. The oocysts is demonstrated in the faeces. Coccidiosis can be treated with sulfadimethoxine (0.5 to 0.7 g/litre drinking water as curative dose and 0.25 g/litre drinking water as preventive dose). The following husbandry measures are important for controlling of coccidiosis.

- Rabbits should be kept in wire cages with sufficient floor space.
- Weaning should be done at proper time.
- Regular cleaning of floor, feeder, nest box and other equipment with proper disinfectants.
- Proper ventilation should be maintained.
- Animal should not be fed any contaminated feed.

Tapeworm infections: Rabbits act as intermediate hosts for *Taenia pisiformis* and *T. serialis*. In the wild, tapeworm infection causes mild diarrhoea with occasionally weight loss in rabbits. Nevertheless, tapeworms are seldom found in domestic rabbits.

Roundworm infections: Round worms such *Trichostrongylus*, *Graphidium* (stomach worm), *Trichuris* and *Strongyloides* have been reported in wild rabbits. *Strongyloides* are found in rabbits living in wet, dark and poorly managed cages. Animals may be infected by consuming green grasses contaminated with round worm larvae.

Ear canker: Ear canker or mange is commonly found in rabbits. The ear mite, *Psoroptes cuniculi* is very common. It is characterized by shaking of head, flapping and scratching of ears. Yellow or brown scabs are also found in the ear canal. The scabs may become waxy. It is frequently complicated by bacterial infections. The middle ear is sometimes affected and the animal shows a symptom of constantly held its head to one side. Insecticides (e.g. malathion) may be applied locally. The hutches must be thoroughly disinfected. Moreover, straw litter must be changed frequently.

Skin mange: It is not very common in domestic rabbits. *Sarcoptes scabiei* and *Notoedres cati* are infrequently found. Lesions start at the edge of lips, eyes and nostrils and then spread to head and forepaws. The affected animals should be culled.

Other diseases: Rabbits have been frequently encountered with many other health problems. They include hutch burn, sore hocks, trichophagy, metritis, dental malocclusion, wet dewlap, torticollis and heat stroke.

Hutch burn or urine burn is a pathological condition in which anus and external genitalia are affected. This noninfectious disease is developed in lagomorphs due to damp and dirty floor conditions. The animal can be treated with antibiotic ointment. As a preventive measure, animals should be kept in clean and dry floors.

Trichophagy or fur eating occurs in domestic rabbits. Overcrowding, unbalanced ration and many other factors may be responsible for this condition.

Rodents:

Physiological parameters: The normal body temperature of rodents usually ranges from 94.6 to 104.0^{0}F. Most rodents, however, have a body temperature below 100.0^{0}F. The respiratory rates vary from 16 to 250 in rodents. The heart rates of rodents range from 96 to 858 beats per minute. The normal rectal temperature, heart rate and respiratory rate of some selected rodents are given in Table 21.4.

Table 21.4 : Physiological data of selected rodents.

Animal	Rectal temperature(^{0}F)	Heart rate(beats/minute)	Respiratory rate(breaths/minute)
Rat	95.2-99.2	261-600	75-115
House mouse	98.0	328-780	84-230
Guinea pig	102.2-104.0	150-400	69-150
Hamster	97.0	300-600	33-250
Chinchilla	100.0	100.0	40-100

Source: The Merck Veterinary Manual, 15th edn.

The composition of milk has been studied in many rodents. Rat milk contains 26.5 per cent total solids and 73.5 per cent water. The other rodents like porcupine and hamster also have high percentage of total solids in milk. The porcupine milk contains 70.0 per cent water and 29.7 per cent total solids, while hamster milk has 73.6 per cent water and 26.4 per cent total solids.

The total blood volume is about 80 ml/kg body weight in common laboratory animals. A maximum blood of 8-12 ml/kg body weight can be drawn without harm. Blood may be collected for various haematological and biochemical studies. The usual sites for blood collections in rodents are ear vessels, tail vessels, jugular veins and femoral veins. Blood may also be collected from heart and orbital sinus. Heart, orbital sinus and jugular vein are selected for collection of large amounts of blood. Small amounts of blood can be collected by clipping a toe nail or by nicking the tip of the tail. It has been reported that about 10-30 per cent nucleated erythrocytes are found in the newborn golden hamster. The salient feature of guinea pig blood is that lymphocytes contain cytoplasmic inclusion bodies like vacuoles called Kurloff bodies. Moreover, guinea pig blood has pseudoeosinophils.

The normal haematological values for some rodents are given in Table 21.5.

Diseases: Many diseases of both infectious and non infectious origins have been reported in feral and captive rodents. Some of the reported diseases have zoonotic importance. Therefore, a disease prevention programme is needed to eliminate

Table 21.5: Haematological values for selected rodents.

Animal	RBC (x10⁶/mm³)	Hb (g %)	PCV (%)	WBC (x10³/mm³)	Neutrophils (%)	Eosinophils (%)	Basophils (%)	Lympho-cytes (%)	Monocytes (%)
Rat	7-10	12-18	35-45	5-23	10-50	0-5	0-1	50-70	0-10
Mouse	7-11	10-20	35.4-40.0	4-12	5-40	0-5	0-1	30-90	0-10
Guinea pig	4-7	11-17	35-45	7-14	20-60	0-5	0-1	30-80	2-20
Hamster	7-8	16.6-18.6	45-49.8	7.02-10.1	18-40	0-1	0-1	56-79.8	2.43
Chinchilla	5.6-8.4	11.8-14.6	27-54	5.4-15.6	39-54	0-5	0-1	45-60	0-5

or restrict infections. A few important diseases found in rodents are described in Table 21.6.

Table 21.6: Some reported diseases in rodents:

Disease	Etiological agent/animal susceptibility	Clinical signs and lesions	Diagnosis	Treatment and control
Salmonellosis	*Salmonella* spp. (e.g. *S. typhimurium* and *S. enteritidis*); mice, rats, hamsters and guinea pigs.	Anorexia, loss of body weight, rough hair coat, conjunctivitis, pale and loose faeces and sporadic death; focal necrosis on spleen and liver.	Clinical signs, faecal examination and serological tests.	Administration of sulfamethazine sodium (2 g/litre of drinking water for 5-10 days); preventive measures against introduction of carrier rodents; prevention of feed and water from faecal contamination.

Tyzzer's disease	*Clostridium piliformis;* a wide variety of laboratory animals such as guinea pigs, rats, mice and gerbils.	Diarrhoea, anorexia, poor body coat, dehydration and sudden death; inflammation of ileum and liver necrosis.	Clinical signs and histological findings of necrotic liver.	Application of oxytetracyclin e (500 mg/litre of water for 15-30 days); isolation of affected animals and strict hygienic measures.
Tularemia	*Francisella (Pasteurella) tularensis);* hamsters, rats, mice, muskrats, voles and beavers.	Anorexia, lethargy, slowed gait and tameness, white pin point lesions in liver, kidney and mesenteric lymph nodes.	Gross and histopathologi cal lesions, and isolation of the causative organism from blood or tissues.	Administratio n of antibiotics e.g. streptomycin; large scale poisoning to reduce susceptible wild rodents and controll of flies and ticks.
Sylvatic plague/wild rodent plague	*Pasteurella pestis;* several species of rodents; occurs mainly through fleas.	Anorexia, listlessness, tameness, and enlarged spleen; nodular necrotic foci in lungs or liver.	Culture of suspected tissue and postmortem findings.	Administratio n of antibiotics or sulfonamides; eradication of fleas; application of insecticides such as DDT and malathion.
Q fever	*Coxiella burnetii;* mice, rats, guinea pigs, hamsters, ground squirrels and cavies.	Fever, splenomegaly and skin rash; nodular lesions in liver, kidney, spleen, adrenal glands and lymph nodes.	Serological tests.	Environment al sanitation.

Murine respiratory mycoplasmosis	*Bordetella bronchiseptica, Mycoplasma pulmonis* etc.; rats, mice and guinea pigs.	Sneezing, nasal discharge, rales, dyspnoea, head tilt, incordination and circling; mucopurulent rhinitis, and bronchopneumonia.	Isolation of causative agent and pathological lesions.	Antibiotic therapy.
Mouse pox	Poxvirus; inbred strains are highly susceptible.	Facial swelling, conjunctivitis with a secondary rash and ulceration of head, tail or extremities; intestinal haemorrhage and focal necrosis in lymph nodes.	Clinical signs, serological tests and postmortem lesions.	Vaccination and strict quarantine measures for newly received mice.
Toxoplasmosis	*Toxoplasma gondi;* many rodents such as rats, mice, guinea pigs and voles.	Fever, enlarged spleen and lymph nodes and visceral granuloma; cysts and trophozoites in lungs, liver, kidney etc.	Serological and histopathological lesions.	Administration of sulfamethazine.
Dermatomycosis	Commonly by *Microsporum gypseum* and *Trichophyton mentagrophytes;* guinea pigs, rats, mice porcupines, chinchilla etc.	Typical ringworm lesions.	Pathological examination.	Administration of griseofulvin and separation of affected animals.

Besides the above mentioned diseases, other ailments like hair chewing, ring tail and pregnancy toxaemia have also been reported in rodents. Gastric spiral bacteria (*Helicobacter* spp) are found in wild rats. Rodents are known to be infested with a number of ectoparasites. They include lice (*Polyplax spinulosa* and *P. serrata*), fleas (e.g. *Xenopsylla cheopis* and *Nosopsyllus fasciatus*) and mites (e.g. *Demodex* spp and

Myocoptes spp). Ectoparasites may be controlled by applying insecticides. The important aspect of antibiotic therapy in rodents is that many of the commonly used antibiotics become toxic to the rodents. Antibiotics may cause disturbances to the gut microflora. Animal may die due to endotoxic shock.

The rodents show the symptoms of diseases arising from deficiency of many nutrients. Commercially prepared feeds, however, are balanced with adequate nutrients. The quality of nutrients may be deteriorated if feeds are stored for long periods. It is generally advised that feeds should be used within three months of the milling date. Laboratory animals are required about fifty nutrients. The following (Table 21.7) information highlights the importance of minerals and vitamins in selected laboratory rodent species.

Table 21.7: The importance of minerals and vitamins and their deficiency symptoms in selected rodent species:

Minerals and vitamins	Dietary requirements (per kg feed)	Deficiency symptoms	Remarks
Calcium	5.0 g for growth and maintenance of non lactating rats and mouse; 8.0 g for guinea pig; 6.0 g for normal bone formation in hamster.	Retarded growth, decreased food consumption, osteoporosis, paralysis of hind legs, internal haemorrhage and reduced activity.	Vitamin-D, lactose and sucrose are essential for absorption of calcium. The deficiency symptoms vary according to the age of the animals affected. Calcium deficiency in young rat, for example, causes tetanus, while in adult, it causes infertility. Milk and fish contain higher amounts of calcium.
Phosphorus	3.0 g for normal growth and maintenance of non lactating rats and mice; 3.5 g for normal bone formation in hamster; and 4.0 g in guinea pigs.	Retarded growth, stiff joints, muscular weakness and low fertility.	Vitamin D and fructose help in absorption of phosphorus.

Chloride	0.5 g for rats and mice.	Reduced poor growth, lowered blood chloride, reduced efficiency of feed utilization and increased blood carbon dioxide content	Rats are relatively insensitive to excess dietary chloride. Hamsters are reluctant to consume chloride spontaneously.
Magnesium	0.5 g for young, adult and non pregnant rats and mice; 0.7 g for lactating strains of mice and 1-3 g for guinea pigs.	Cardiac arrhythmias, hyperirritability, vasodilatation and deposition of mineral especially in renal tissues.	Phytate reduces the absorption of magnesium.
Potassium	3.6 g minimal requirement for rats and 5 g for lactating rats; 2.0 g for mice and 5 g for guinea pigs.	Reduced appetite, stunted growth, diarrhoea, lethargic, abnormally short hair and ascites.	Requirements vary according to different strains of rats.
Sodium	0.5 g for growth, maintenance, gestation, and lactation in rats and mice.	Retarded growth, reduced appetite, increased heat production and corneal lesions.	Animals are relatively insensitive to excess sodium.
Copper	5.0 mg for growth and maintenance in rats and 6.0 mg in mice. 8.0 mg for pregnant and lactating mice and rats; 6.0 mg for normal growth and development of guinea pig.	Abnormalities in skeletal, nervous, vascular, immune and reproductive systems of young rats; enlarged hearts and atrophied spleen.	Copper deficiency symptoms develop more rapidly when rats are fed sucrose and fructose. Mice are reported to be relatively resistant to copper toxicosis.
Iodine	100-200 µg in rats and mice; 150 µg for guinea pigs and 0.15 mg for hamsters.	Goiter, stunted growth, impaired reproduction and still births.	Most commercially available natural ingredient diets have adequate amounts of iodine.

Iron	75 mg for pregnant and lactating rats; 25-100 mg may support normal growth and haematopoiesis in male mice and 50 mg for reproduction, growth and development of guinea pigs.	Anaemia, hyperlipidemia, elevated metabolic rate; low maternal weight gain; and frequently neonatal mortality in hamsters.	Important for formation of haemoglobin.
Manganese	About 5.0 mg for normal growth and development of rats and 10.0 mg for mice; 40.0 mg for normal growth and development of guinea pigs.	Poor growth, bone abnormalities, reduced food consumption and early mortality.	It is an important trace element for laboratory animals.
Molybdenum	150 µg for rats, mice and guinea pigs; 0.10 mg for hamsters.	Poor growth.	Toxicity depends on the dietary concentrations of copper and sulfate.
Selenium	150 µg is the minimal requirement for growing rats, mice and guinea pigs; 400 µg for pregnant and lactating rats and mice; 150-200 µg for growth and maintenance of hamster and 400 µg for pregnant and lactating hamster.	Poor growth, cataracts, sparse hair coats and reproductive failure.	Selenium may be obtained form organic (selenocystine) and inorganic (sclenite) sources.

Zinc	10-12 mg for growing and adult rats and mice; 25-30 mg for pregnant and lactating rats and mice.	Anorexia, retarded growth, alopecia, hyperirritability, thickened epidermis and abortion in guinea pigs.	Galvanized cages with solid floor require less amounts of zinc.
Vitamin A	2400 IU for rats and mice.	Defects in vision, retarded growth, corneal opacity, abortion, infertility and hyperkeratosis.	Large amounts can store in liver. Protein deficient diet causes lower serum concentrations of Vitamin A. cod liver oil, green plants and yellow maize contain high amounts of vitamin A.
Vitamin D	1000 IU cholecalciferol for rats, mice and guinea pigs.	Rickets, enlarged bones and joints, and hypocalcemic tetany.	Amounts required depend on calcium and phosphorous intake.
Vitamin E	32 IU for mice if dietary lipid contains less than 10 per cent; 27 IU for rats when dietary lipid contains less than 10 per cent.	Haemolysis, muscular dystrophy and accumulation of yellow pigment in smooth muscles.	Relatively non toxic. Green plants and grain contain high amounts of Vitamin-E.
Vitamin K	1 mg phylloquinone for rats and 5 mg phylloquinone for guinea pigs.	Increased clotting time, haemorrhages and decreased prothrombin level.	It is orally non toxic. Green plants and fish meal have high amounts of vitamin K.
Vitamin B_6	6.0 mg for growth, maintenance and reproduction in rats; 2-3 mg for guinea pigs.	Retarded growth, renal atrophy, alopecia, paraplegia and degeneration of tail.	Important for many biochemical activities; cod liver oil, milk and grains contain more vitamin-B_6.

Biotin	0.2 mg for rats, mice, guinea pigs and hamsters.	Weight loss, alopecia, depigmentation of hair, rough coat and decreased reproduction.	Raw egg white impedes biotin absorption. Biotin is produced by gut flora.
Folic acid	1.0 mg for rats, 0.5 mg for mice; 3-6 mg for guinea pigs; 2 mg for golden hamsters.	Reduced growth rate, anaemia, leucopenia and diarrhoea.	Soybean meal and fishmeal have high folic acid contents.
Thiamin	4.0 mg for rats; .0 mg for mice; 2.0 mg for guinea pigs.	Polyneuritis, weight loss, vomiting, reduced food intake, muscular fatigue and early death.	It is needed for many biochemical functions. They are abundantly found in grains, oil seeds and milk products.

Source: NRC: Nutrient Requirement of Laboratory animals, 1995, USA.

Carnivores

Physiological parameters: The normal rectal temperature in canids ranges from 100.5 to 102.5^{0}F. Felids have a rectal temperature of 98.0 to 102.0^{0}F. The normal body temperature in procyonids has been recorded 99.0 to 101.5^{0}F. The respiratory rate in canids ranges from 11 to 20 breaths per minute. The normal respiratory rates in felids vary between 25 and 30 breaths per minute. Respiration rates in procyonids range from 15 to 25 per minutes. The heart rate of felids ranges from 120 to 140 beats per minute. However, the large members in the family Felidae have a heart rate of 40 to 50 per minute. Canids have a heart rate of 75 to 185 beats per minute. Bears have a normal heart rate of 60 to 90 beats per minute depending on the physiological status of the animals.

Blood is collected for various studies. Femoral, jugular, cephalic, saphenous and tail veins are the usual sites for collection of blood in carnivores. Blood may by collected by heart puncture in minks and ferrets under anaesthesia. Howell-jolly bodies are occasionally observed in small numbers in the red blood cells of domestic cats, leopards and cheetahs. The RBCs of dogs and cats sometimes form rouleaux (roll of red blood cells like a pile of coins). Heinz bodies within the erythrocytes are naturally found in cats. Examination of blood in spotted hyena is important for determination of sex. The sufficient numbers of the "drumstick" nuclear appendage in neutrophils are found in the female. Normal haematological values of some selected carnivores are presented in Table 21.8.

Table 21.8: Normal haematological values of selected carnivores:

Animal	RBC (x10⁶ /µl)	Hb (g/d l)	PC V (%)	WBC (x10³/µl)	Neutro phils(%)	Eosino phils(%)	Baso phils (%)	Lymph o cytes(%)	Mono cytes(%)
Tiger	6-8	9-14	35-45	10-15	63	2	0	30	5
Leopard	6-8	8-13	35-45	10-14	63	2	0	30	5
Snow leopard	9.06	11.1	34	12.5	73	1.0	0	21.5	4.5
Domestic cat	5-10	8-15	24-45	5.5-19.5	35-75	0-12	0-1	15-55	0-7
Asian lion	6.9-10.9	-	-	8.2-19.2	54-97	0.6	0	7-37	0-2
Striped hyena	6.5	14.7	45	11-19	-	-	-	-	-
Sloth bear	6.31	17.5	49	9.5	81	1	-	17	1
Giant panda	5.2-8.0	-	32-49	4-14.5	61-84	0-14	-	6-27	0-22
Red panda	4.8-12.8	9.6-17.4	29-54	3.1-14.2	13-67	0-8	0-11	6-11.1	0-13
Wolf	5.48	11.8	35	7.9	85	3	-	11	1

Source: Schalm's Veterinary Hematology by Nemi C Jain; Zoo and Wild Animal Medicine by Murray E Fowler ; Diseases of Exotic animals by J.D.Wallach & William J. Boever.

The dog milk contains 75.4 per cent water, 11.2 per cent protein, 3.1 per cent carbohydrate, 9.6 per cent fat and 0.7 per cent ash. The leopard milk contains 22.6 per cent solids.

Diseases: A large number of infectious and noninfectious diseases have been reported in carnivores. Some of the well mentioned diseases are rabies, brucellosis, feline panleukopenia, infectious canine hepatitis, canine distemper and leptospirosis. Helminthiasis in wild carnivores causes high morbidity and mortality especially in young carnivores. Wild animals tend to mask clinical symptoms of infectious diseases. Therefore, it is very difficult to diagnose a particular disease in free living carnivores. However, it is a great challenge to the veterinarians who are involved in handling and diagnosing of diseases in wild and zoo animals.

Mammary gland cancer has been reported in captive exotic felids. Feline spongiform encephalopathy has been reported in cheetah, mountain lion and domestic cat. Gastritis associated with spiral bacteria has been reported in cheetah, ferrets, domestic dogs and cats. A brief note on various diseases of carnivores is documented in Table 21.9.

Table 21.9 : Some reported diseases in carnivores.

Disease	Host	Etiology	Clinical signs	Diagnosis	Treatment and control
Feline panleukope nia (FPL)	Lions, tigers, leopards, cheetahs, wild cats etc.	Caused by parvo virus. It is also called feline distemper, feline infectious enteritis, and cat plague.	Depression, anorexia, diarrhoea, vomiting, dehydration, haemorrhagic enteritis, marked leucopenia and anaemia.	Diagnosis is based on clinical signs, necropsy findings and isolation of virus. Advanced techniques like monoclonal antibodies and polymerase chain reaction are important in diagnosing FPL.	A hyper immune antiserum is useful (2-4 ml/kg body weight). Treatment includes fluid therapy (20 ml/kg body weight), whole blood transfusions (10-20 ml/kg body wt. on every alternate day) and administration of antibiotics to prevent secondary infections. Inactivated and modified live vaccines advocated for prophylactic measures. Kittens are vaccinated at 8, 10, 12 and 16 weeks intervals. Control of vectors (flies and fleas) is important.
Rabies	All members are susceptible. Common in foxes, dogs, raccoons, skunks etc. Wild carnivores are more susceptible.	Caused by rhabdo virus.	Profuse salivation, posterior ataxia, paresis and paralysis are the symptoms of rabies.	Diagnosis by clinical symptoms and serological tests.	Young animal (3–4 months) is vaccinated by killed virus vaccine. The minimum exposure of captive animals to stray dogs and cats should be strictly followed. Modified live virus rabies vaccine recommended for domestic animals should be avoided in free living animals.
Feline viral rhinotrache itis(FVR)	Exotic felids are susceptible to FVR.	Caused by herpes virus.	Sneezing, dyspnoea, fever, bronchopneu monia, depression, eye discharge and abortion.	Clinical signs, tissue culture and serological tests.	Killed or modified live virus vaccines (initial dose at 12 weeks of age followed by annual booster dose) is recommended. Supportive therapy should be followed.
Feline infectious peritonitis (FIP)	Commonly found in 6 months to 2 years of age wild felids. Domestic cats are also susceptible.	Caused by corona virus.	Clinical signs include anorexia, fever, ataxia, ascites, jaundice, nasal discharge and leukocytosis.	Diagnosis is based on history, clinical signs and serological tests.	The removal of abdominal fluid and supportive therapy are the important measures of FIP.

Disease	Host	Etiology	Clinical signs	Diagnosis	Treatment and control
Canine distemper	Occurs in canids, mustelids, procyonids, viverrids, hyenids and ursids.	Caused by paramyxovirus.	Exhibit the clinical signs of anorexia, diarrhoea, emaciation, lethargy, purulent discharge from eyes and nose, and neurological signs.	Diagnosis can be made by clinical pictures, histopathological lesions and serological tests.	Treatment includes administration of antibiotics, antipyretics, anticonvulsants and electrolyte therapy. Immunization can be done by killed or modified live vaccine.
Anthrax	Reported in all free living carnivores.	Causes by *Bacillus anthracis.*	Fever, depression, weakness, oozing of blood from mouth, nostrils and anus, septicaemia and subcutaneous oedema.	Microscopic examination of blood and culture of the organism.	Administration of parenteral antibiotics is essential. Enclosure should be disinfected with formalin. Dead animal should be deeply buried with opening the body. However, it is advisable to burn the carcass. Anthrax spore vaccine (1 ml for adult and 0.5 ml for young) is advisable.
Brucellosis	Exotic canids, procyonids and mustelids.	Caused by *Brucella* sp.	Clinical signs include infertility, abortion and enlarged lymph nodes.	History and postmortem lesions are important for diagnosis the disease.	Affected animals should be treated with antibiotics. Suspected animals should be isolated.
Valley fever	Reported in tiger.	Causal agent is *Coccidioides immitis.*	Loss of body weight, anorexia, depression and respiratory problems.	History, postmortem findings and serological tests.	Amphotericin B may be given.
Salmon poisoning	Reported in canids.	Caused by *Neorickettisia helminthaeca.* Animals are infected by eating salmon and trout that harbour the vector fluke, *Nanophyetas salmincola*	Lethargy, rise of body temperature, anorexia, vomiting, diarrhoea and dehydration.	Clinical signs, history and examination of faecal matter.	Administration of antibiotics with supportive fluid therapy is recommended. Fish should be well cooked before giving to the animals.

Disease	Host	Etiology	Clinical signs	Diagnosis	Treatment and control
Trypanoso miasis	Occurs in felids, (e.g. lion, and tiger),canids (e.g. fox and dog), procyonids (e.g. pandas) and mustelids (e.g. ferrets).	Caused by *Trypanosoma evansi* (common in felids), *T. brucie* and *T. congolense* (common in canids) and *T. cruzi*.	Anorexia, loss of body weight, intermittent fever, constipation, oedema and diarrhea.	Clinical findings, demonstration of parasites in blood films and serological tests.	Trypanosomiasis can be treated by barenil (3 mg/kg body weight- sc). Control of tse tse fly or other biting flies and supportive therapy are important.
Toxoplasm osis	Cheetah, lion, snow leopard, wolf, fox, weasel etc.	Caused by *Toxoplasma gondii*.	Anorexia, fever, depression, cough, dyspnoea, icterus and neurological disorders.	Clinical signs, histopathologic al examination and serological tests.	It can be treated by sulphadiazine (60 mg/kg body weight) with pyrimethamine (0.4 mg/kg body weight).
Feline Infectious anaemia (FIA).	Small cats, tigers etc.	Caused by *Haemobartone lla felis*.	Depression, anorexia, weakness, fever, jaundice and splenomegaly.	Demonstration of parasites in blood and clinical signs.	Administration of antibiotics (oxy tetracycline 20 mg/kg body weight orally).
Toxocara	Wild felids, canids and procyonids.	Caused by *Toxocara cati*, *T. canis*, *T.leolina* etc.	Rough body coat, poor growth, diarrhoea, anaemia and pneumonic signs.	Observing the eggs in faeces and postmortem finding.	May be treated with piperazine (100 mg/kg body weight orally in single dose).
Chastek paralysis	Occurs in fox and mink.	Caused by thiamine deficiency in the animals. Induced by feeding of certain types of raw fishes containing thiamine-splitting enzyme that destroys dietary thiamine.	Loss of appetite, emaciation, salivation, ataxia, abnormal gait, convulsion and death.	History, clinical signs, and postmortem findings.	Parenteral administration of thiamine is important. Fish containing enzyme thiaminase should be cooked.

Disease	Host	Etiology	Clinical signs	Diagnosis	Treatment and control
Steatites (yellow fat disease)	Normally occurs in young and growing mink. It is also reported in felids.	Caused by either deficiency of vitamin E or feeding of excessive unsaturated fatty acids.	Clinical signs are lethargy and lumpy subcutaneous fat.	Clinical symptoms, history and fat biopsy.	Corrected by parenteral injection of vitamin E (15-20 mg/IM for 3-4 days.). Rancid feed should not be fed to the animals.

Elephants:

Physiological parameters: The body temperature of elephant ranges from 97.5 to 99.0°F and it may be recorded from the rectum or faecal bolus. The normal heart rate is reported to be 29-57 beats per minute and it can be measured by palpation of arteries in the ears. The normal respiration rate in elephant is reported to be 4-12 breaths per minute. The average volume of urine produced by an adult elephant is about 50 litres per day. It urinates on an average eight to twelve times a day.

Blood is collected from the auricular veins of elephant. The blood can be easily collected if the back of the ear is washed with hot water. The haematological values for Indian elephant are as follows: RBC $(x10^6/\mu l)$:1.98-4.0, Hb (%):12.0-15.5, PCV (%):30.0-43.3, WBC $(x10^3/\mu l)$:6.4-14.0, neutrophils (%): 22.0-50.0, eosinophils (%): 6.0-15.0, basophils (%):0.0-2.0, lymphocytes (%):40.0-60.0, and monocytes (%): 0.0-5.0.

The elephant's milk is white and watery and the milk composition of Indian elephant is as follows: water (78.1%), fat (11.6%), protein (4.9%), lactose (4.7%) and ash (0.7%).

Diseases: The information on elephant diseases is mostly based on the findings from the Asian elephants. Some reported diseases in elephants are pasteurellosis, anthrax, foot and mouth disease, salmonellosis, tetanus, elephant pox, enterotoxaemia trypanosomiasis, babesiosis, rinderpest, tuberculosis and rabies. The other ailments like cracked sole, heat stroke, colic, trunk injury, gastritis, impaction, subcutaneous abscesses and dermatitis have also been reported in elephants.

Regular deworming (once in 6-12 months) with albendazole @ 2.5 mg/kg body weight is reported to be effective against helminthic infection in elephants.

Colic in elephant is due to overfeeding and changing in diet. The animal shows abnormal postures (e.g. crossing the hind legs), restless, lying down and getting up. Soapy water enemas and feeding of good quality of feeds are suggested in order to relief from colic.

Trunk paralysis is found in elephants. Trauma, bacterial infections, migration of parasites and nutrient deficiencies are the factors responsible for trunk paralysis.

Cracked sole is commonly seen in elephants. This is due to poor sanitation, inadequate exercise, nutritional and traumatic factors. Cracked sole causes pain, lameness, ulceration and granulation. The treatment includes debridement of the crack and antibiotic therapy.

Perissodactyls:

Physiological parameters: The normal rectal temperature of rhinoceros varies from 98.6 to 102.2^0 F. The heart rate of these animals is recorded as 64 - 67 beats per minute, while the respiratory rate is reported to be 20-40 breaths per minute. The normal body temperature of domestic horse is 100.5^0F.

Blood may be collected from the jugular vein in horse, donkey, ass and zebra. In rhinoceros, blood may be collected from large ear veins. The haemogram of Asiatic wild ass (male) is as follows: RBC $(x10^6/\mu l)$: 6.14, Hb (%): 10.47, PCV (%): 38.50, WBC $(x10^3/\mu l)$:7.62, neutrophils (%):54.45, eosinophils (%):4.25, lymphocytes (%): 40.25 and monocytes (%):0.95.

The milk of the African wild ass contains 88.3 per cent water, 1.4 per cent fat, 2.0 per cent protein, 7.4 per cent lactose and 0.5 per cent ash. The composition of mare's milk is as follows: 88.8 per cent water, 1.9 per cent fat, 2.5 per cent protein, 6.2 per cent lactose and 0.5 per cent ash.

Diseases: The followings are the important diseases found in perissodactyls.

African horse sickness: This is a highly fatal viral disease of horses, asses and mules. It is transmitted by insects. The acute pulmonary form is characterized by elevated body temperature, intense dyspnoea, coughing, pulmonary oedema and frothy nasal discharge. Subacute cardiac form is characterized by oedema on the head, neck, lips and eyelids. Postmortem lesions of African horse sickness include hydrothorax, severe pulmonary oedema, hydropericardium and haemorrhages on the endocardium. Diagnosis is based on the clinical symptoms and isolation of virus. Annual vaccination of the animals and destruction of insect vectors are important in controlling this disease.

Equine infectious anaemia: It occurs in horses, mules and asses. Equine infectious anaemia is caused by lentivirus. The acute form is characterized by high rise of temperature (108^0F), extreme weakness, excessive thirst, anorexia, depression, oedema of the lower abdomen and nasal haemorrhages. A presumptive diagnosis can be made from the clinical signs and characteristic lesions during acute case. The control measures include eradication of infected animals, vaccination, restriction of the movements of horse and sanitary measures.

Tuberculosis: Tuberculosis has been reported in rhinoceros, tapirs and zebras. Clinical signs include dyspnoea, coughing and depression. Postmortem lesions are found in spleen, lymph nodes and lungs. Affected animals are usually discarded.

Anthrax: This disease has been reported in rhinoceros and zebra. Anthrax is characterized by septicaemia, lameness and discharge of blood from nostril and anus. Peracute and acute forms cause in rapid death. Diagnosis can be made on the basis of clinical signs and examination of blood. Animal is vaccinated by virulent spore vaccine as a prophylactic measure.

Rinderpest: This dreaded disease has been reported in zebra. It is caused by myxovirus. Rinderpest is characterized by high fever, eye discharge, diarrhoea, ulcers on the mucosa of buccal cavity and profuse salivation. Animals may be vaccinated by attenuated virus vaccines.

Big head: Big head caused by *Clostridium sordelli* in the black rhinoceros, is characterized by anorexia, ataxia and oedematous swellings on the head and neck.

Babesiosis: Babesiosis has been reported. in rhinoceros and zebra. It is caused by blood protozoan parasites *Babesia* spp. Haemolytic anaemia, dyspnoea, diarrhoea and jaundice are the clinical findings of babesiosis.

Other diseases: Red maple *(Acer rubrum)* toxicosis has been reported in horse and grevy's zebra. It is clinically characterized by anorexia, anaemia, polypnoea and hematuria. Ingestion of wilted or dead leaves of red maple is toxic to the animal. Equine herpes viruses (EHV-1, 2, 3 and 4) have been reported in free-ranging mountain zebras. EHV-1 infection in horses causes abortion, neonatal foal disease and or paralysis. Leucoencephalomalacia has been reported in rhinoceros. The aetiology is unknown. This fatal disease has no treatment.

Artiodactyls:

Physiological parameters: The normal rectal temperature varies among the members of the order Artiodactyla. Normally, the very large and the very small mammals have low body temperatures. The normal heart rates have been recorded in various artiodactyls and it can be measured from the middle coccygeal, facial, or femoral arteries. The middle coccygeal or facial arteries, for instance, are the sites for recording pulse in cattle and buffaloes, whereas, the femoral artery is chosen for recoding pulse in sheep and goats. The normal respiratory rates are counted by observation of rib movements or by feeling the nasal air movements. Biotelemetry transmitters can be used to collect the physiological data in wild artiodactyls. Table 21.10 shows the normal rectal temperatures, heart rates and respiratory rates of some selected artiodactyls.

Table 21.10: The normal temperature, heart rate and respiratory rate of some selected artiodactyls.

Animal	Body temperature (oF)	Heart rate (beats/minute)	Respiratory Rates (breaths/minute)
Barking deer	101.4-102.0	-	-
Bison	102.3	45-50	-
Yak	98.0-104.0	45-68	38-86
Wild pig	101.0-102.5	60-80	12-20
Giraffe	99.6-101.5	40-66	12-20
Camel	99.6-100.2	40-50	5-12

Camel can conserve body fluids during heat stress. The giraffe has a unique circulatory system in which the animal is able to regulate and maintain cerebral circulation despite a range of head elevations from 4.2 m above the heart to 3.0 m below the heart. The rate of water loss from the skin in hot weather is several times greater in hippopotamus than other mammals. So, it is difficult for these animals to maintain body water balance in dry air condition.

The composition of milk of some artiodactyls is presented in Table 21.11.

Table 21.11: Milk composition of selected artiodactyls:

Animal	Water (%)	Fat (%)	Protein (%)	Lactose (%)	Ash (%)
Red deer	78.0	9.5	7.4	4.5	0.6
Bison	86.9	1.7	4.8	5.7	0.9
Giraffe	77.0	12.5	5.7	3.4	0.9
Yak	82.7	6.5	5.8	4.6	0.9
Camel	86.5	4.0	3.6	5.0	0.8
Cow (domestic)	87.3	3.9	3.2	4.6	0.7
Buffalo (domestic)	82.8	7.4	3.8	4.8	0.8
Goat(domestic)	87.0	4.2	3.5	4.2	0.8
Pig(domestic)	81.2	6.8	4.8	5.5	1.0
Sheep(domestic)	80.7	7.9	5.2	4.8	0.9

The blood is collected for various studies. It may be collected from jugular vein, coccygeal vein, anterior vena cava, internal thoracic vein or femoral vein. The jugular vein is the common site for collection of blood in cattle, sheep, goat and antelopes. The coccygeal vein may also be used for taking blood samples in cattle. The haematological values of some members of the order Artiodactyls are given in Table 21.12. Erythrocytes of camelids are typically elliptical in shape and are more resistant to haemolysis in hypotonic saline solution. The classic sickling phenomenon (semicircular shaped erythrocytes) is an important physiological characteristic of certain deer. It has been observed that the normal erythrocytes in deer are not sickle shaped when first removed from the body. The sickling phenomenon is observed when the blood sample is kept at either refrigerator or room temperature. This phenomenon, however, can be prevented by acidification of the blood samples. The sickling can also be enhanced by supplying oxygen gas in the blood samples. The "drumstick" nuclear appendage has been occasionally observed in nuetrophils of giraffe's blood.

Table 21.12: Haematological values of various artiodactyls.

Animal	RBC(x 10^6/ µl)	Hb (g/ dl)	PC V (%)	WB C (x 10^3/ µl)	Neutro phils (%)	Eosino phils (%)	Baso phils (%)	Lympho cytes (%)	Mono cytes (%)
Blackbuck	13.28	17.4	49.0	5.9	77.0	0.0	0.0	20.0	3.0
Yak	4.36-7.75	8.0-12.4	20-32	8.02-13.35	23.2-42.4	2.44-16.32	0-1.52	42.14-66.86	3.20-8.87
Nilgai	5.62-8.03	11.9 - 15.8	35.9 - 47.7	3.2-6.9	17.4-44.4	0.0-0.9	-	54.9-73.6	0.0-0.6
Camel	6.1-9.3	10.0 - 15.1	20-33	10.5-28.3	30-60	2.0-17.5	0-0.5	33-58	1.5-6.0
Chevrotain	14-17	14.4	65-74	8.4-11.0	14-24	0-11.0	0-2	70.0	0-3
Cheetal	10.5	19.0	53.0	6.0	-	-	-	-	-
Cattle (domestic)	5-10	8-15	24-46	4-12	15-45	2-20	0-2	45-75	2-7
Sheep (domestic)	8-16	8-16	24-50	4-12	10-50	0-10	0-3	40-75	0-6
Goat (domestic)	8-18	8-14	19-38	4-13	10-59	0-10	0-2	40-75	0-6
Pig (domestic)	5-8	10-16	32-50	11-22	28-47	1-11	0-2	39-62	2-10
Giraffe	12.82	15.1	43.0	14.4	65.0	2.5	0.5	30.5	1.5

Source: Schalm's Veterinary Hematology by Nemi C Jain; Zoo and Wild Animal Medicine by Murray E Fowler..

Diseases: Many important diseases found in domestic ruminants have also been reported in their wild counterparts. Some widely known infectious diseases such as foot and mouth disease, rinderpest, anthrax, haemorrhagic septicaemia, brucellosis, blue tongue, tuberculosis and vibriosis have been reported in many wild ruminants. African swine fever has been reported in wild pigs and peccaries. Like wild ruminants, hippopotamuses are susceptible to anthrax, rinderpest, haemorrhagic septicaemia and tuberculosis. Infectious keratoconjunctivitis has been documented in some wild artiodactyls such as deer and big horn sheep. Deer are frequently associated with antler anomalies. Genetic factor, trauma, hormonal and nutritional factors are responsible for anomalies of antlers.

Aflatoxicosis has been reported in many artiodactyls. The toxic effects of aflatoxin ingestion include decreased feed intake, decreased weight gain, liver damage and suppression of immune systems. Aflatoxins are more toxic to young animals than adults.

Capture myopathy has been commonly found in deer, wild goats, antelopes and many other artiodactyls. It normally occurs during transportation and restraining. Capture myopathy is characterized by weakness, depression, stiffness, incoordination of movements, paralysis and occasionally haemoglobinuria. Fascioliasis, neurofilariasis and lungworm infections have been reported in many members (e.g. deer, blackbuck and mountain goat) of the order Artiodactyla.

A large number of ectoparasites have also been documented in exotic artiodactyls. A short description of some selected diseases of artiodactyls is presented in Table 21.13.

Table 21.13: Reported diseases of artiodactyls.

Disease	Host	Aetiology	Clinical findings	Diagnosis	Treatment and control
Rinderpest	All artiodactyls	Caused by virus belonging to paramyxo group.	Pyrexia, aggressiveness, polydipsia, gastroenteritis and ulcerative stomatitis.	History of outbreak, clinical findings, gross lesions, postmortem findings and serological tests.	Secondary treatment includes administration of antibiotics, fluid and electrolytes. Inoculation of vaccine as a protective measure.
Foot and mouth disease	All artiodactyls	Caused by various serotypes of epitheliotropic virus.	Vesicular eruptions in the epithelium of oral cavity, tongue, teats, udder and feet.	History, clinical signs and laboratory tests.	Antiseptic mouthwash (e.g. boric acid and glycerin) apply over mouth and foot lesions. Immunization is important. Restrict the animal movement.

Mucosal disease	Many artiodactyls such as gaur, banteng, deer, antelopes and pigs.	Caused by virus (genus *Pestivirus*)	High rise of temperature (106^{0}F), Anorexia, depression, polydipsia, nasal discharge, corneal opacity, diarrhoea and abortion.	Typical clinical findings and gross and microscopical lesions.	Supportive therapy includes administration of antibiotics, fluid and electrolytes. Isolation of affected animals and proper hygienic measures.
Malignant catarrhal fever	Deer, antelopes, gaur, wild goats and other wild ruminants.	Caused by herpes virus.	High rise of temperature, dyspnoea, diarrhoea, catarrhal inflammation of upper respiratory tract, lymph adenopathy and keratoconjunct ivitis.	Clinical findings, histopatholog ical lesions and isolation of virus.	Prevent secondary bacterial infection. Vitamin A may be given as supportive treatments. Separation of affected animals.
African swine fever	Wild pigs.	Caused by virus.	High rise of temperature (105-108^{0}F), anorexia, incoordination of gait, nasolacrymal secretion, dyspnoea and coughing.	Clinical findings, characteristic lesions and isolation of virus.	Inoculation of vaccine, restriction on movements of animal. Dead animal should be buried. Quarantine measures are important.

Anthrax	All artiodactyls such as deer, antelope, hippopotam us, giraffe and bison.	Caused by *Bacillus anthracis.*	Signs of elevated body temperature, dyspnoea and oedema. Blood oozes from the natural orifices after death.	Clinical findings, microscopic examination of blood films, Ascoli's test etc.	Administration of antibiotics. Annual vaccination is recommended. The carcass should not be opened and it should be deeply buried or burnt. Restriction of animal movement.
Haemorrha gic septicaemia	Deer, antelope, bison, wild sheep, wild pigs etc.	Caused by *Pasteurella multocida.*	Elevated body temperature, depression, salivation, mucopurule nt discharge, pneumonia and septicaemia.	Clinical findings, identification of organism, culture of organism and animal inoculation test.	Administration of antibiotics and vaccination. Sulphadimidin e @ 100 mg/kg body weight for five days in spotted deer is effective.
Black quarter	Deer, bison, antelope, wild sheep etc.	Caused by *Clostridium chauvoei.*	High rise of body temperature, stiffness, oedematous swellings, development of focal gangrenous and emphysemat ous myositis and lameness.	Microscopic examination of blood smear, biological test and fluorescent antibody technique.	Administration of antibiotics, proper burning of dead animal and vaccination are suggested.

Contagious bovine pleuropneumonia	Yak, bison, antelope etc.	Caused by *Mycoplasma mycoides*.	High rise of temperature (105°F), depression, anorexia, arched back, panting, dilated nostrils, nasal discharge, and oedematous swellings on throat and dewlap.	History, clinical manifestation, culture of organisms and other tests.	Slaughtered of affected animals, strict quarantine measures, restriction on animal movements are important.
Anaplasmosis	Bison, deer, antelopes and other wild ruminants.	Caused by *Anaplasma marginale*.	Fever, nasal discharge, lacrymation, coughing, dehydration, anaemia, rough body coat and enlarged lymph nodes.	Clinical findings, examination of blood films, animal inoculation test and serological tests.	Tetracycline (10-15 mg/kg body weight) maybe administered. Supportive therapy including liver extract, mineral mixture. Isolation of carrier animals and control of insect populations are important.
Theileriosis	Deer, antelopes, and other wild ruminants.	Caused by *Theilaria* spp.	High rise of temperature, enlargement of superficial lymph nodes, nasal discharge, coughing, anaemia, oedema and diarrhoea.	Clinical findings, lesions, examination of blood films and serological tests.	Control of tick population, vaccination of animals in enzootic areas and other supportive therapy.

Babesiosis	Deer, bison, antelopes, wild sheep etc.	Caused by *Babesia* spp.	High fever, anorexia, depression, weakness, anaemia, jaundice and haemoglobinuria.	Clinical signs, examination of blood films and serological tests.	Barenil (4 mg/kg body weight, im), supportive therapy includes haematinics, vitamin B complex injections, and control of tick population.
Heartwater	Wild buffaloes, antelope, giraffe, deer and Himalayan tahr etc.	Caused by *Cowdria ruminantium*.	Fever, inappetence, diarrhoea and neurological signs. Postmortem findings include hydrothorax, pulmonary oedema, ascites, oedema of lymph nodes and swollen spleen.	Clinical signs, postmortem findings and serological tests.	Control measures include administration of oxytetracycline (6-10 mg/kg body weight.i.v,b.i.d. for 3-4 days), vaccination and control of ticks.
Coccidiosis	Blackbuck, gazelle, deer, etc.	Caused by *Eimeria* spp.	Anorexia, depression, diarrhoea, haemorrhages and emaciation.	Presence of oocysts in the faecal samples.	Treatments include administration of sulphonamides, and strict sanitary measures.

Chapter 22

Health Care of Wild and Zoo Reptiles

Physiological parameters: Most reptiles are exothermic animals. The heat generated by metabolic activity is not sufficient enough to raise their body temperature. However, many reptiles can increase or decrease their body temperature through behavioural means. Basking, for example, is an important behaviour in which some reptiles raise their body temperature. The preferred body temperature (PBT- the temperature in which a reptile is physiologically adapted) varies according to species. In most reptiles, it ranges from 68 to 103.1^0F. Shedding of skin is an important physiological phenomenon of many reptiles. This is commonly found in snakes. The pulse rate of most active reptiles ranges from 20 to 35 beats per minute.

Haematological studies in reptiles are very important for diagnosis of diseases. A few drops to about 1.5 ml of blood can be collected from reptiles. However, this depends on the size of the animal. The jugular and the lateral caudal veins are the usual sites for the collection of blood in crocodilians. Cardiocentesis is the suitable method for collection of blood in young crocodilians as well as young and many adult chelonians. Small amount of blood may be collected by toe clippings in chelonians. Blood in marine turtles may be collected from the flipper and the jugular veins. Several drops of blood may be obtained by clipping a posterior toe in small lizards

Like amphibians, all reptiles have nucleated erythrocytes, leucocytes and thrombocytes. The reptilian erythrocyte is oval shaped with somewhat centrally placed nucleus. Howell Jolly bodies are occasionally found in the reptile erythrocytes. The haematological values of some reptiles are presented in Table 22.1.

Diseases: The diseases caused by both pathogenic and nonpathogenic organisms have been reported in reptiles. Some bacterial diseases that are reported in reptiles include salmonellosis, tuberculosis, cutaneous ulcerative disease, septicaemia and stomatitis. Reptiles which are unable to adapt to captive environment result in anorexia, lethargy, weight loss, dehydration and death. It has been reported that the high incidence of mortality in chelonians is due to the problems of digestive tract. Malnutrition, poor husbandry and lack of sanitary and hygienic measures

Table 22.1: The haematological values of selected reptiles.

Animal	RBC (x10^6/mm^3)	Hb (g/dl)	PCV (%)	WBC (x10^3/mm^3)	Neutrophils (%)	Eosinophils (%)	Basophils (%)	Lymphocytes (%)	Monocytes (%)
Python	1.5-2.5	-	25-40	6-12	0-20	-	0-10	10-60	0-3
Tortoise	1.2-3	-	23-37	3.8	0-3	-	2-15	25-50	0-4
Juvenile hawksbill turtle	5.5	12.8	47.0	5.2	43.0	0	1.0	54.0	4.0

have a prominent role in causing diseases to the captive reptiles. However, this section is aimed at highlighting the overall important of reptile diseases.

Ulcerative stomatitis or mouth rot is a common infectious disease of snakes. It has also been reported in turtles and lizards. *Aeromonas, Pseudomonas* and *Klebsiella* spp are the normal causal agents of ulcerative stomatitis. Poor nutrition and trauma are also the important factors causing this disease. It is clinically characterized by ulceration and formation of caseous exudates. Treatment includes debridement of necrotic tissue, topical application of antiseptics and supportive therapy. Systemic antibiotic therapy (e.g. enrofloxacin at the dose rate of 10 mg/kg body weight I/M for four days) may also be advocated. Supportive therapy with multivitamins (e.g. vitamin A injection at 11,000 units/kg body wt I/M once and vitamin C injection @ 200mg/kg body wt twice a week) and parenteral fluids is also important. The affected animal may be fed through tubes.

Septicaemic cutaneous ulcerative disease has been reported in turtles. It is caused by a gram negative bacterium *Citrobacter freundii*. Turtles are believed to be infected through skin abrasions. The disease is clinically characterized by anorexia, lethargy, paralysis of limbs, loss of claws or digits, ulceration and petechial haemorrhages on the shell and skin. Multiple necrotic foci are found in internal organs. Systemic antibiotic therapy and good hygienic measures are suggested.

Ulcerative shell disease or shell rot found in turtles is caused by *Beneckea chitinovora*. This contagious and chronic bacterial disease is clinically characterized by erythema and pitting of the shell with ulceration. Treatment consists of parenterally administration of antibiotics and topical application of iodine. The newly acquired animal should be kept in a separate enclosure for 15 days. The disease can be prevented by isolation of affected animal and avoidance of overcrowding. Crabs, lobsters and cray fish should not be fed or housed with turtles. The infected animal should be treated immediately.

Salmonellosis has been reported in tortoises, crocodiles, turtles, snakes and lizards. Reptiles are important reservoirs of salmonellosis. The disease is normally asymptomatic in these animals. However, diarrhoea, anorexia and listlessness may be observed in affected animals. It has the major zoonotic importance. Therefore, the animal with salmonellosis infection should be handled with great care. Hygienic measures that include sanitary disposal of faeces, disinfecting potential fomites and washing of cages are important. The other factors such as overcrowding, drinking of stagnant water and ingestion of contaminated food should be avoided in order to prevent salmonellosis. Zoo reptiles may harbour many serotypes of *Salmonella*.

Viral encephalitis has been reported in wild turtles and snakes. Herpes virus like infection has been reported in iguana lizards and pond turtles. It is characterized by anorexia, lethargy and lymphoid hyperplasia of spleen. Chronic rhinitis has been reported in tortoises. The main clinical signs include rhinitis, stomatitis and glossitis.

Mycotic diseases reported in reptiles may be caused by *Beauveria bassiana*, *Paecilomyces fumoosroseus*, *Geotrichum candidum* and *Fusarium solani*. Malnutrition, dampness, cold, stress and debilitation are the predisposing factors causing fungal infections to the reptiles.

However, a summary of infectious diseases with their treatment and control in reptiles is given in Table 22.2.

Parasitic diseases: Wild and captive reptiles harbour a wide variety of ectoparasites as well as endoparasites. Protozoan diseases such as amoebiasis, coccidiosis and trypanosomiasis occur in reptiles. *Entamoeba invadens* causing amoebiasis is the most commonly found protozoan parasite in reptiles. It causes high morbidity and mortality among snakes and lizards. Turtles are occasionally infected with this parasite. Amoebiasis is clinically characterized by emaciation, loss of body weight, vomiting and mucoidal or haemorrhagic diarrhoea. Ulcerative gastritis, colitis and enlarged liver are the postmortem findings. Affected animal can be treated with metronidazole (160 mg/kg body weight for 3 days). Paromomycin at a dose rate of 25-100 mg per kg body weight may be administered daily for four weeks. Amoebiasis can be prevented by providing adequate quarantine and sanitation measures. The common enclosure for keeping snakes and turtles should be avoided.

Coccidiosis in reptiles is caused by *Isospora, Eimeria, Caryospora* and *Cryptosporidium* spp. Eimeria is commonly found in reptiles. Reptiles can be infected through contaminated soil or faeces that contain oocysts. Coccidiosis causes anorexia, restlessness, regurgitation, haemorrhagic enteritis and intussusceptions. Clinical diagnosis is made by demonstration of oocysts in faecal matter. Sulfamethoxydiazine (90 mg/kg body weight on first day followed by 45 mg/kg body weight daily for 5 days) may be orally administered.

Trypanosomiasis occurs in reptiles. *Trypanosoma enhydris* has been reported in fresh water snakes. The important flagellates found in reptiles include *Leishmania*,

Table 22.2 : A summary on infectious diseases in reptiles and their control.

Disease	Pathogens	Treatment/ control	Remarks
Common bacterial diseases in reptiles include glossitis, stomatitis, osteomyelitis, intermandibular or pharyngeal cellulites and cutaneous abscesses.	Common pathogens are *Aeromonas, Pseudomonas, Proteus, Klebsiella* and *Acinetobacter. Citrobacter* is the primary causal agent of septicaemic cutaneous ulcerative disease of soft shelled turtles. *Staphylococcus, Mycoplasma* and *Streptococcus* are occasionally found. Rarely found pathogens in reptiles include *Pasteurella, Salmonella, Mycobacterium, Coxiella* and *Haemobartonella.*	Localized lesions can be corrected with debridement, topical application of antimicrobials and systemic antibiotic therapy.	Bacterial diseases are primarily due to gram negative pathogens. Poor sanitation, inadequate husbandry and trauma are mainly responsible for most diseases. Oral medication through mixing with food should be avoided whenever possible. Oral drug can be given through a feeding needle. Injection is to be given in the front half of the body as the reptiles have a renal portal system. Intramuscular injection is done one third of the snake's total length caudal to the head in the lateral muscle group midway between the spine and true lateral surface of the body. Intramuscular injections in crocodilians, chelonians and lizards are given in the caudal muscle groups of the front legs or the epaxial musculature.
Commonly encountered viral diseases include hepatic necrosis, papillomas, dermal lesions, and neurological disease.	Aetiologic agents include Adenovirus, Herpes virus, Poxvirus, Paramyxovirus and boid inclusion body disease virus.	Treatment includes supportive care, treatment of secondary pathogens and good husbandry.	Infected animal should be kept in quarantine.

Common fungal diseases in reptiles include superficial dermal and shell mycosis, cutaneous ulcerative disease, respiratory disease and chelonian scute infections.	Aetiologic agents include *Dermatophyton, Tricophyton, Phycomycosis, Chromomycosis, Aspergillus, Candida, Penicillium* and *Fusarium.*	Fungal diseases can be treated with topical or systemic antifungal agents. Griseofulvin and ketoconazole are widely used as systemic treatments.	Hygienic and sanitary measures are important.

Trichomonas, Hexamita, Monocercomonas and *Tritichomonas.*

Trematodes found in reptiles are mostly nonpathogenic parasites. Flukes have been isolated from the lungs and digestive tract of snakes. *Ochetosoma aniarum, Pneumatophilus* spp and *Dasymetra* spp are the common trematodes found in reptiles. *Bilorchis mehari* has been reported in the gall bladder of freshwater tortoise. *Lechriorchis stomatrema* is commonly seen in mouth, pharynx, oesophagus, trachea and lungs of some snakes. Animal infected by trematodes may be treated with praziquantel at a dose rate of 7 mg/kg body weigh intramuscularly.

A large number of nematodes occur in reptiles. Ascarids (e.g. *Polydelphis* and *Ophidascaris* spp) have been reported in reptiles. The adult parasites harbour in oesophagus, stomach or small intestine of snakes. Infected larvae will cause purulent lesions in the various organs of the affected animals. Lizards and turtles are commonly infected with oxyurids. It can be diagnosed by demonstration of eggs or intact adults in the faecal samples. Filariasis caused by *Macdonaldius oschei* has been reported in reticulated python. The clinical signs of filariasis include cutaneous swellings and necrotic dermal lesions. The diagnosis of filariasis can be made by demonstration of microfilariae in the blood or by observation of adult filariae in the lesion.

Strongyloides have been reported in snakes. The affected animal shows the symptoms of diarrhoea and respiratory distress. The infective larvae migrate through the host's lungs. Hookworms like *Kalicephalus* and *Oswaldocruzia* occur in reptiles. It causes lethargy, anorexia, general debilitation, anaemia, ulceration, intestinal obstruction and peritonitis. Ivermectin (0.2 mg/kg body I/M) or pyrantel pamoate (8 mg/kg body weight) can be used for treatment of nematode infestations. Ivermectin is contraindicated in chelonians and crocodilians.

Cestodes or tapeworms have been found in reptiles. Crocodilians, however, are rarely infected with cestodes. Adult tapeworms compete for nutrients or they may cause intestinal obstruction. Transmission occurs through ingestion of an intermediate host. The diagnosis of tape worm infections can be made by observing eggs or segments of the parasites in the faecal samples. Praziquantel (7mg/kg body weight) and bunamidine HCL (50 mg/kg body weight) may be suggested for treatment of adult cestodes.

Mites and ticks are known to occur in wild reptiles. Ticks such as *Hyalomma, Amblyoma* and *Aponomma* are found in reptiles. Both soft and hard ticks are responsible for the transmission of blood parasites. Ticks can be manually removed from the infested animal. *Ophionyssus* mites are commonly found in snakes. Ivermectin is used for the treatment of mites. Pyrethroid insecticide should be used carefully. Turtle may be infested by fly larvae. Leeches are commonly found in aquatic turtles. Leeches also cause cutaneous fibroepithelioma in green sea turtles.

Hypoglycaemic shock has been reported in crocodilians. It is clinically characterized by opisthotonos, mydriasis and torticollis. The animal may be treated with glucose solutions.

Duodenal volvulus has been reported in hawksbill turtles and green turtles. It may be due to dietary factors. Volvulus is induced by physically turning the animal over for restraint purpose. Therefore, restrainer should avoid 360^0 rotation when overturning turtle onto its carapace.

Intestinal foreign bodies such as rocks, plastic bags, coins, corals and bones are occasionally found in marine turtles and tortoises. Metabolic bone disease occurs due to prolonged deficiencies of calcium and phosphorus, improper ratio of calcium and phosphorus in the diet and deficiency of vitamin D.

Wild tortoises are commonly exhibited with traumatic injuries.

Chapter 23

Health Care of Wild and Zoo Birds

Physiological parameters: The study of avian haemogram is important for diagnosis of various diseases. The total blood volume in avians is estimated at about ten per cent of animal's body weight. Blood samples may be collected from the jugular vein, the brachial vein, and the claw. However, blood is sometimes collected from the occipital venous sinus as well as through cardiac puncture. The maximum amount of blood that can be collected from a small bird is about 0.5 ml. However, 2 ml of blood can be safely obtained from an average-sized parrot. The blood should be stored in a tube containing the anticoagulant EDTA.

Birds have nucleated erythrocytes. In general, the values for total red blood cell counts vary from 2.5 to 4.5 million mm^3. Normally the smallest birds tend to have higher values for red blood cells. The values of blood also depend on many factors such as age, sex, nutrition, season, altitude and hormones. Wild birds, for example, tend to have higher numbers of red blood cells during the seasons of migratory activity. The life span of avian erythrocytes is relatively short as compared to mammal. The life span of pigeon erythrocytes, for example, ranges from 35 to 45 days. The erythrocytes of domestic fowl have a life span of 20-35 days. Normally one to two per cent of avian erythrocytes are polychromatophilic and their numbers may increase due to inflammation and haemorrhage. Normal values for haemoglobin vary from 10 to 20 g per 100 ml of blood. The packed cell volume in birds ranges from 37 to 53 per cent. The haemogram of some selected birds is given in Table 23.1.

The normal body temperature, respiratory and heart rates of some selected birds are as follows (Table 23.2):

Disease: Wild and captive birds are susceptible to a wide variety of diseases which are caused by infectious or noninfectious agents. The infectious agents include virus, bacteria, fungus and parasites. The common causes for death of wild and captive birds are reportedly due to non infectious origins like starvation, bad weather, pollution of water, poisoning, nutrient deficiencies, collision with electrical wires and aeroplanes.

Table 23.1 : Haematological values of some selected birds.

Birds	RBC (x 10⁶/m m³)	Hb (g /dl)	PCV (%)	WBC (x 10³/m m³)	Hetero phils (%)	Eosino phils (%)	Baso phils (%)	Lympho cytes (%)	Mono cytes (%)
Ostrich	2.4	15.6	48.0	21.0	-	-	-	-	-
Turkey	2.3-2.7	12.5-14.0	38.0	-	43.4	0.9	3.2	50.6	1.9
Pigeon	3.2-3.4	16.0-20.0	42.0	13.0-18.5	23.0-42.8	1.9	2.4	47.8	5.1
Mallard duck	2.7-3.6	16.0	46.0	23.4-24.8	24.3-48.0	2.2-7.0	1.5-5.0	32.0-61.7	8.0-10.8
Barn owl	2.6-3.4	11.7	36.0-52.0	13.2-20.0	36.5	17.5	2.7	43.3	-
Golden eagle	1.8	-	36.0	27.0	21.2-39.0	-	-	7.1-13.7	0.0-5.1
Amazon parrot	2.4-3.5	19.0-25.0	40.0-64.0	4.0-8.4	8.0-30.0	1.1	0.0	69.0-92.0	0.0
Pheasant	2.2-3.6	8.0-11.2	28.0-42.0	-	48.0	1.0	10.0	34.0	8.0

Egg binding problem is commonly seen in game birds, doves and finches. This condition is characterized by the inability to expel the egg from the oviduct. Other reproductive disorders such as salpingitis, prolapse of the oviduct and rupture of egg yolks have also been reported. The gout has been reported in wild birds (e.g. vulture). This is characterized by the deposition of urate in the viscera and joints. Obesity is commonly seen in many pet birds. Cannibalism occurs in captive birds. Poor management, nutrient deficiencies and improper feeding may

Table 23.2 : Normal body temperature, respiratory and heart rates of selected birds.

Bird	Temperature (⁰F)	Respiration rates (breaths /minute)	Heart rates (beats/minute)
Ostrich	106.0	-	-
Pigeon	110.0	25.0-30.0	-
Duck	106.5	42.0-110.0	210.0-220.0
Goose	106.0	14.0- 40.0	-
Turkey	105.4	28.0- 49.0	93.0
Fowl	106.5	17.0- 27.0	250-450
Crane	105.5	-	-
Owl	104.5	-	-
Pelican	105.5	-	-

be the reasons for this problem. Dehydration, emphysema, feather loss and crop impaction have been reported in captive birds. Fractured on legs, toes, feet and wings are occasionally encountered in captive birds. Feather eating or feather plucking is commonly found among caged parrots. It can be corrected by providing adequate minerals and vitamins. Captive seed-eater birds are occasionally shown abnormal growth of beaks and claws. This condition can be prevented by providing grit, calcium blocks etc. Captive and wild birds often exhibit mounting problem. Poisoning is an acute problem of wild birds nowadays. Contaminated food, poisonous plants and pesticides used in agricultural crops are the common sources of poisons causing lower fertility. Caged birds may be exposed to lead poisoning. Leaves, twigs and bark of many plants may be toxic to birds. The widely reported diseases found both in wild and captive birds are documented in Table 23.3.

Table 23.3 : Diseases of captive and wild birds.

Disease	Host	Aetiology	Clinical signs and pathological lesions	Diagnosis	Treatment and control
Infectious coryza	Cranes, pigeons, guinea fowl, pheasants, turkeys, ducks, and sparrows.	This bacterial disease is caused by *Haemophilus para gallinarum*	The clinical signs include facial oedema, mucoid nasal discharge, dyspnoea and conjunctivitis.	Diagnosis is based on examination of smears prepared from exudates, history, clinical signs and haemagglutination inhibition test.	Sulphachloropyridazine and trimethoprim are used. Depopulation, cleanliness, good management, vaccination and isolation of sick birds are important.
Pullorum disease	Turkeys, pigeons and sparrows.	Causal agent is *Salmonella pullorum.*	General malaise, loss of appetite and white diarrhoea. Lesions include grayish-white necrotic foci in liver, lungs and heart.	History, clinical signs, lesions, and isolation and identification of the organism.	Nitrofuran and sulphamerazine may be given. The control measures include proper disinfection, isolation of infected birds and cleaning of infected equipment.

Fowl typhoid	Ostrich, peacock, guinea fowl, swan, pheasant and other wild birds.	Caused by *Salmonella gallinarum*.	Dullness, fever, loss of appetite, bluish combs and wattles, and yellowish diarrhoea. Enlarged liver (copper colour), spleen and discoloured ova.	Same as pullorum disease.	Captive birds are to be vaccinated with killed vaccine.
paratyphoid	Pigeons, hornbills, cockatoos, parrots, guineafowl, ducks, swans etc.	Caused by *Salmonella typhimurium*.	Depression, poor growth, weakness, severe gastroenteritis and dehydration. Lesions include enlarged liver, necrotic foci in liver and unabsorbed yolk.	Clinical signs, isolation of the organism and serological tests.	Antibiotics can be administered. Control measures include good hygiene, separation of infected birds and control of predators such as rats.
Fowl cholera	Most species of wild and captive birds.	Caused by *Pasteurella multocida*.	Dyspnoea, mucoid oral discharge, diarrhoea, fever, prostration, and cyanosis of wattles and combs. Lesions include purulent facial oedema and conjunctivitis.	Clinical signs, postmortem lesions, presence of organisms in specimens of blood and serological tests.	Sulpha drugs may be given for treatment of fowl cholera. The dead animal should be buried. Stress should be avoided. Vaccination should be routinely practised.
tuberculosis	All captive and wild birds. Commonly found among the members of the order Anseriformes.	Caused by *Mycobacterium avium*.	Clinical signs include debility, diarrhoea and emaciation. Yellowish caseous nodules are seen in liver.	Demonstration of acid fast bacilli in smears taken from most lesions.	Depopulation is important. Proper disinfection and sanitation of cages are essential. The infected bird should be buried or burnt.

Botulism or western duck sickness	Water fowl, raptors and gulls.	*Clostridium botulinum.*	Anorexia, greenish diarrhoea, flaccid paralysis of neck, legs and wings are clinical signs.	Clinical symptoms and animal test.	Dead birds should be removed. Regular supply of fresh water is vital.
Mycoplasmosis	Commonly found in pheasants, turkeys, peacocks, parrots and pigeons.	Caused by several *Mycoplasma spp.*	Coryza, respiratory tract infection, coughing are the clinical signs.	Clinical symptoms, postmortem lesions, culture of the organisms and serological tests.	Administration of antibiotics.
Avian pox	Many wild and captive birds.	Caused by avian pox virus.	Wart like nodules on the skin, conjunctivitis, erythema, and oedema.	Clinical signs, lesions and laboratory tests.	Symptomatic treatment is suggested. Pigeon pox vaccine or the egg embryo adopted vaccine is recommended.
Infectious laryngotracheitis	Pheasants, peafowl, pigeons, guinea fowl, cranes, sparrows etc.	Caused by Herpes virus.	Clinical signs include dyspnoea, coughing, gasping, conjunctivitis and cyanosis of wattles.	Clinical symptoms, characteristic lesions, isolation of the organism and serological tests.	Antibiotics are to be administered in order to prevent secondary bacterial infections. Birds should be vaccinated against this disease.
Newcastle disease	Found in all species of birds. Galliformes are very susceptible.	Caused by avian paramyxo type I group.	Loss of appetite, closed eyes, prostration, greenish or yellowish diarrhoea, rhinitis, torticollis and paralysis.	Clinical signs, isolation of the organisms and serological tests.	Symptomatic treatment is advised. Vaccination of birds is essential.

327

Duck plague	Ducks, swans and geese. Wild fowl serve as reservoirs.	This contagious disease is caused by herpes virus.	Ruffled feathers, dullness, nasal discharge, haemorrhagic diarrhoea and tremors.	Clinical signs, isolation of the virus and serological tests.	Modified live vaccine is used as prophylactic measure. Environmental decontamination is vital. Treatment of open water with calcium hypochlorite is important.
Pacheco's parrot disease	All psittacines are susceptible.	Caused by herpes virus.	Vomiting, diarrhoea and neurological signs are the clinical symptoms.	Diagnosis is based on clinical signs, postmortem lesions and culture of the organism.	Treatment is symptomatic.
Aspergill osis	All wild and captive birds.	Causal agent is *Aspergillus fumigatus*.	Loss of appetite, dyspnoea, depression, gasping, emaciation, increased thirst and neurological signs. Air sacs get thickened.	Gross lesions, demonstratio n of fungi in the lungs and culture of the organism.	Infected litter and feed should be disposed off. Good hygienic measures are vital. Newly birds should be kept in clean and dust free rooms.
Coccidio sis	Peafowl, quail, parrot, owl, duck, goose, crane, and many other birds. Commonly found in game birds.	Caused by *Eimeria* and *Isospora* spp. *E. gruis* and *E. rechenowi* are commonly found in whooping and sandhill cranes.	Depression, loss of appetite, diarrhoea, emaciation and death.	Diagnosis based on characteristic gross lesions, examination of faecal contents and clinical symptoms.	Sulphamezathine may be used. Hygienic measures like disinfection, use of proper waterers and feeders are important.

Aspergill osis	All wild and captive birds.	Causal agent is *Aspergillus fumigatus.*	Loss of appetite, dyspnoea, depression, gasping, emaciation, increased thirst and neurological signs. Air sacs get thickened.	Gross lesions, demonstratio n of fungi in the lungs and culture of the organism.	Infected litter and feed should be disposed off. Good hygienic measures are vital. Newly birds should be kept in clean and dust free rooms.
Coccidio sis	Peafowl, quail, parrot, owl, duck, goose, crane, and many other birds. Commonly found in game birds.	Caused by *Eimeria* and *Isospora* spp. *E. gruis* and *E. rechenowi* are commonly found in whooping and sandhill cranes.	Depression, loss of appetite, diarrhoea, emaciation and death.	Diagnosis based on characteristic gross lesions, examination of faecal contents and clinical symptoms.	Sulphamezathine may be used. Hygienic measures like disinfection, use of proper waterers and feeders are important.
Histomo niasis	Turkey, peafowl, grouse, pheasant, guineafowl etc.	The causal agent is *Histomonas meleagridis.*	Clinical signs include dullness, cyanosis and yellowish droppings. Enlarged caeca may exhibit necrotic ulcers.	Demonstrati on of parasites in the faecal contents and postmortem lesions.	Strict hygienic measures are vital.

References and Further Reading

Abbasi, S.A.1997. Wetlands of India: Ecology and Threats. Discovery Publishing Hou; e, New Delhi.

Ahmed, A. 1997. Live Bird Trade in Northern India. TRAFFIC –India. WWF for Nature – India, New Delhi.

Alderton, D. 1991. Crocodiles and Alligators of the World. Blandford, London.

Ali, S. and Ripley, S. D.1987. Birds of India and Pakistan (Second ed.), Oxford University Press.

Amstrup, S.C. and Wiig, Y.1991. Polar Bears. Proceedings of the Tenth Working Meeting of IUCN/SSC Polar Bear Specialist Group, IUCN, Gland, Switzerland.

Anderson. I. R. 1976. Animals in Danger. Frederick Warne & Co. Ltd, London.

Animal Management- Working with Zoo Animals. National Extension College, London.

Arora, B.M. 1994. Wildlife Diseases in India. Periodical Expert Book Agency, New Delhi.

Bailey, J.A. 1984. Principles of Wildlife Management, John Willey and Sons, Inc. USA.

Baer, C. K (ed). 1996. Proceedings- American Association of Zoo Veterinarian's Annual Conference.

Bedi, R. and Bedi, R. 1984. Indian Wildlife, Brijbasi Printers Private Ltd., New Delhi.

Bellairs, A. 1969. The Life of Reptiles. Vols. 1 and 2. London, Weidenfeld and Nicolson.

Benton, M. J. 1990. The Reign of the Reptiles. Crescent Books, New York.

Bere. R. 1970. Antelopes. Arco Publishing Company, Inc. New York and Arthur Barker Limited, London,

Berger, J. 1986. Wild Horses of the Great Basin. The University of Chicago Press,

Berwick, S. H. and Saharia, V.B. (eds). 1995. Wildlife Research and Management. Oxford University Press.

Biosphere Reserves. 1986. Proceedings of the First National Symposium, Government of India, Ministry of Environment and Forests, New Delhi.

Birds: The Wealth of India.1990. Publication and Information Directorate, CSIR, New Delhi.

Boyd,I.L.(ed.).1993. Marine Mammals- Advances in Behavioral and Population Biology. The Zoological Society of London, Clarenden Press, Oxford.

Bramwell, Martyn (ed). 1973. The Atlas of World Wildlife. Portland House, New York.

Brown, R. D. (ed.). 1992. The Biology of Deer. Springer Verlag, New York Inc. USA.

Bubenik, G. A. and Bubenik, A. B. (eds). 1992. Horns, Pronghorns and Antlers. Springer-Verlag, New Your, Inc, USA.

Burton, M.1977. How Mammals Live. Galley Press, England.

Campbell, Б.1983.The Dictionary of Birds in Color. Exeter Books, New York.

Campbell, T.W. 1988. Avian Haematology and Cytology. Iowa State University Press, Iowa, USA.

Carr, A. 1968. The turtle. A Natural History. Cassell, London: Natural History Press.

Caro, T.M.1994. Cheetahs of the Serengeti Plains. The University Chicago Press, Chicago.

Catton, C. 1990. Pandas. Christopher Helm (Publishers) Ltd, London.

Chadha, S.K. (ed). 1992. Conserving Wildlife in India. Vinod Publishers & Distributors. Jammu & Kashmir, India.

Chadwick, D.H. 1993. The fate of the elephant. Viking Penguin Books.

Chakravarty, Kalyan. 1991. Man, Plant & Animal Interaction. Darbari Prakashan, Calcutta.

Chandler, E.A., Gaskell, C. J. and Gaskell, R.M. (eds). 1994. Feline Medicine and Therapeutics, Second edition, Blackwell Scientific Publications.

Chapman, J. A. and John, E.C.F.1990. Rabbits, Hare and Pika: Status Survey and Conservation Action Plan. IUCN/SSC Lagomorph Specialist Group. Switzerland.

Chauhan H.V.S and Roy, S. 1996. Poultry Diseases, Diagnosis and Treatment, Second edition. New Age International (P) Limited, New Delhi.

Choudhury B.C. and Bhupathy, S. 1993. Turtle Trade in India: A study of Tortoises and Freshwater Turtles. WWF- India, New Delhi.

Churchfield, S. 1990. The Natural History of Shrews. Christopher Helm (Publishers) Ltd., London.

Coles, B.H. 1985. Avian Medicine and Surgery. Blackwell Scientific Publications.

Compendium of Environment Statistics, Central Statistical Organization, Ministry of Planning and Programme Implementation, Government of India, New Delhi.

Cooper, J. E. and Jackson Oliphant F. (eds). 1981. Diseases of Reptilia, Vol 1 & 11, Academic Press, London.

Crandall, L. S. 1964. Management of Wild Animals in Captivity. Chicago, Chicago University Press, USA.

Dagg, A. l. and Foster, J. B.1976. The Giraffe: Its Biology, Behavior, and Ecology. Van Nostrand Reinhold Company.

Daniel.J.C. 1983. The Book of Indian Reptiles. Bombay Natural History Society, Mumbai, India.

Daniel,J.C. 1996. The Leopard in India -A Natural History. Natraj Publishers, Dehra Dun.

Darlington, P.J. Jr. and Robert, E. 1982. Zoo Geography: The Geographical Distribution of Animals. Krieger Publishing Company, Inc, Florida, USA.

Das, A.K. and Dev Roy, M.K. 1984. A General Account of the Mangrove Fauna of Andaman and Nicobar Islands. Zoological Survey of India, Calcutta.

Das, I. 1995. Turtle and Tortoises of India. Oxford University Press.

De Vos, Antoon. 1982. A Manual on Crocodile Conservation and Management in India. FAO.

Dein, F. Joshua., Brunson, D. B. and Malik Pradeep. 1991. Wildlife Immobilization Workshop, Kanha National Park and Tiger Reserve, Wildlife Institute of India and U.S. Fish and Wildlife Service.

Desai, A. The Indian Elephant. Vigyan Prasar & Sanctuary Magazine.

Dhyani, S.N. 1994. Wildlife Management. Rawat Publication, Jaipur, Rajasthan, India.

Ditmars, R.L. 1937. Snakes of the World. Macmillan, New York.

Dobson, A. P. 1996. Conservation and Biodiversity. Scientific American Library.

Donkin, R.A. 1985. The Peccary – With Observation on 'he Introduction of Pigs to the New World. The American Philosophical Society.

Dugar, P.J. (ed). 1990. Wetland Conservation – A Review of Current Issues and Required Action. IUCN, Switzerland.

Dunbar, B. A.A. 1982 (Indian Edition). Wild Animals in Central India. Natraj Publishers, Dehradun, India.

Duncan, P. 1992. Zebras, Asses and Horses- An Action Plan for the Conservation of Wild Equids. IUCN/SSC Equid Specialist Group, IUCN, Gland, Switzerland.

Dutta, A.K. 1991.Unicornis: the Great Indian one- horned Rhinoceros. Konark Publishers Pvt. Ltd, New Delhi.

Eaton, L.R. and R. E. Eaton. 1982. The Cheetah: The Biology, Ecology and Behaviour of an Endangered Species. Krieger Publishing Company, Florida, USA.

Edgoankar, A. and Chellam, Ravi. 1998. A Preliminary Study of the Ecology of the Leopard, *P.p. fusca* in Sanjay Gandhi NP. Maharashtra, WII, Dehradun.

Edney, A.T.B. and Hughes, I.B. 1986. Pet care. Blackwell Scientific Publication.

Edward, D and Clive Turner. 1980. Bats. The Pitman Press Ltd, Sussex, U.K.

Encyclopedia of the Animal World. 1972. Bay Books, Sydney.

Fauna of Andhra Pradesh, Part 1. 1993. Zoological Survey of India.

Fauna of Orissa. 1991. Part IV, Zoological Survey of India.

Fauna of West Bengal- Part 1. 1992. Zoological Survey of India. Kolkata.

Fauna of Western Himalayas (U.P.), Part 1. 1995. Zoological Survey of India. Kolkata.

Favre, D.S. 1989. International Trade in Endangered Species: A Guide to CITES, Martinus Nijhoff Publishers, Netherlands.

Findley, J.S. 1993. Bats: A Community Perspective. Cambridge University Press, Cambridge.

Finlayson, C. M. and Vander Valuk, A.G. (eds). 1995. Classifications and Inventory of the World Wetlands. Kluwer Academic Publishers, Netherlands.

Fowler, M. E. 1986. Zoo and Wild Animal Medicine (Second ed). W.B. Saunders Company, West Washington Square, Philadelphia, USA.

Fowler, M. E. and Miller, R.E.(eds.). 1999. Zoo and Wild Animal Medicine: Current Therapy. 4th ed. W.B Saunders Company, Piladelphia, USA.

Freeman, H.(ed.).1988. Proceedings of the Fifth International Snow Leopard Symposium. Srinagar, India,

Friend, M. (ed). 1987. Field Guide to Wildlife Diseases, Vol. 1 (General Field Procedures and Diseases of Migratory Birds). United States Department of the Interior fish and Wildlife Service, Washington, USA.

Fyre, F. L. 1978. Husbandry, Medicine and Surgery in Captive Reptiles, V. M. Publishing

Inc, Kansas, USA.

Gandhi, M., Husain, O. and Panjwani, Raj .1996. Animal Laws of India. Universal Law Publishing Co. Pvt. Ltd, Delhi.

Gangstad, E.O. 1990. Natural Resource Management of Water and Land. Van Nostrand Reinbold, New York.

Gaston, K. J. and Spicer, J. I.1998. Biodiversity-An Introduction. Blackwell Science Ltd.

Gee, E.P . 1964. The Wildlife of India. Collins, London.

Gogate, M.G., Thosure P.G. and S.B. Banubakode (eds). 1992. Two Decades of Project Tiger, Melghat (1973- 1993): Papers and Proceedings. Government of India.

Gopal, B.(ed. & compiled). 1995. Hand Book of Wetland Management. World Wildlife Fund-India, New Delhi, India.

Grenard, Steve. 1991. Handbook of Alligators and Crocodiles. Krieger Publishing Company Florida, USA.

Groombridge, B.(ed.). The 1994. IUCN Red List of Threatened Animals. Gland, (Switzerland) and Cambridge (UK), IUCN.

Grzimek, B(ed.). 1972. Grzimek's Animal Life Encyclopedia. New York, Van Nostrand Reinhold.

Gupta, S.C., Gupta, N. and Nivsarkar, A.E. 1999. Mithun . ICAR, New Delhi.

Hand Book of Animal Husbandry. 1990. Publications and Information Division, ICAR, New Delhi (2nd revised ed).

Hand Book of Procedures. Vol 11, 1990-93. Import and Export Promotion (Act and Orders), Ministry of Commerce, Government of India.

Hanfee, F. 1998. Wildlife Trade- A Handbook for Enforcement Staff: TRAFFIC – India and WWF – India, New Delhi.

Harrington, F.H. and Paquet, P.C.1982. Wolves of the World- Perspectives of Behaviour, Ecology and Conservation. Noyes Publications, New Jersey, USA.

Hawkey, C.M. and Dennett T.B. 1989. Comparative Veterinary Haematology. Wolfe Publishing Limited, England.

Haynes. Gary. 1991. Mammoths, Mastodons and Elephants -Biology, Behaviour and the Fossil Record. Cambridge University Press, Cambridge, New York.

Hoff, G. L. and Davis, J. W. Noninfectious Diseases of Wildlife. The Iowa State University Press, USA.

Howard, E. B. (ed). 1983. Pathology of Marine Mammal Disease. Vol. 1 & 11: CRC Press Inc. Florida, USA.

Identification Manual for Indian Wildlife Species and Derivatives in Trade: TRAFFIC India. 1996. WWF- India and Ministry of Environment and Forests, Government of India.

Indian Board for Wildlife. 1972. Project Tiger: A Planning Proposal for Preservation of tiger (*Panthera tigris tigris* Linn.) in India. New Delhi, Government of India.

Jain, N. C. 1986. Schalm's Veterinary Haematology (4th ed). Lea & Febiger, Washington Square, Philadelphia, USA

Jenkins, M.D. Tortoises and Freshwater Turtles. The Trade in South East Asia. TRAFFIC – Southeast Asia.

Jha, L.K. and Sen Sarma, P.K. 1993. Agroforestry: Indian perspective. Ashish Publishing House, New Delhi.

John ,M.C., Jayathangaraj, M.G. and Kalaimathi, R. 1995. Wildlife Student's Manual. Department of Wildlife Science, Madras Veterinary College, Tamil Nadu Veterinary And Animal Science University, Madras.

Johnston, A.M. 1986. Equine Medical Disorders. Blackwell Scientific Publications.

Jordan, W.J. and Hughes, John. Care of the Wildlife. Macdonald & Co. London & Sydney.

Joshi B.P. 1991. Wild Animal Medicine, Oxford & IBH Publishing Co. Pvt. Ltd, New Delhi.

Junge, R. E. (ed). 1994. Proceedings of the Association of Reptilian and Amphibian Veterinarians and American Association of Zoo Veterinarians.

Junge, R.E.(ed.). 1993. Proceedings: American Association of Zoo Veterinarian's Annual Conference.

Junge, R.E.(ed.).1995. Proceedings: American Association of Zoo Veterinarians, Wildlife Disease Association and American Association of Wildlife Veterinarians.

Kamra,D.N.,Agarwal,N.,Chaudhary,L.C.,Pathak,N.N., Garg,A.K. and Sastry,V.R.B.(eds.). 1999-2000. Centre of Advanced studies in Animal Nutrition: Nutrition and Feeding of wild animals. IVRI, Uttar Pradesh.

Khanna, L.S. 1993. Principles and Practice of Silviculture. Khanna Bandhu, Dehradun, India.

Kinne, Otto (ed). 1980. Diseases of Marine Animals, Vol. 1, John Wiley and Sons Ltd.

Kinne,Otto. Disease of Marine Animals, Vol. 11, Biologische Anstalt Helgoland, Hamburg, Germany.

Kirk, R. W. (ed.). 1989. Current Veterinary Therapy: Small Animal Practice. W.B. Saunders Company, Philadelphia, USA.

Kotwal, P.C. and Mukherjee, S. (eds). 1998. Biodiversity Conservation in Managed Forest and Protected Areas. Agro Botanica, Bikaner, Rajasthan.

Kumar, H.D. 1995. General Ecology. Vikas Publishing House Pvt. Ltd., New Delhi.

Lair.R. C. 1997. Gone Astray: The care and Management of the Asian Elephant in Domesticity. FAO Regional Office for Asia and the Pacific (RAP), Thailand.

Lanworn, R..A. 1972. The Book of Reptiles. The Hamlyn Publishing Group Limited.

Laws, R.M., Parke, I.S.C. and Johnstone., R.C.B.1975. Elephants and Their Habitats. Clarenden Press, Oxford.

Lyon, J. G. 1993. Practical Handbook for Wetland Identification and Delineation. Lewis Publishers.

Lyubashenko, S. Y. 1983. Disease of Fur bearing Animals. Oxonian Press Pvt. Ltd., New Delhi.

Macdonald, D.The Encyclopaedia of Mammals, Vol 1 & 11, George Allen & Unwin, London & Sydney.

Mandal A.K. and Nandi. N.C. 1989. Fauna of Sundarban Mangrove Ecosystem, West

Bengal, India. Zoological Survey of India, Calcutta (Kolkata).

Manfredi, P. In Danger -Indian Wildlife and Habitat. Local Colour Private Limited and Ranthambhore Foundation, New Delhi.

Martin, B. P. 1987. World Birds. Guinness Superlatives Ltd, London Road, Great Britain.

Martin, E.B. Rhino Exploitation. World Wildlife Fund, Hong Kong.

Martin, E B. Rhino Exploitation- the Trade in Rhino Products in India, Indonesia, Malayasia, Burma, Japan and South Korea: World Wildlife Fund, Hong Kong.

Maulton, M.P. Wildlife Issues in a Changing World, St. Luice Press, Florida, USA.

McCurnin, D. M. 1990. Clinical Textbook for Veterinary Technicians. W.B. Saunders Company, USA, (2nd ed).

McDiarmid A. (ed). 1969. Diseases in Free–living Wild Animals: Symposia of the Zoological Society of London. Academic Press Inc. Ltd, London, U.K.

McNeely,J., Harrison, J. and Dingwall, P. (eds). 1994. Protecting Nature – Regional Reviews of Protected Areas. IUCN, Switzerland.

Miller, S. Douglas and Daniel, D.E.(eds). 1982, Cats of the World: Biology, Conservation and Management. Proceedings of the Second International Symposium, Texas, USA.

Mills,Gus. and Hofer, H. 1998. Hyaenas: Status Survey and Conservation Action Plan. IUCN /SSC Hyaena Specialist Group.

Murphy, J. B. and Collins, J. T. 1980. Reproductive Biology and Diseases of Captive Reptiles, Society for the Study of Amphibians and Reptiles.

Nahlik A.J. de. Deer Management. David & Charles Inc. North Pomfret, USA.

Nair, S.M. 1992. Endangered Animals of India and their Conservation. National Book Trust, New Delhi, India.

National Commission on Agriculture (NCA), 1976, Government of India, New Delhi.

National Geographic Book of Mammals. 1983. Vol 1 & 11, National Geographic Society, Washington , USA,

Negi, S.S.1992. Himalayan Wildlife–Habitat and Conservation. Indus Publishing Company, New Delhi.

Nilgiri Tahr.1992-93.North American Regional Studbook, Minnesota Zoological Garden.

Nivsarkar, A.E., Gupta, S.Ç. and Gupta, N. 1998. Yak Production. ICAR, New Delhi.

Nowell, K. and Jackson, P. (ed). 1996. Status, Survey and Conservation Action Plan: Wild Cats. IUCN/SSC Cat Specialist Group, IUCN, Switzerland.

Oliver, W. L. R. (ed). 1993. Status Survey and Conservation Action Plan – Pigs, Peccaries and Hippos. IUCN/ SSC Pigs and Peccaries Specialist Group and IUCN/SSC Hippo Specialist Group, IUCN, Switzerland.

Olivier, R.C. 1978. On the ecology of the Asian Elephant *Elephas maximus* Linn. With Particular Reference to Malaya and Sri Lanka, Ph.D dissertation, University of Cambridge.

Osmaslon, B.B. and Sale, J.B. 1989. Wildlife of Dehradun and Adjacent Hills. Natraj Publishers, Dehradun, Uttaranchal.

Pandey, P., Kothari, A. and Singh, S. (eds.). 1991. Directory of National Parks and Sanctuaries in Andaman and Nicobar Islands- Management, Status and Profiles, Indian Institute of Public Administration, New Delhi.

Pandey, V. (ed). 1992. Himalaya's Encyclopaedic Dictionary of Environmental Pollution Vol.2 (Water and Marine Pollution). Himalaya Publishing House, Bombay.

Parker, H.W. 1965. Natural History of Snakes. Brit. Mus. (Nat. Hist.) London.

Patnaik S.K. and Acharjo, L.N. (eds). 1996. Indian Zoo Year Book, Vol. 1. Indian Zoo Director's Association and Central Zoo Authority, New Delhi.

Patnaik, S.K. and Acharjo, L.N. (eds). 1997. Indian Zoo Year Book, Vol. 11. Indian Zoo Director's Association & Central Zoo Authority, New Delhi.

Pedersen , Niels. 1991. Feline Husbandry: Diseases and Management in the Multiple Cat Environment. American Veterinary Publications, Inc. USA.

Penny. M. 1987. Rhinos- Endangered Species. Christopher Helm Publishers Ltd., London.

Perrin,W. F. L., Brownell, R.L., Kaiya, Z. Jr. and Jiankang, L. 1986. Biology and Conservation of the River Dolphins. Proceedings of the Workshop on Biology and Conservation of the Platanistoid Dolphins. Wuhan, China, IUCN, Switzerland.

Phull, A. and Phull, R. K.1994. Rabbit Farming and its Economics. International Book Distributing Co., Lucknow, India.

Population and Habitat Viability Assessment Workshop Briefing Book- Manipur Brow-antlered Deer. 1992. Zoological Garden, Mysore. Karnataka.

Pope, C.H. 1956. The reptile World. A Natural History of the Snakes, Lizards, Turtles and Crocodilians. Routledge & Kegan Paul, London.

Porritt, J. 1990. Where on Earth Are We Going. BBC Books, London.

Prater, S.H. 1980. The Book of Indian Animals. Bombay Natural History Society. Bombay.

Proceeding of the National Symposium on the Asian Elephant. 1989. The Asian Elephant-Ecology, Biology, Diseases, Conservation and Management. Trichur, Kerala Agriculture University, India.

Proceedings of the symposium on Biosphere Reserves, September 14-17, 1988, Colorado, USA.

Project Elephant. 1993. New Delhi, Government of India, Ministry of Environment and Forests.

Przewalsky Horse and Restoration to its Natural Habitat in Mongolia. 1986. FAO, Rome.

Putman, R. 1988. The Natural History of Deer. Christopher Helm (Publishers) Ltd. London, U.K.

Rabbit: Husbandry, Health and Production. 1986. Food and Agriculture Organization of the United Nations. Rome.

Radostits, O.M., Blood, D.C. and Gay, C.C. 1994.Veterinary Medicine (8th ed.). ELBS, Bailliere Tindall, London.

Ramona and Marris, D.1996. Men and Pandas. McGraw- Hill Book Company.

Randall, C.J. 1985. A Colour Atlas of Diseases of the Domestic Fowl & Turkey. Wolfe Medical Publication Ltd. England.

Randall, C.J. 1985. Diseases of the Domestic Fowl and Turkey. Wolfe Medical Publications Ltd. England.

Ranjitsingh, M.K. 1989. The Indian Blackbuck. Natraj Publisher, Dehradun, Uttaranchal, India.

Ranjitsingh, M.K. 1997. Beyond the Tiger-Portraits of Asian Wildlife. Brijbasi Printers Pvt. Ltd., New Delhi.

Rathore, G.S. 1986. Camels and their Management. Publication and Information Division, Indian Council of Agricultural Research, New Delhi.

Raymond, L. D. 1989. Reptiles of the World. Cosmo Publications, New Delhi. India.

Recheigl, M. Jr. (editor in chief). 1977. CRC Handbook Series in Nutrition and Food. Vol 1 (Diets for Mammals), CRC Press, Inc. Ohio, USA.

Rekoff, March.1978. Coyotes–Biology, Behavior and Management. Academic Press Inc, New York, USA.

Rementsova,M..M. 1987. Brucellosis in Wild Animals. Oxonian Press Pvt. Ltd., New Delhi.

Richard J (ed). 1976. Mycobacterial Infections of Zoo Animals. The Symposia of the National Zoological Park, Smithsonian Institution Press, Washington, USA.

Ridgway, S. H. and Harrison, S. Richard.(eds). 1981. Hand Book of Marine Mammals. Vol.1, 2, 3 and 4. Academic Press Ltd. London, U.K.

Robert Stroud. 1964. Stroud's Digest on the Diseases of Birds. T.F.H. Publications, Inc,

Rodgers, W.A. and Panwar, H.S. 1988. Planning a Wildlife Protected Area Network in India. Dehradun, India.

Saharia, V.B. 1982. Wildlife in India. Natraj Publishers, Dehra Dun.

Salim, A. 2003.The Book of Indian Birds. Bombay Natural History Society & Oxford University Press, Bombay, Thirteenth edition.

Santiapillai, C. and Jackson, P. 1990. The Asian Elephant – An Action Plan for its Conservation. IUCN/SSC Asian Elephant Specialist Group, Switzerland.

Santiapillai, C. and Peter J. 1996. The Asian Elephant: An Action Plan for its Conservation. IUCN, Gland, Switzerland.

Santra, A.K. 2003. Studies on Elephant (*Elephas maximus*) Movement in South West Bengal and its Impact. Unpublished Ph.D Thesis, West Bengal University of Animal and Fishery Sciences, Kolkata.

Sanyal, R. B. 1995 (Reprinted). Handbook of the Management of Animals in Captivity. Natraj Publishers, Dehradun.

Schaller, G. B., Jinchu Hu, Wenshi Pan and Jing, Zhu Jing. 1985. The Giant Pandas of Wolong. The University of Chicago Press, Chicago and London.

Schaller, G.B.1972. The Serengeti Lion- A Study of Predator- Prey Relations. The University of Chicago Press, Chicago and London.

Schaller, G.B.1998. The Deer and the Tiger- A Study of Wildlife in India. Natraj Publishers, Dehradun, Uttaranchal, India.

Schenkel, R., S.Lottle – Hulliger. and P. Verlag Paul. 1984. Ecology and Behaviour of the Black Rhinoceros. A field study. Hamburg and Berlin.

Scott, D.A. 1989. A Directory of Asian Wetlands. Gland (Switzerland) and Cambridge (U.K.), IUCN.

Sharma B.D. (ed.). 1998. Wildlife and Diseases in India. Asiatic Publishing House, New Delhi.

Sharma, B.D.1994. High Altitude Wildlife of India. Oxford and IBH Publishing Co. Pvt. Ltd., New Delhi.

Sheldon,.J.W.1992. Wild Dogs: The Natural History of the Nondomestic Canidae. Academic Press, Inc. San Diego, California, USA.

Sigurd, Raethel Heinz. Bird Disease.T.F.H. Publications, Inc. Ltd.

Silver, C.1983.World of Horses. Omega Books Ltd., England.

Singh, S. and Rao, K. 1984. India's Rhino Re-introduction Programme. Department of Environment and Forests, Govt. of India.

Sukumar, R. 1985. Ecology of the Asian Elephant (*Elephas maximus*) and its Interaction with Man in South India, Ph.D Thesis, Indian Institute of Science, Bangalore.

Sukumar, R. 2003. The living elephants. Evolutionary Ecology, Behaviours and Conservation. Oxford University Press.

Svendsen Per and Hau, Jann(eds). 1994. Hand Book of Laboratory Animal Science, Vol I & II, CRC Press, Florida, USA.

Swenson, M. J. and Reece, W.O. (eds). 1996. Dukes Physiology of Domestic Animals. 11th ed., Panima Publishing Corporation, New Delhi, (First Indian Reprint).

Tejwani, K. G. 1994. Agroforestry in India. Oxford & IBH Publishing Co. Pvt. Ltd, New Delhi.

The Ecological Studies of Snow Leopard and its Associated Prey Species in Hemis High Altitude National Park, Ladakh. 1990. Wildlife Institute of India, Dehradun, Uttaranchal.

The IUCN Amphibia- Reptilia Red Data Book. Part F (Testudines, Crocodylia and Rhynchocephalia, 1982, IUCN, Gland, Switzerland.

The IUCN Red List of Threatened Animals. 1990. World Conservation Monitoring Centre, Cambridge, UK.

The IUCN Red List of Threatened Animals. 1994. World Conservation Monitoring Center, Cambridge, UK.

The Merck Veterinary Manual. 15th and 16th eds. 1987 & 1998. Merck & Co. Inc. New Jersey, USA.

The New Encyclopaedia Britannica, 15th ed. 1987.

The Red Data Book on Indian Animals. Part 1- Vertebrate (Mammalia, Aves, Reptilia and Amphibian). 1994. Director, Zoological Surgery of India, Calcutta.

The Wild Animals of India (Reprinted from the Journal of the Bombay Natural History Society). 1988. Daya Publishing House, Delhi.

The Wildlife (Protection) Act, 1972 (As amended up to 1991), Natraj Publishers, Dehra Dun.

The Wildlife (Protection) Act, 1972 (As amended up to 2003). 2003. Natraj Publishers, Dehra Dun.

Tilson, R. L. and Seal, U.S.(eds).1987. Tigers of the World – the Biology, Biopolitics, Management and Conservation of an Endangered Species. Noyes Publication, New Jersey, USA.

Tiwari, S.K. 1985. Zoo Geography of India and South East Asia. CBS Publishers and Distributors, Delhi.

Tiwari, S.K. 1997. Wildlife Sanctuary of Madhya Pradesh. APH Publishing Corporation, New Delhi.

Tortoises and Freshwater Turtles- An Action Plan for their Conservation. 1989. IUCN/ SSC Tortoise and Fresh Water Turtle Specialist Group, IUCN, Switzerland.

Trivedi, P.R. and Singh U.K., Environmental Laws on Wildlife, Commonwealth Publishers, New Delhi.

Various Scientific Journals (e.g. Sanctuary Asia, Zoo's Print, Zoo Zen, Journal of Zoo and Wildlife Medicine, Cheetal, Oryx etc.) and Leading Daily News Papers (The Times of India, The Telegraph, etc.).

Walker, E.P. 1978. Mammals of the World. Vol. I & II, the Johns Hopkins University Press.

Wallace, R. A. 1990. Biology- The World of Life. Harper Collins Publishers USA, 5th ed.

Wallach, J. D. and Boever,W. J. Diseases of Exotic Animals W.B. Sounders Company, Philadelphia, USA.

Webb, B.H., Johnson, H. A. and Alford, J. A. 1987. Fundamentals of Dairy Chemistry. CBS Publishers and Distributors, Delhi.

Wemmer,C., Teare J.A. and Pickett, Charles. A Zoo Biologist's Manual. National Zoological Park, Smithsonian Institution, Washington, USA.

Whitaker, R. 1978. Common Indian Snakes. Macmillan Co. of India Ltd., New Delhi.

Wilson, D. E. 1997. Bats in Questions- The Smithsonian Answer Book. Smithsonian Institute Press, Washington and London.

Wintzer, Hanns –Jurgen. 1986. Equine Diseases: A Text Book for Students and Practitioners. Springer Verlag, New York.

Woodroffe, Gordon. Wildlife Conservation and the Modern Zoo. Saiga Publishing Co. Ltd., Surrey, England.

Worden, A. N. 1989. Hand Book of Laboratory Animals. Anmol Publication, New Delhi.

Workshop on Control of Illegal Wildlife Trade in India. 1998. Wildlife Institute of India, Dehradun, 1996, India.

Young, E. (ed). The Capture and Care of Wild Animals. Human and Rousseau Publishers (Pvt.) Ltd, Cape Town, Pretoria, South Africa.

Zuber, C. 1978. Animals in Danger. Barrons / Woodbury, New York.

Index